"With a compassionate eye and a generous heart, Brenda Paik Sunoo has beautifully captured the fascinating, dwindling world of the *haenyeo*. I am grateful she has given voice to these 'grannies of the sea,' and their powerful, moving stories. Dive into this book and prepare to be amazed."

GAYLE BRANDEIS

Author of *The Book of Dead Birds, Delta Girls,* and *Fruitflesh: Seeds of Inspiration for Women Who Write*

"*Moon Tides* is a magnificent picture and storybook of the aging woman divers of Jeju Island, off the southwestern coast of South Korea. It draws you into the lives of these stalwart women. It's an inspiring story of aging women's productivity and survival in an entrenched patriarchal society."

SETSUKO MATSUNAGA NISHI, Ph.D.

Professor Emerita of Sociology, Brooklyn College and the Graduate Center, The City University of New York

"Imagine visiting Jeju Island, where Korean grandmothers free-dive to harvest seaweed and shellfish. Even at age 70, many plunge 50 feet and hold their breath underwater for two minutes. Wouldn't you like to talk with these daring women? Photojournalist Brenda Paik Sunoo, who lived for a while in her grandmother's homeland to research this book, invited the locals to share their honest, intimate stories. *Moon Tides* amplifies their voices with Brenda's engaging, award-winning photographs, plus woodcuts and poetry. You'll discover details about the women's work, relationships, traditions, community, and spiritual lives. *Moon Tides* is a beautiful, compelling book; I loved it."

PAOLA GIANTURCO

Author of *Celebrating Women* and *Women Who Light the Dark*

"Jeju grannies show the spirit of never, ever giving up in their pursuit of being the best that they could possibly be in their chosen life's field. I like to think that I must have inherited that spirit during my earlier sports career."

SAMMY LEE, M.D.

First Asian American to win the gold medal for the United States in diving at the 1948 Olympics

"A fascinating and pictorially stunning study of a little known culture—the diving women of Jeju Island, known as *haenyeo*, who maintain their shamanist traditions and diving skills despite age, suffering, and personal privation. Their resiliency and sense of purpose are an inspiration to us all."

SUSAN K. GOLANT

Co-author of *Within Our Reach: Ending the Mental Health Crisis* (with former First Lady Rosalynn Carter)

MITCH GOLANT, Ph.D.

Health psychologist and Senior VP of Research & Training for the Cancer Support Community (CSC)

"*Moon Tides* is a truly welcome book in the way it informs the world about the *haenyeo*—a unique and valuable cultural legacy of Korea, but one that is on the wane. Brenda Paik Sunoo's meticulous eye and prose delve deep to provide a vivid picture of the *haenyeo*'s lives, realizing an accomplishment as yet unachieved by anyone else. I hope that this book will allow many to share in this accomplishment and the profound emotions encountered therein."

SEO KANG-SOO

Director, Korean Culture and Information Service

"Brenda Paik Sunoo's *Moon Tides: Jeju Island Grannies of the Sea* offers an oblique meditation on aging through her extended encounter with the diving women of rocky, windy Jeju Island, off the southwestern coast of South Korea. Likely the last generation of Jeju women to seek their livelihoods from deep sea diving, most of Paik Sunoo's subjects are now in their sixties or older; their daughters have access to easier lives, and the ocean around Jeju no longer provides an abundant harvest, but some of these grannies still plunge into the sea, and others are only recently retired. These women speak of lives of poverty, hardship, and loss, but with a resilience that neither romanticizes the past nor asks for pity. Their words reveal the satisfactions of their own competence in the sea and their knowledge of having navigated difficult lives. Paik Sunoo's stunning photographs capture the strength and animation of her subjects, the joy they find in each other's company, and the profound beauty of these weathered faces. *Moon Tides* would make a lovely birthday gift for any active woman approaching granny status."

LAUREL KENDALL

Anthropologist and curator at the American Museum of Natural History in New York
Author of *Shamans, Nostalgias, and the IMF: South Korean Popular Religion in Motion*

"*Moon Tides* is a vivid exploration of the untold stories of the female sea divers of South Korea's Jeju Island. Brenda Paik Sunoo's powerful photography and first-person interviews bring a hidden world to life and connect these incredible women for the first time with a wide global audience."

CLARE WINTERTON

Executive Director of the San Francisco-based International Museum of Women

MOON TIDES

MOON TIDES

JEJU ISLAND GRANNIES OF THE SEA

Brenda Paik Sunoo

with Youngsook Han

For Jennie + Jenna
May you both live joyfully
+ remain in good health like
Jeju's grannies of the sea. I
enjoyed meeting Cody!

Best,
Brenda
Paik
Su

Seoul Selection

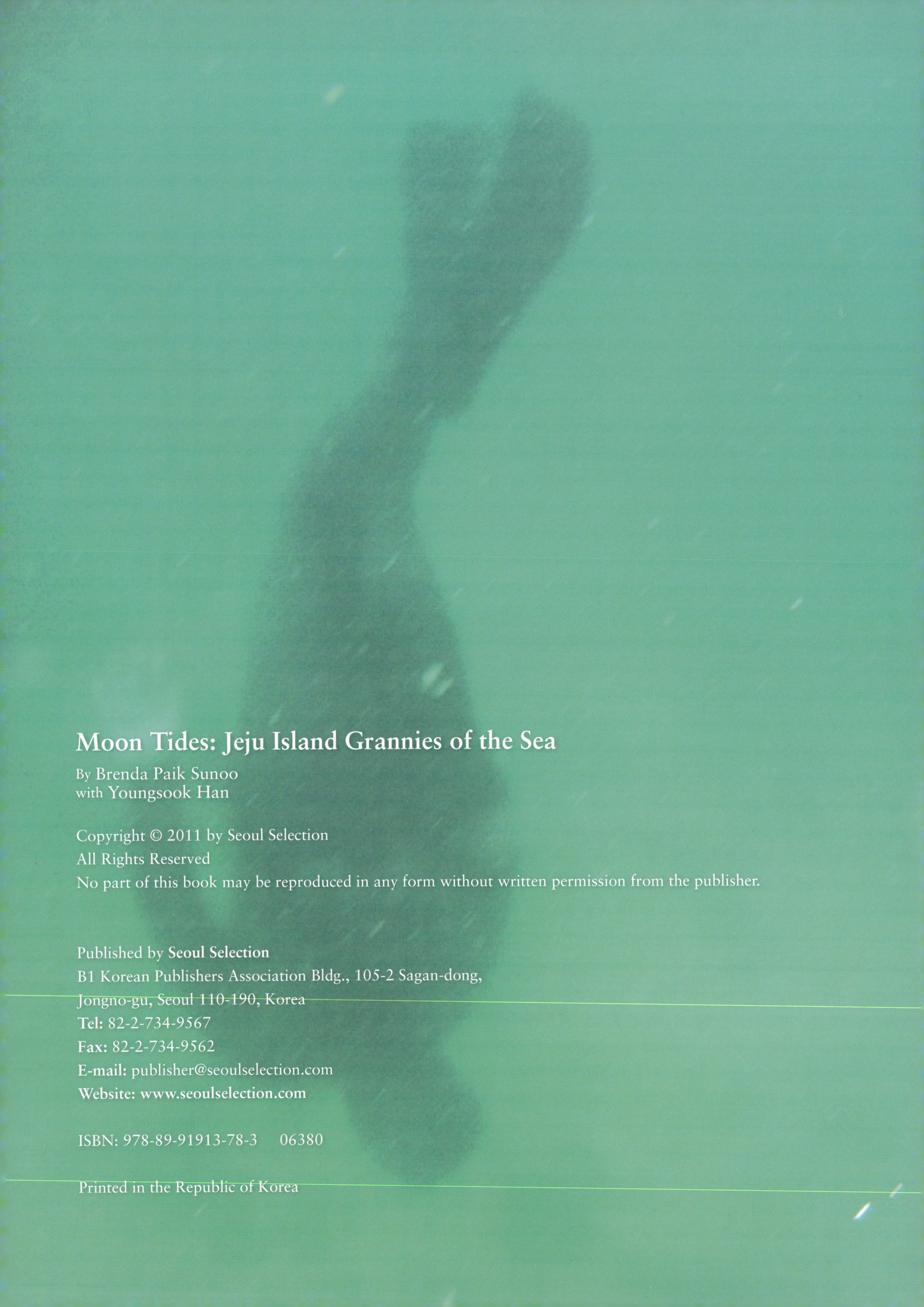

Moon Tides: Jeju Island Grannies of the Sea

By Brenda Paik Sunoo
with Youngsook Han

Published by **Seoul Selection**
B1 Korean Publishers Association Bldg., 105-2 Sagan-dong,
Jongno-gu, Seoul 110-190, Korea
Tel: 82-2-734-9567
Fax: 82-2-734-9562
E-mail: publisher@seoulselection.com
Website: www.seoulselection.com

ISBN: 978-89-91913-78-3 06380

Printed in the Republic of Korea

In honor of Song Kuang-do and Whang Ai-sung,
my grannies who crossed the seas to America in 1905 and 1913

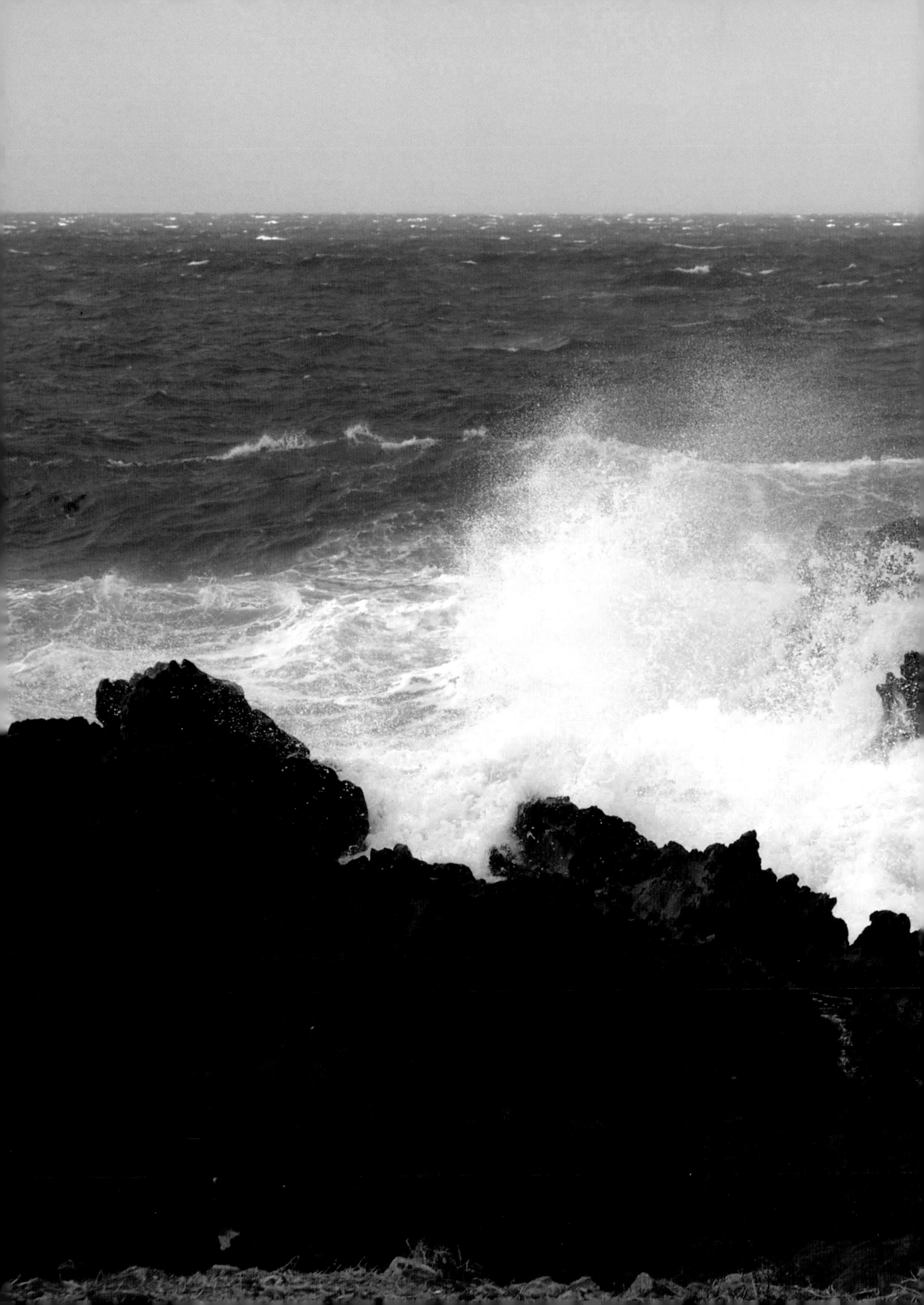

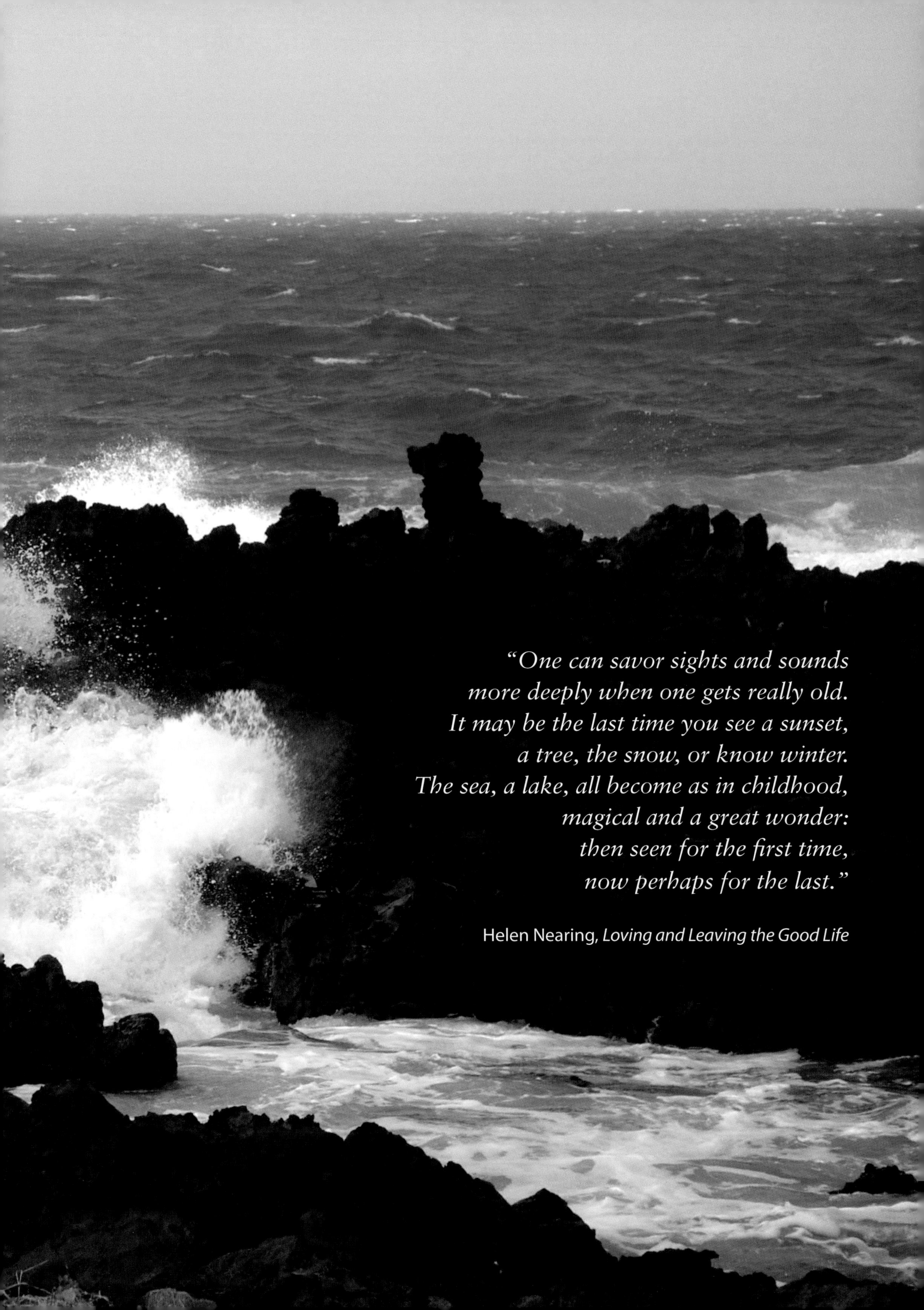

"One can savor sights and sounds
more deeply when one gets really old.
It may be the last time you see a sunset,
a tree, the snow, or know winter.
The sea, a lake, all become as in childhood,
magical and a great wonder:
then seen for the first time,
now perhaps for the last."

Helen Nearing, *Loving and Leaving the Good Life*

CONTENTS

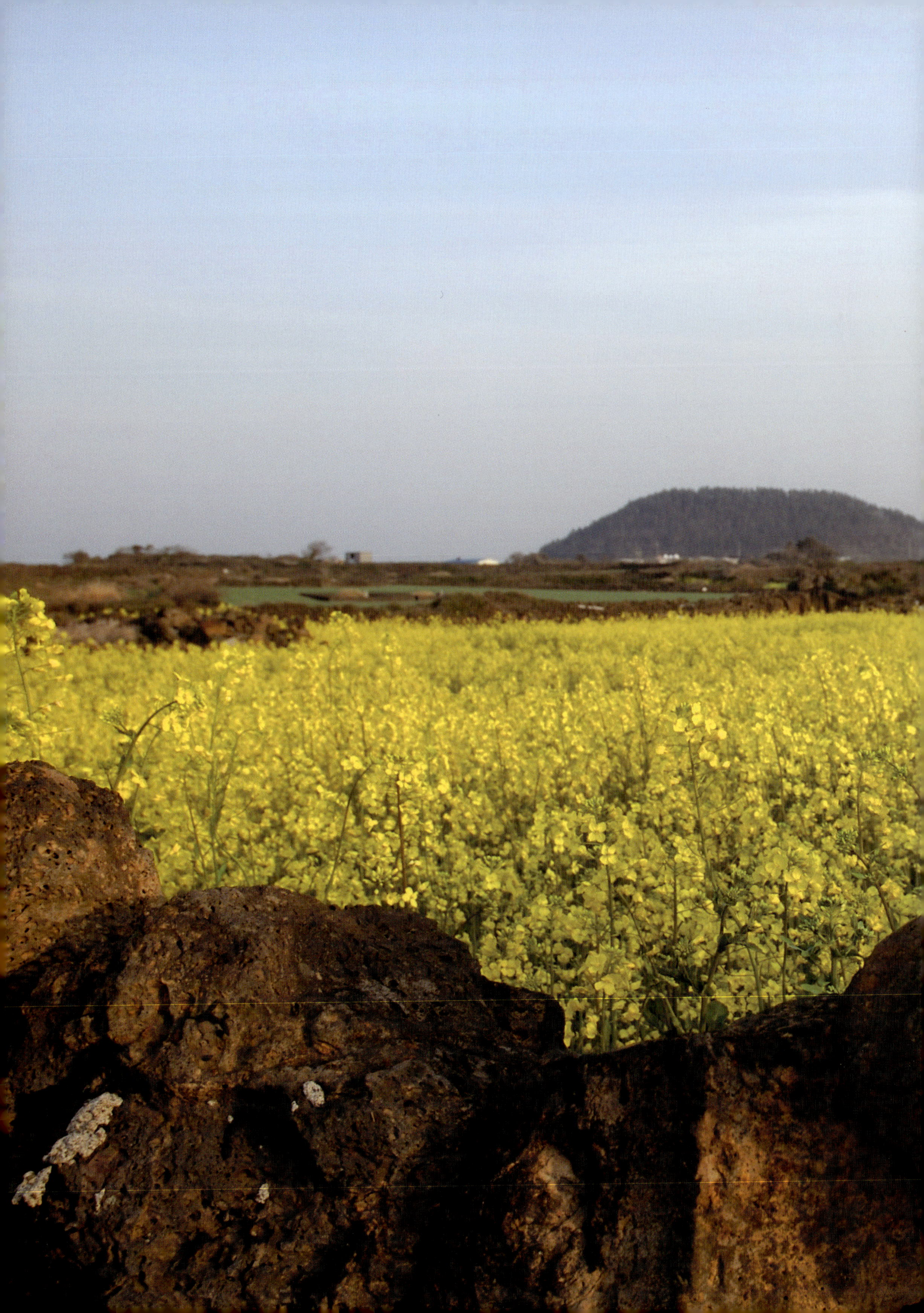

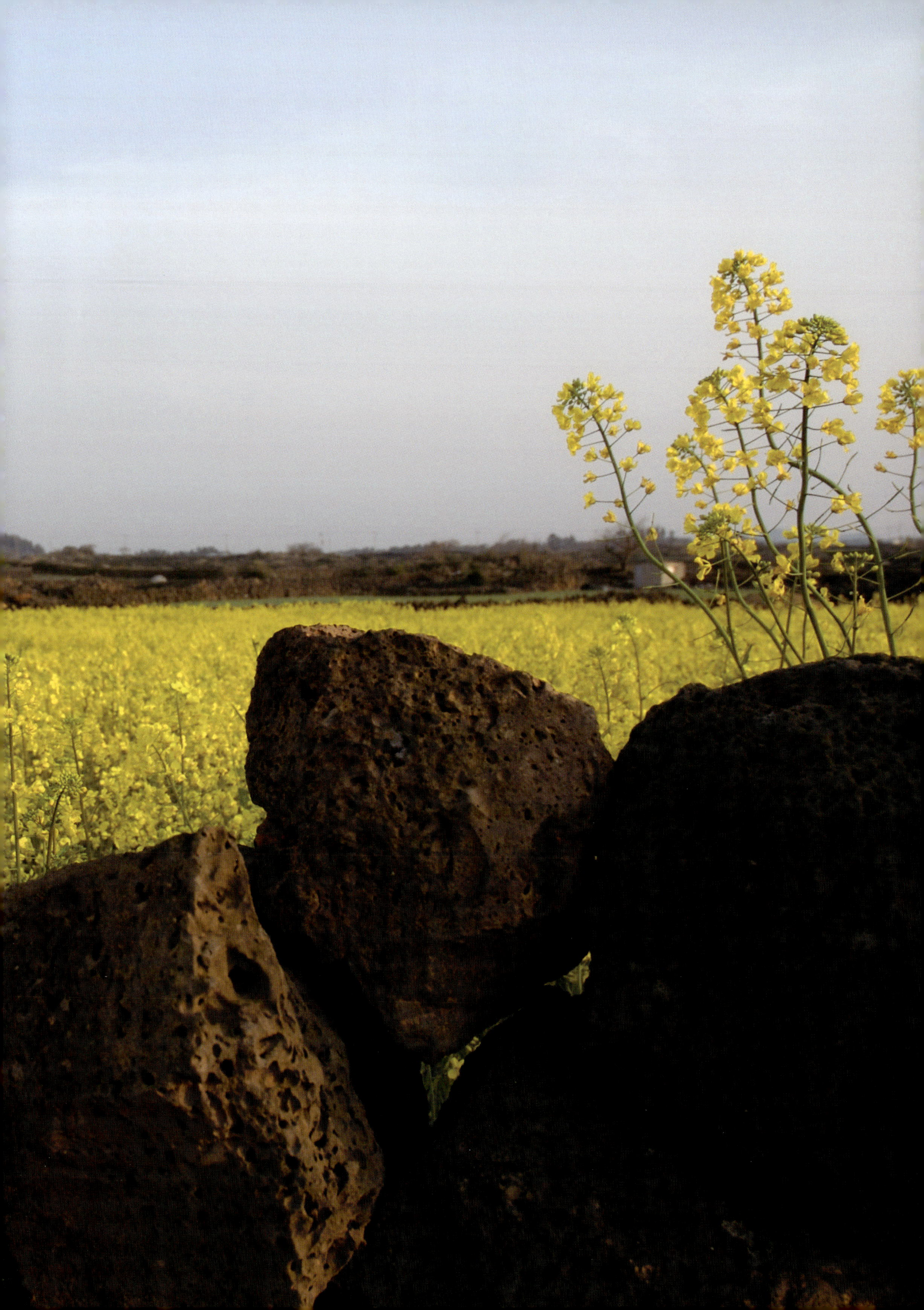

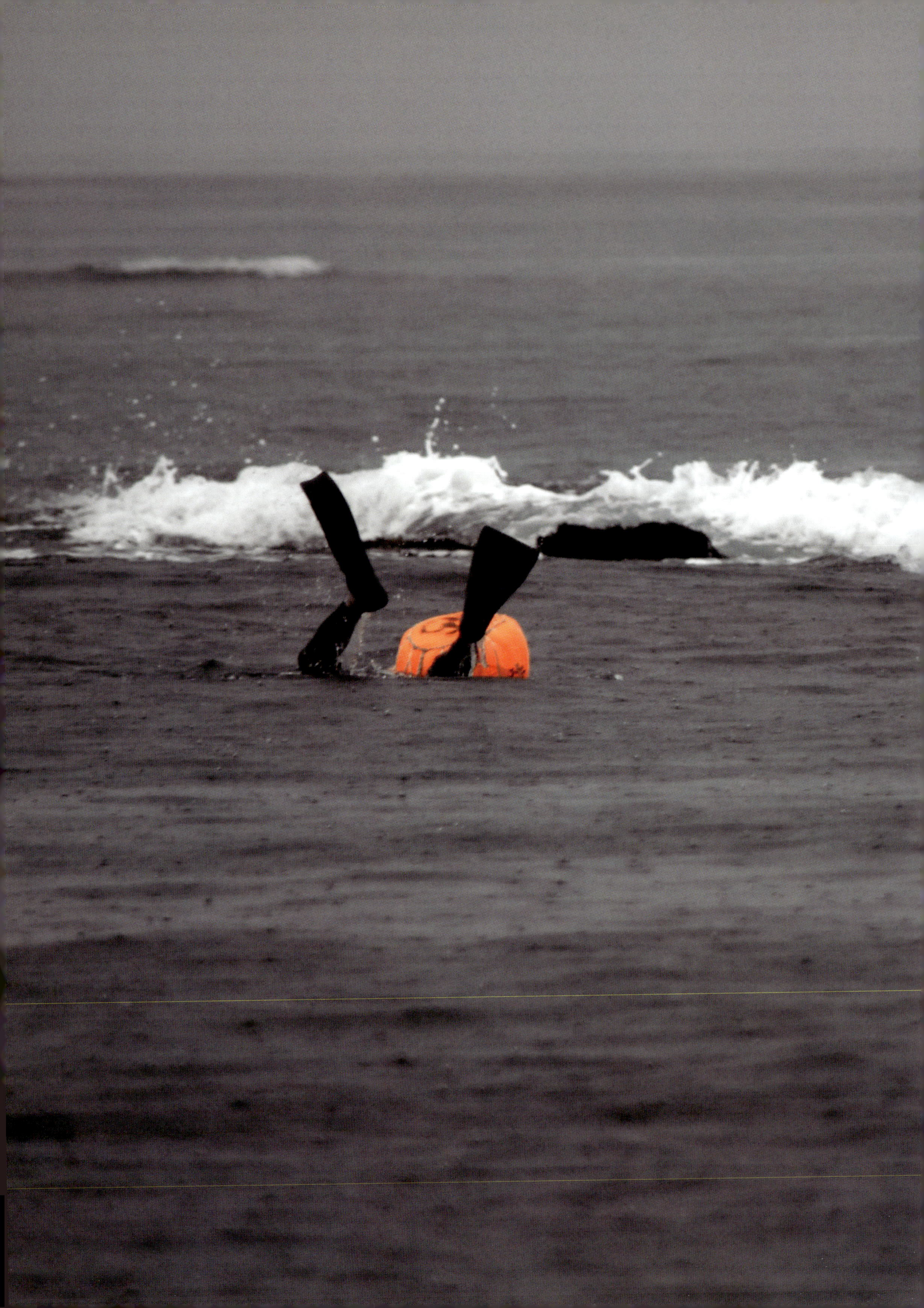

INTRODUCTION

I'm newly arrived to the shores of aging. And yes, a bit fearful of the unknown. After all, the eldest of 75 million American baby boomers began turning 60 in 2004.

In some ways, aging is like free diving. In the sea, one's mental and physical endurance is tested with a single breath. On land, one's endurance is tested within a single lifetime. Quite simply, there's no guaranteed safety net for either phenomenon. Aging and free diving both invite risks into mysterious and magical realms.

I was born on February 13, 1948. During my last few physical exams, I have repeatedly been warned to watch my cholesterol, supplement my diet with vitamin D, monitor uterine fibroids, irrigate my sinuses, and practice Pilates due to a herniated disc. It's no wonder that I've never dreamt of diving into the sea. That is, until I encountered the sea women of Jeju Island, Korea—known as *haenyeo*.

For centuries, these divers have faced the tempestuous tides of history and struggle for survival. Their intimate relationship to the land and sea, their shamanist beliefs, and their communal village life have protected them throughout their entire lives. In return, many have sustained a continuous life of purpose and resiliency well into their 90s. To a graying baby boomer like myself, their stories are gleaming beacons—illuminating a wise, practical and fearless course.

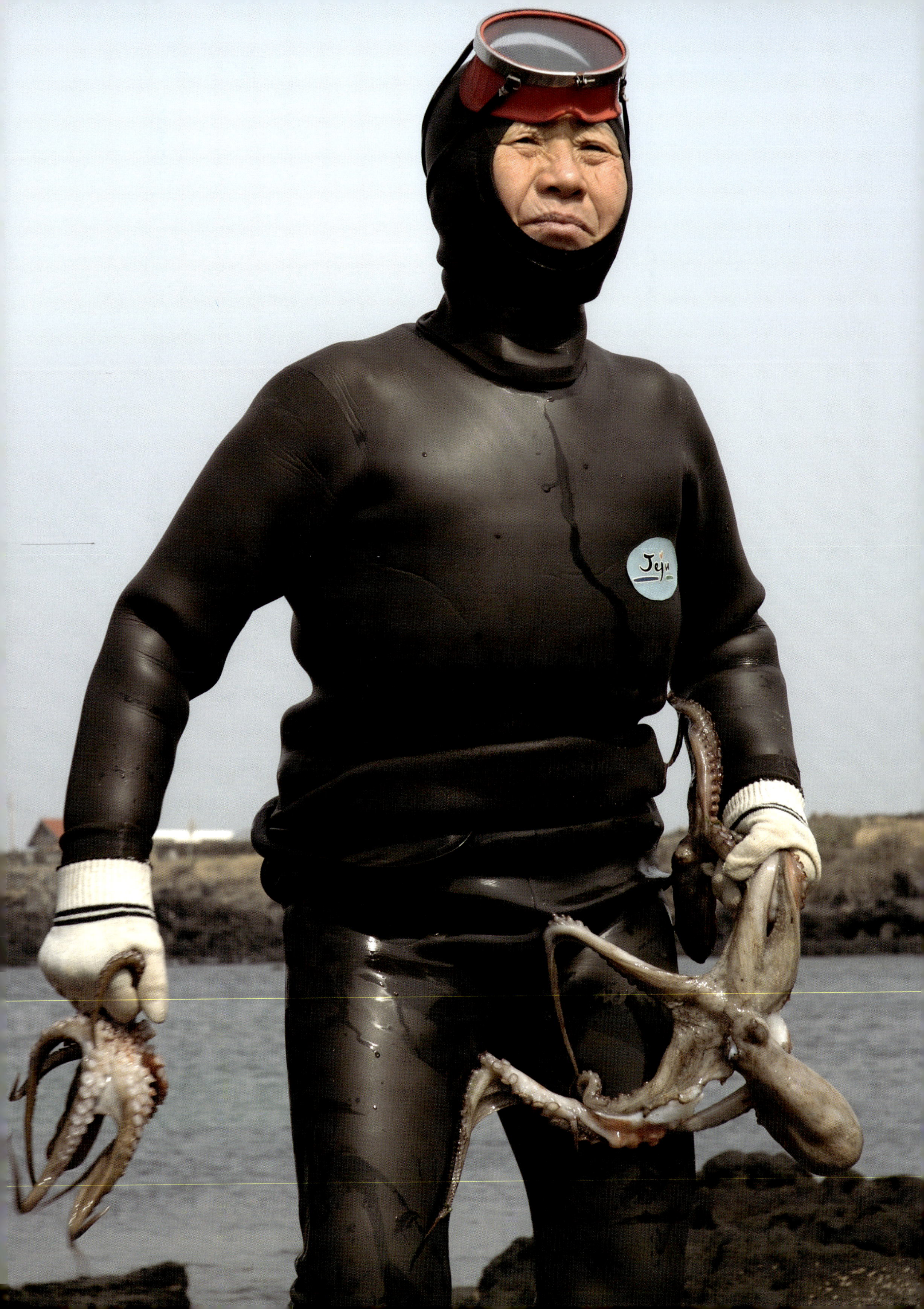
Jeju

I first visited Jeju Island in the 1980s, and learned that Jeju locals praise the island's "three abundances," or *samda*: the wind, black volcanic rocks, and women (mythical goddesses, shamans and divers). I came away learning very little about the "abundant" women during that first trip. What impressed me the most, though, was their uniqueness.

In the United States, women generally do not dive as a full-time occupation. The *ama* divers of Japan dive for pearls. But the *haenyeo* harvest seaweed and shellfish, such as the abalone, turban shell, conch, sea urchin, and sea slug. Diving and farming have always defined their livelihood. Although I was unaware of it at the time, the images of these robust women had carved a niche in my subconscious, popping up periodically whenever I thought about the sea.

By the time I revisited Jeju in the mid-90s, I had become a journalist. My instincts told me to pay closer attention to these women. I was told that during the 1970s there were an estimated 15,000 *haenyeo* working as divers. By 2002, there were only 5,600 female divers, and more than half were 60 years or older. In ten years, the total number of divers will likely be only half of that number.

Scattered throughout the more than 100 villages around the island, these women work year-round in groups organized

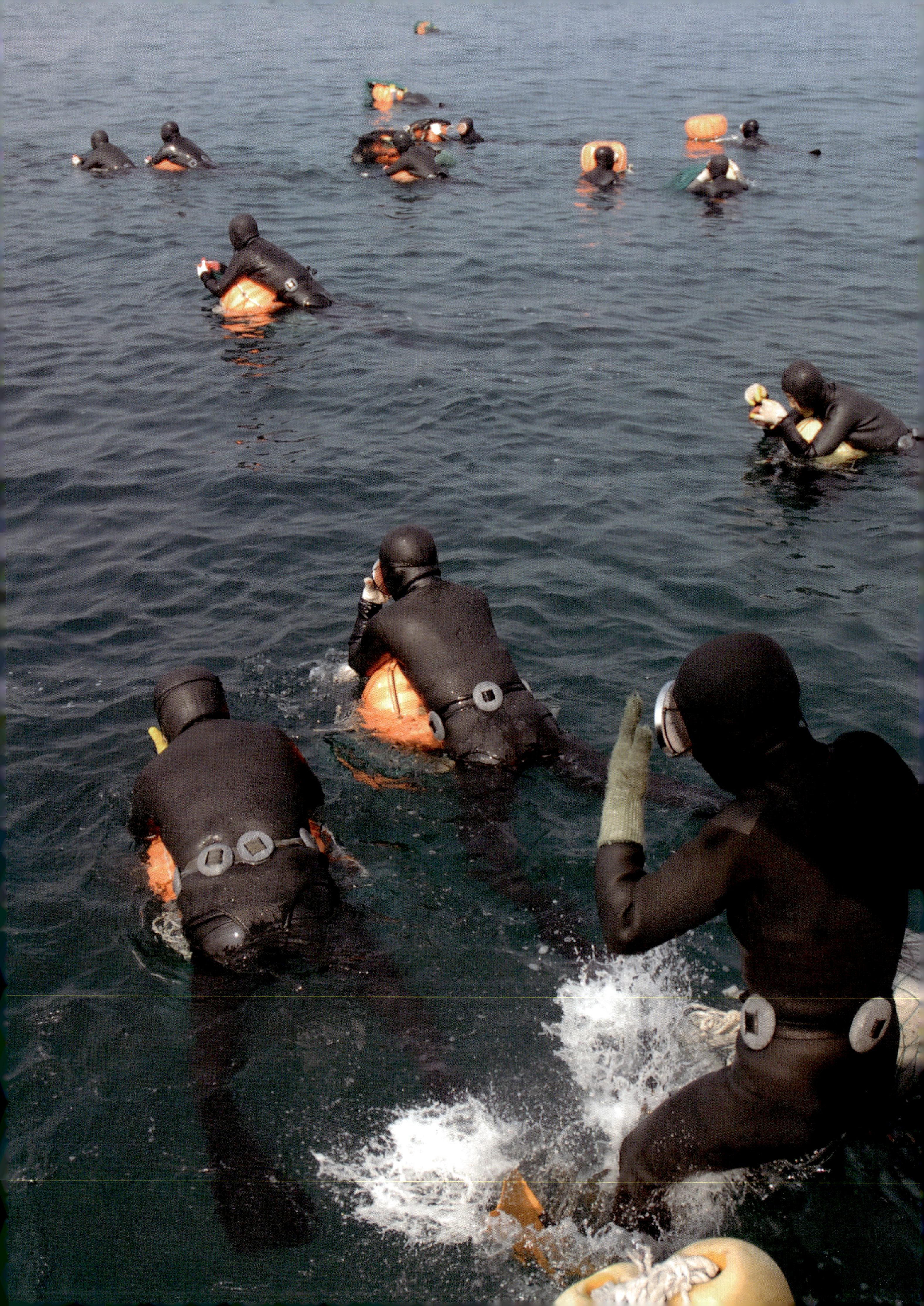

by their respective community diving associations, known as *eochongye*. Many are known to dive 65 feet under the sea while holding their breath for two minutes or longer. Most will tell you that they never took formal diving lessons. They taught themselves how to swim as children after following their mothers and grandmothers to the sea. The job is dangerous, especially if a diver's greed surpasses the "critical line"—when the concentration of carbon dioxide in the blood forces the breathing impulse before one safely reaches the surface.

Their work periods in the sea are determined by the lunar calendar. For the most part, it accurately predicts the timing and height of the tides. *Haenyeo* dive year-round (usually 15 days per month) during "neap" tides—when the vertical distance and gravitational pull between high and low tides (each occurring twice a day) is the least. This knowledge became quite pertinent, as it also determined when I could or could not photograph them or arrange appointments for interviews. Their schedule and activities were ruled by the moon, and therefore so were mine. Nothing could be guaranteed. As I was often told, "Perhaps we can meet. It depends on the tides."

For reasons of tradition, culture, comfort, and regulations, the *haenyeo* have never used scuba diving equipment. They dive in three-piece rubber suits, aided by a glass mask, fins, a floating

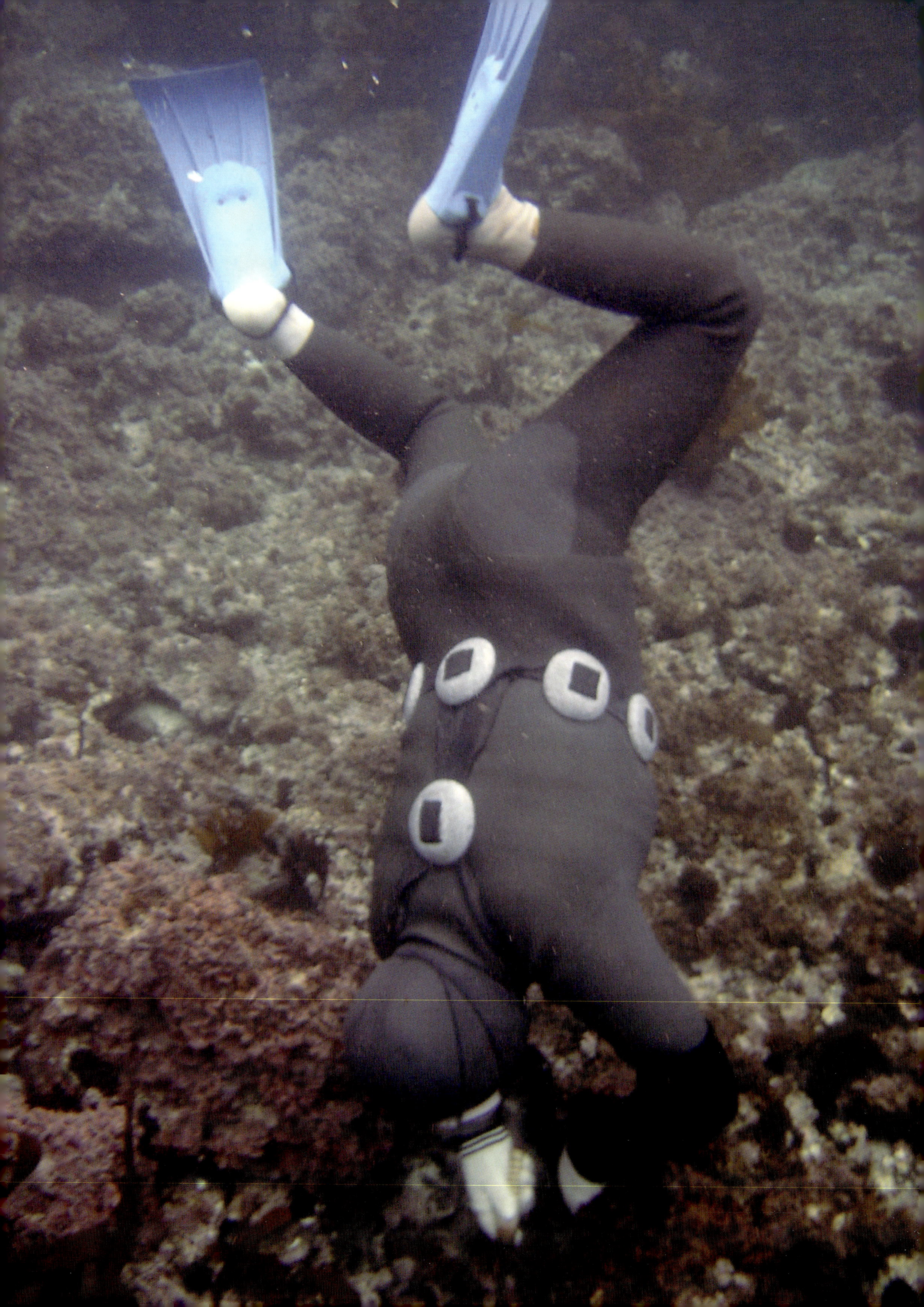

device with a net (called *taewak*), and metal implements for plucking abalone or cutting seaweed.

Depending upon skill and circumstances, women divers work along the shallow coasts, in deeper waters, and on the mainland or overseas as they did in the past—travelling by boat as far away as Japan, China, and Vladivostok in Russia. One *haenyeo* I met worked in Japan up until seven years ago. The most highly-skilled *haenyeo* can earn as much as $30,000 per year, which goes a long way in Jeju. But their earnings are diminishing, they say, as quickly as the sea products are being destroyed by environmental pollution.

People often ask why Jeju women have dominated the diving profession since the 16th century. For one thing, women are believed to withstand high water pressure better than men. Also, with more body fat, women can endure cold water for longer periods.

There are other historical reasons, too. Once considered an abandoned island, Jeju was used as a banishment destination for male scholars who fell out of political favor with the king during the Joseon Dynasty (1392–1910). The men were regarded for their cultural literacy, not proletarian aptitude. Subsequent tides of history—Japanese occupation from 1910 to 1945; the April 3 Massacre (1948); the Korean War (1950–

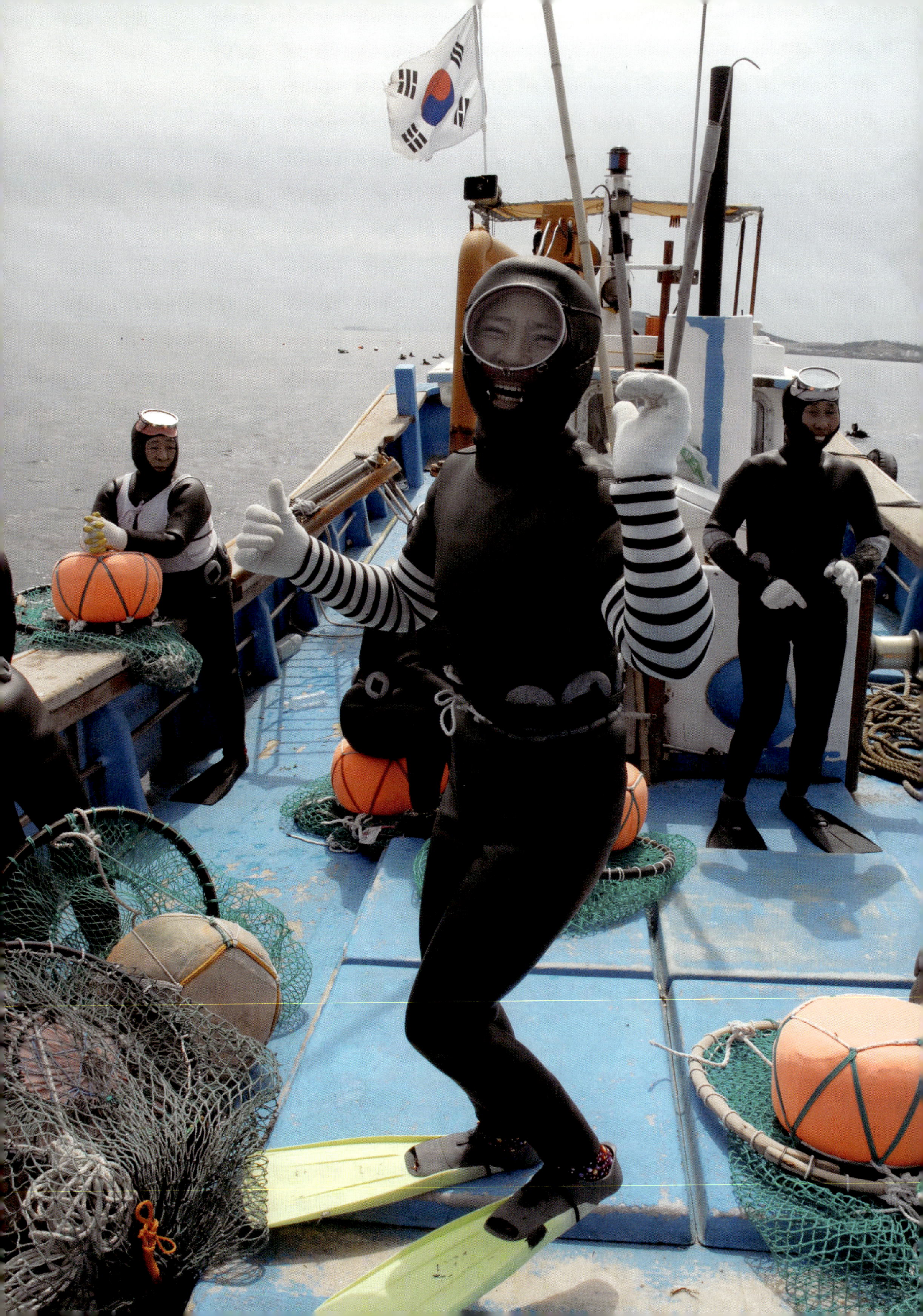

53), and other events—greatly reduced the population of male citizens, casting the women as primary economic providers. Under the Confucian order, however, they were still exploited and perceived as low-level workers, although engaged in non-traditional jobs.

One cannot move around Jeju Island without sensing the women's presence everywhere. All along the coast, you'll see large stone statues that represent the divers carrying their floating devices and filled nets. When you hear whistling (*sumbisori*) emanating from the sea, it is the sound of *haenyeo* rising to the surface to release carbon dioxide from their lungs. If ordering *jeonbokjuk* (abalone rice gruel), a local specialty, chances are the restaurant is owned and operated by local *haenyeo*. When you spot a granny whizzing by on her three-wheeled motorcycle, she, too, is likely a *haenyeo*. On warmer days, the women assemble together at seaside pavilions (*jeongja*) to abandon their loneliness. Many of them are retired divers and widows. Sharing their woes, they seek comfort from familiar villagers.

I felt a strong pull to document these women's stories before they died. By 2002, my husband Jan and I had experienced a shift in our personal and professional circumstances. We were eager to re-engage life after eight years of surviving the

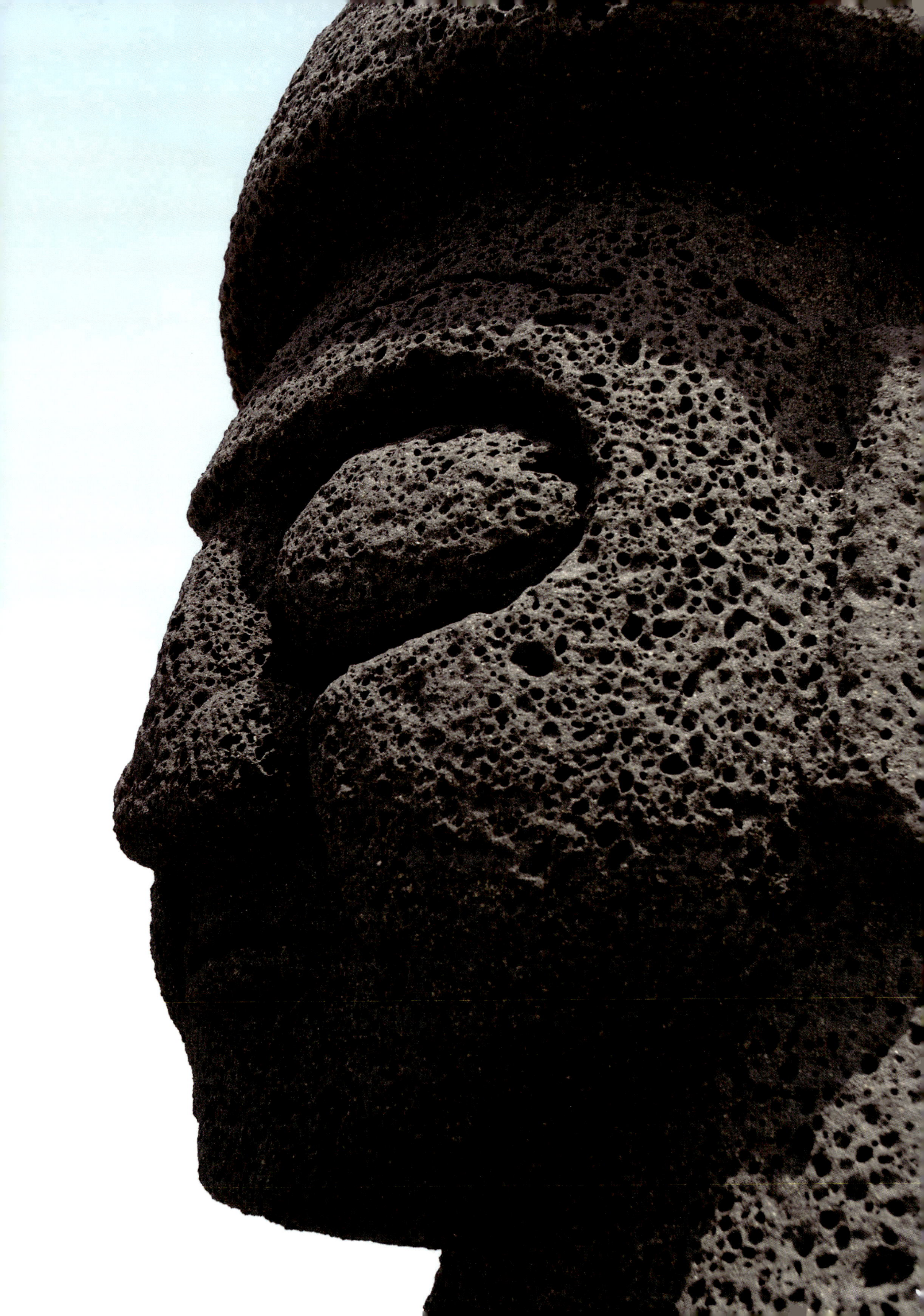

sudden death of our younger son in 1994. Jan accepted a three-year overseas assignment in Vietnam as an industrial relations specialist for a United Nations agency. When the project ended in 2005, he decided to continue his project for three more years in Southeast Asia. Finally, the opportune moment had arrived to turn my vision into reality. As it turned out, I was able to arrange three stays in Jeju (2007, 2008, and 2009) for longer periods, not as a tourist but as a sojourner lured by the whistles blowing from the sea.

I lived in Gimnyeong-ri on the northern side of the island in 2007 and 2008. This village of 3,000 residents is known to be 2,000 to 3,000 years old. The whole island, locals believe, is inhabited by 18,000 gods and goddesses. Its myths are integral to Jeju culture: Seolmundae, goddess of creation, created Jeju Island, including Mt. Halla and over 360 *oreum*, or secondary volcanic cones. Samseung is in charge of pregnancy and birth, and Gusamseung is in charge of death. Jacheongbi, or Sekyeong, governs agriculture. Madam Yeosan governs kitchens, and the daughter of Noiljeodaegwiil manages bathrooms! One of the most revered goddesses is Yeongdeung, goddess of wind and the sea, who watches over the safety of fishermen and seaweed.

Special shamanist ceremonies are conducted every year on the

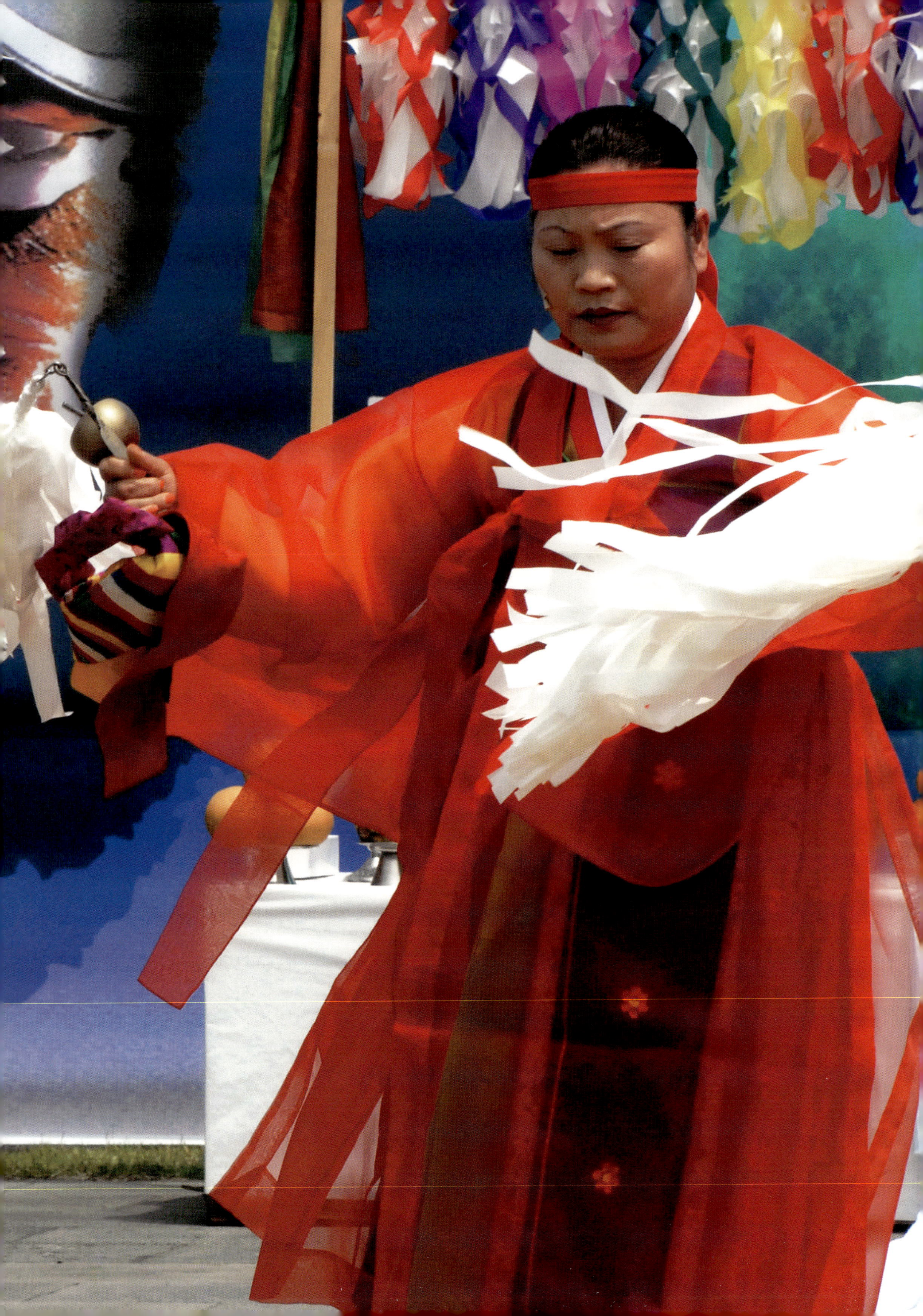

second day of the first lunar month to welcome Yeongdeung. Two weeks later, after she has scattered five grains on land and turban shells and seaweed in the ocean, she is sent back to the sea. Shamans conduct an elaborate ritual that bids farewell to the goddess until she is welcomed back the following year.

During the course of my three stays, I was very fortunate to meet numerous *haenyeo* and other Jeju locals through personal referrals and serendipity. I was continually reminded how curiosity and persistence—and a genuine effort to speak Korean—could establish initial rapport. Still, I could not have interviewed and learned as much as I did without the professional support and friendship of so many who are cited and thanked in my acknowledgements.

For most of my field research, I was always accompanied by Prof. Youngsook Han, my translator and interpreter. One day, while viewing exhibits at the Haenyeo Museum in Hado-ri, Youngsook remarked: "I had almost forgotten that I was raised among the *haenyeo* in my mother's village. I sat with them at the *bulteok* (fire pit). I even remember playing with their tools." She also said that a *haenyeo* midwife had delivered her at birth. As I watched her skillful swimming in the sea one time, I teased her: "You can take the girl out of the sea, but you can't take the sea out of the girl."

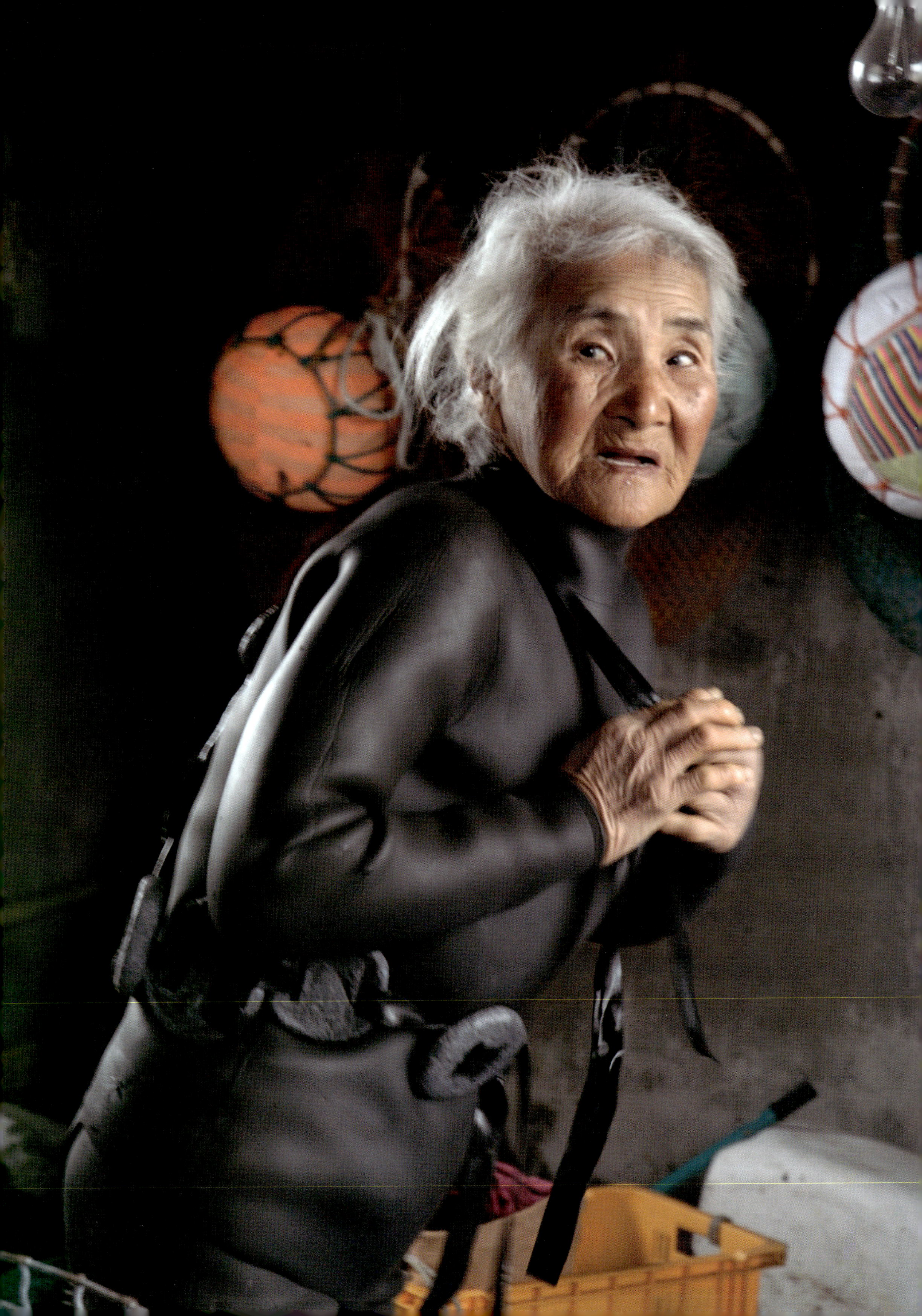

By the end of my second visit in 2008, something else significant occurred in my life: I turned 60. My husband and I were still living in Vietnam, where more than 60 percent of the population was born after 1975. That put us in the category of elders among a nation of youth. Even though I've always felt young at heart, one's body can't defy the gravity of aging. Or the psychological vulnerability attached to it. It dawned on me that I also was emotionally drawn to Jeju and the *haenyeo* because they reminded me of my maternal *halmeoni*, Whang Ai-sung. My grandmother, who was born in north Pyongyang on August 4, 1890, and died in Los Angeles in 1967, lived with my family during most of my childhood.

I missed the presence of Korean grannies like her. The *haenyeo*'s wrinkled faces, thick knuckles, voluptuous breasts, gray hair, and robust personalities all reminded me of her. Even their temper and humor rang with familiarity. Had my grandmother been born in Jeju, I believe she, too, may have become a *haenyeo*. She possessed the same survival-driven ethos. As it was, my grandparents and earlier Korean-American immigrants became farmers, barbers, small grocers, tailors, and restaurant owners.

When I compare the circumstances of the *haenyeo* and of my grandmother during the same time period, I realize both

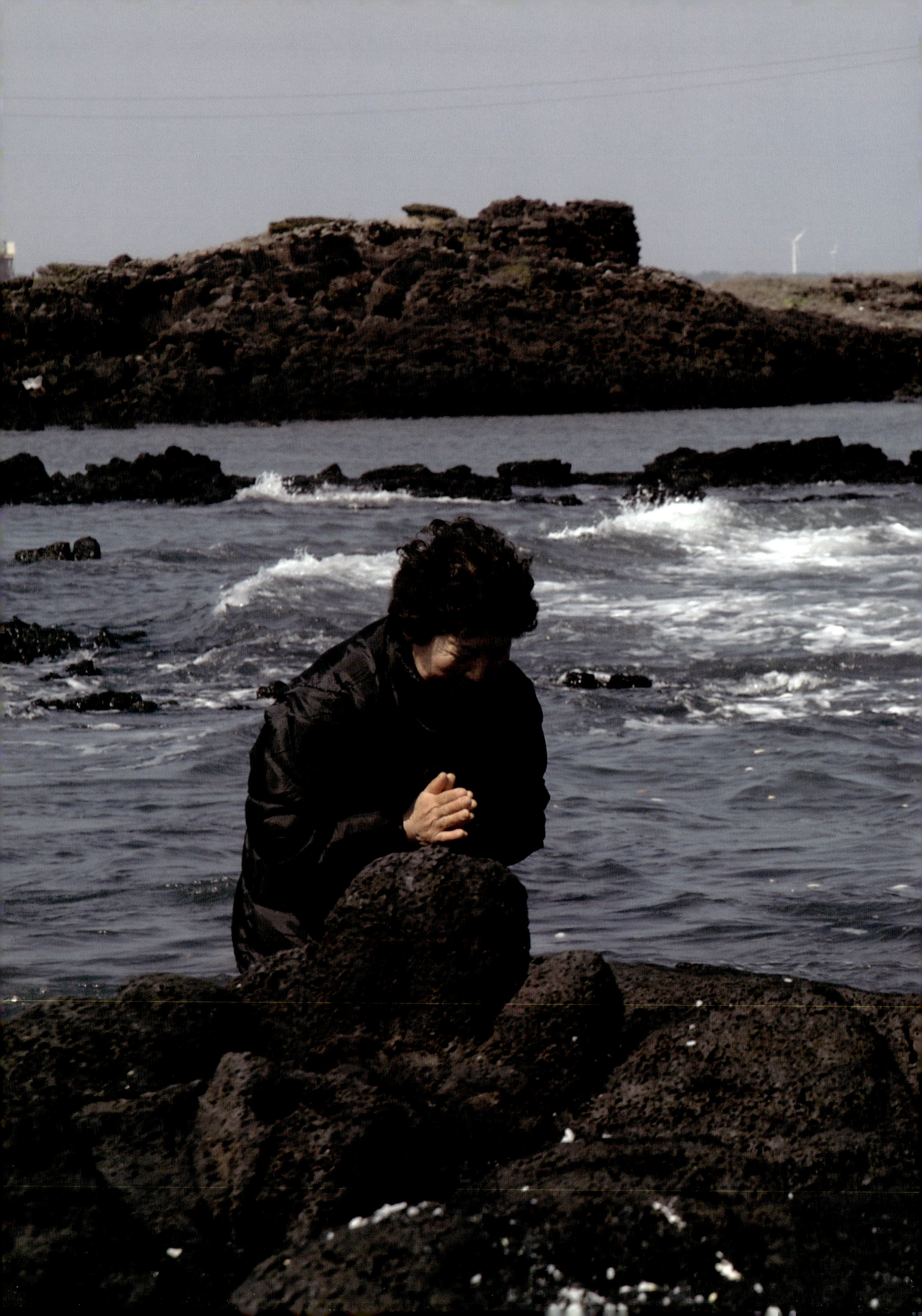

contexts were informed by patriotism. My grandmother—and many of her Korean-American peers and fellow Christians—also devoted much of their time, energy, and money to the cause of Korea independence during Japanese occupation. Through their community associations and churches, they would raise money to support the overseas patriots—some of whom lived in exile in Manchuria. One elderly *haenyeo* surprised me with her understanding of Korean-American history: "Aaah yes, your grandparents suffered a lot in those days. We know their immigrant stories." Although an ocean apart, those who departed were still remembered by those who remained—even as far away as Jeju.

Moon Tides: Jeju Island Grannies of the Sea is my homage to the *haenyeo*. I have organized the book according to seven aspects of their multi-dimensional lives. Each one reveals how the *haenyeo* have been "lifted by the wind and tides" of: Survival, Shamans, Suffering, Aging, Compassion, Family, and the Future. Through photos and translated interviews, my intention has been to represent the women (ages 39–93) in their own voices and work/life environments—not only in their rubber suits. They are, after all, grannies, wives, mothers, sisters, community volunteers, patriots, and social activists. And to each other, lifelong girlfriends. Through their examples,

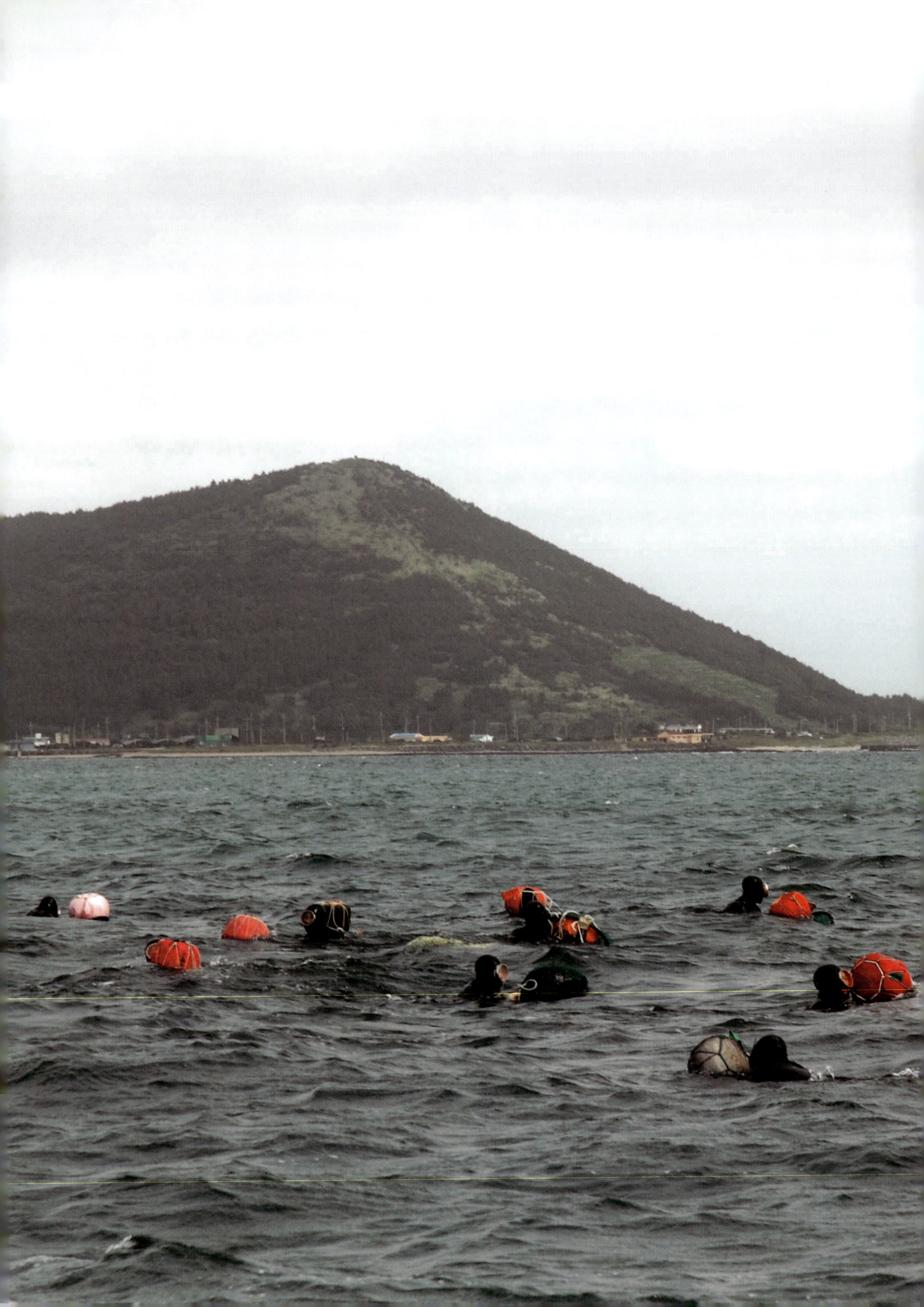

I believe that one's life purpose can be as continuous and infinite as the sea. For centuries, these Jeju divers have kept their lives afloat—in spite of risks and danger. As one granny told me, "If we didn't work, we couldn't survive."

Having been blessed to drift among them for a while, I feel more comfortable with aging. Thankfully, one is never too old to be inspired by others. I have even begun to imagine my life as a septuagenarian and octogenarian. Granted, I will never become a working woman of the sea like the *haenyeo*. But if I do live well into my 90s, I hope to lead a continuous life of purpose as they have. Gray-haired, wrinkled, and polyp-free.

Brenda Paik Sunoo

2011

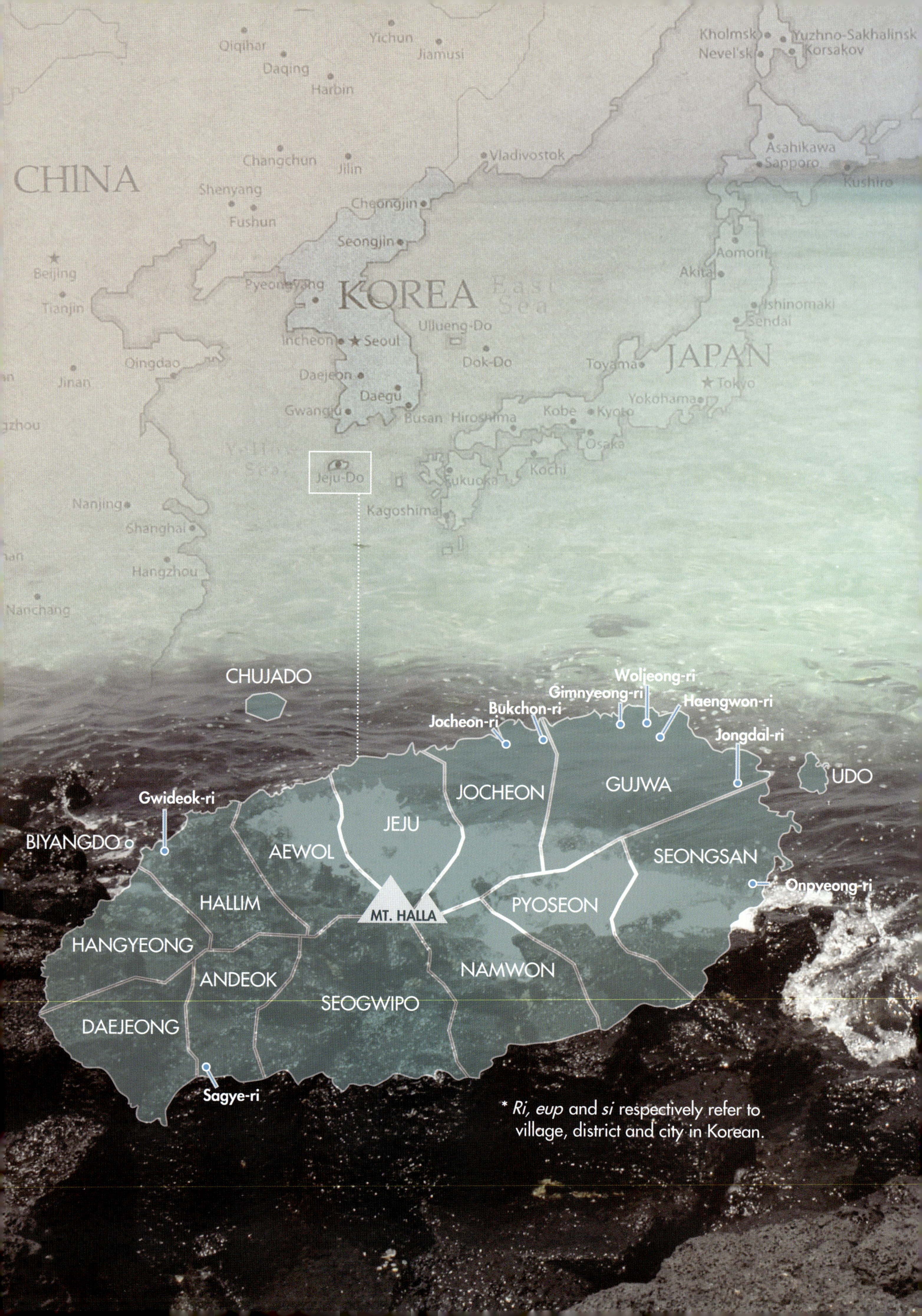

* *Ri*, *eup* and *si* respectively refer to village, district and city in Korean.

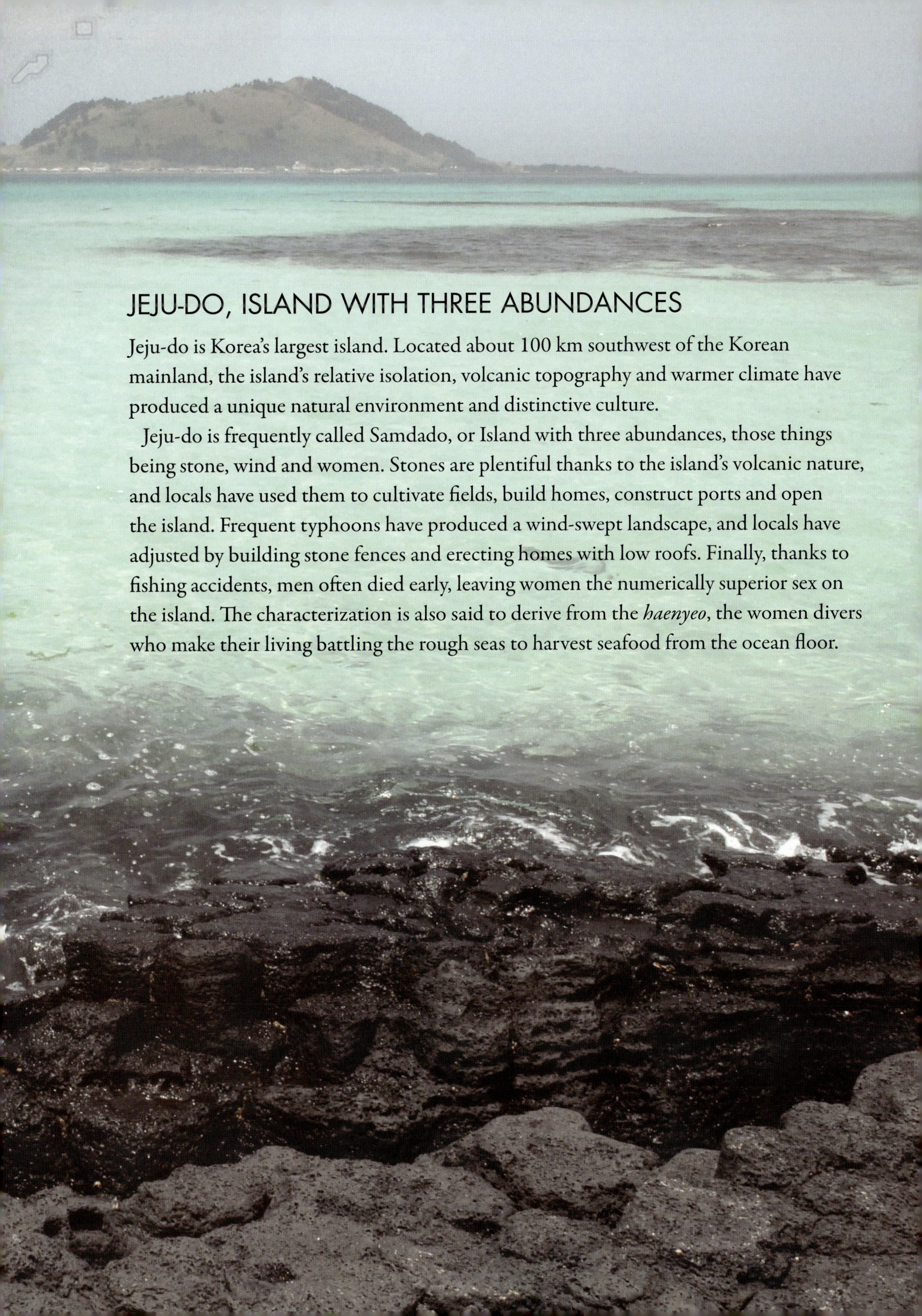

JEJU-DO, ISLAND WITH THREE ABUNDANCES

Jeju-do is Korea's largest island. Located about 100 km southwest of the Korean mainland, the island's relative isolation, volcanic topography and warmer climate have produced a unique natural environment and distinctive culture.

Jeju-do is frequently called Samdado, or Island with three abundances, those things being stone, wind and women. Stones are plentiful thanks to the island's volcanic nature, and locals have used them to cultivate fields, build homes, construct ports and open the island. Frequent typhoons have produced a wind-swept landscape, and locals have adjusted by building stone fences and erecting homes with low roofs. Finally, thanks to fishing accidents, men often died early, leaving women the numerically superior sex on the island. The characterization is also said to derive from the *haenyeo*, the women divers who make their living battling the rough seas to harvest seafood from the ocean floor.

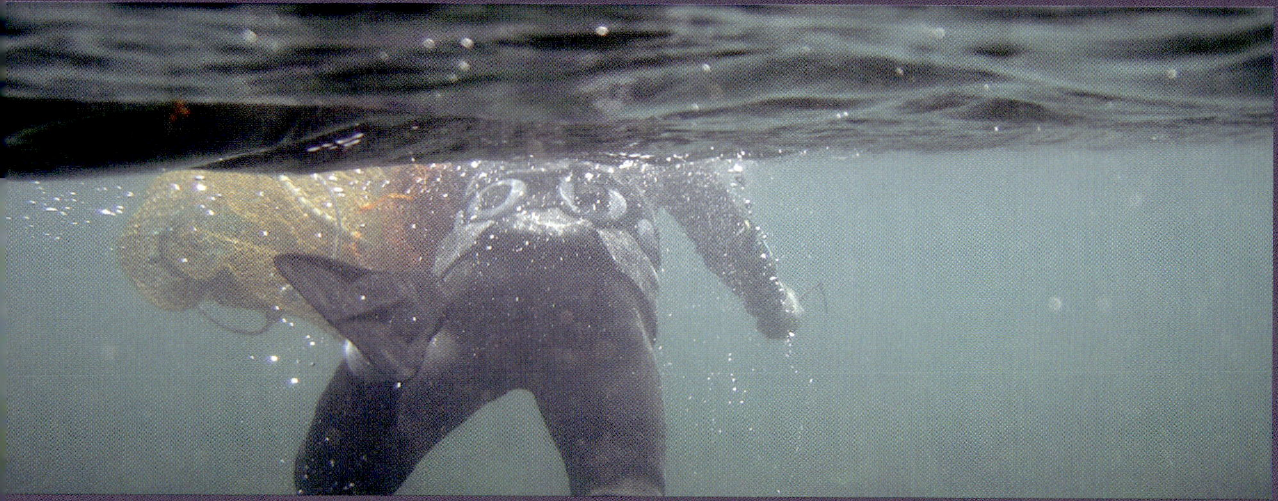

Chapter One

LIFTED BY THE WIND & TIDES OF SURVIVAL

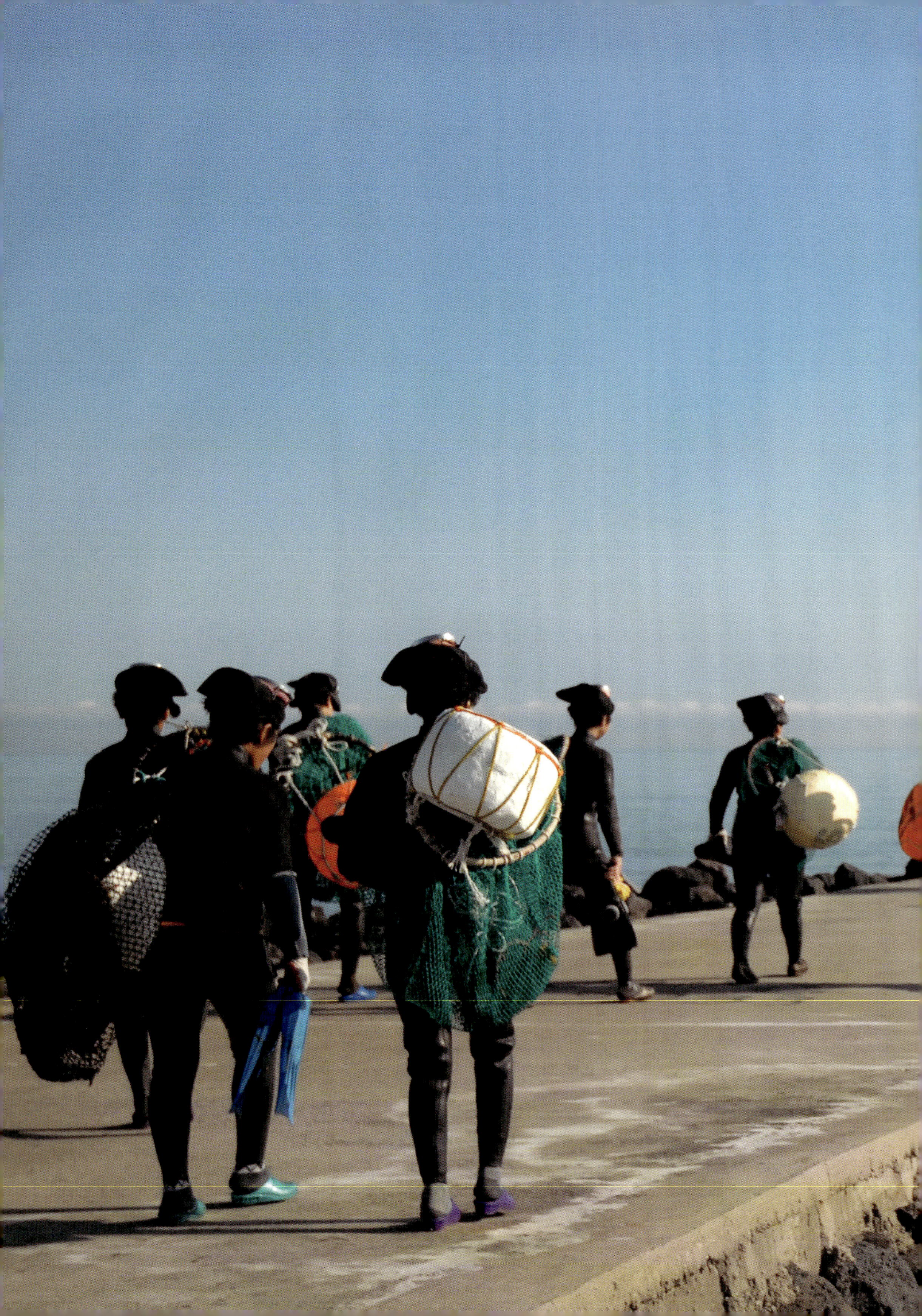

Over the past decade, the *haenyeo* population has decreased from 5,789 in 2000 to an estimated population of 2,500. Many believe this may be the last generation of Jeju Island's woman divers. Aside from the *ama* divers of Japan, there is no other place on earth where women free dive for seaweed and shellfish. Ask them if they want their daughters to carry on the tradition, and most will say that the work is much too dangerous. They prefer that their daughters find jobs on the mainland or jobs that don't require them to endanger their lives.

Given the pressures of economic survival and political circumstances, most *haenyeo* I met felt they had no choice but to follow the job, even if far away from home. They have always regarded the sea as a working field. Few, if any, ever swam in the water for simple pleasure. If they weren't scouring the sea for seaweed, they were plowing the ground for tangerines, onions, garlic, cabbage and flowers. Although the women worked both on land and sea, the latter working field has taken the greatest toll on their physical wellbeing.

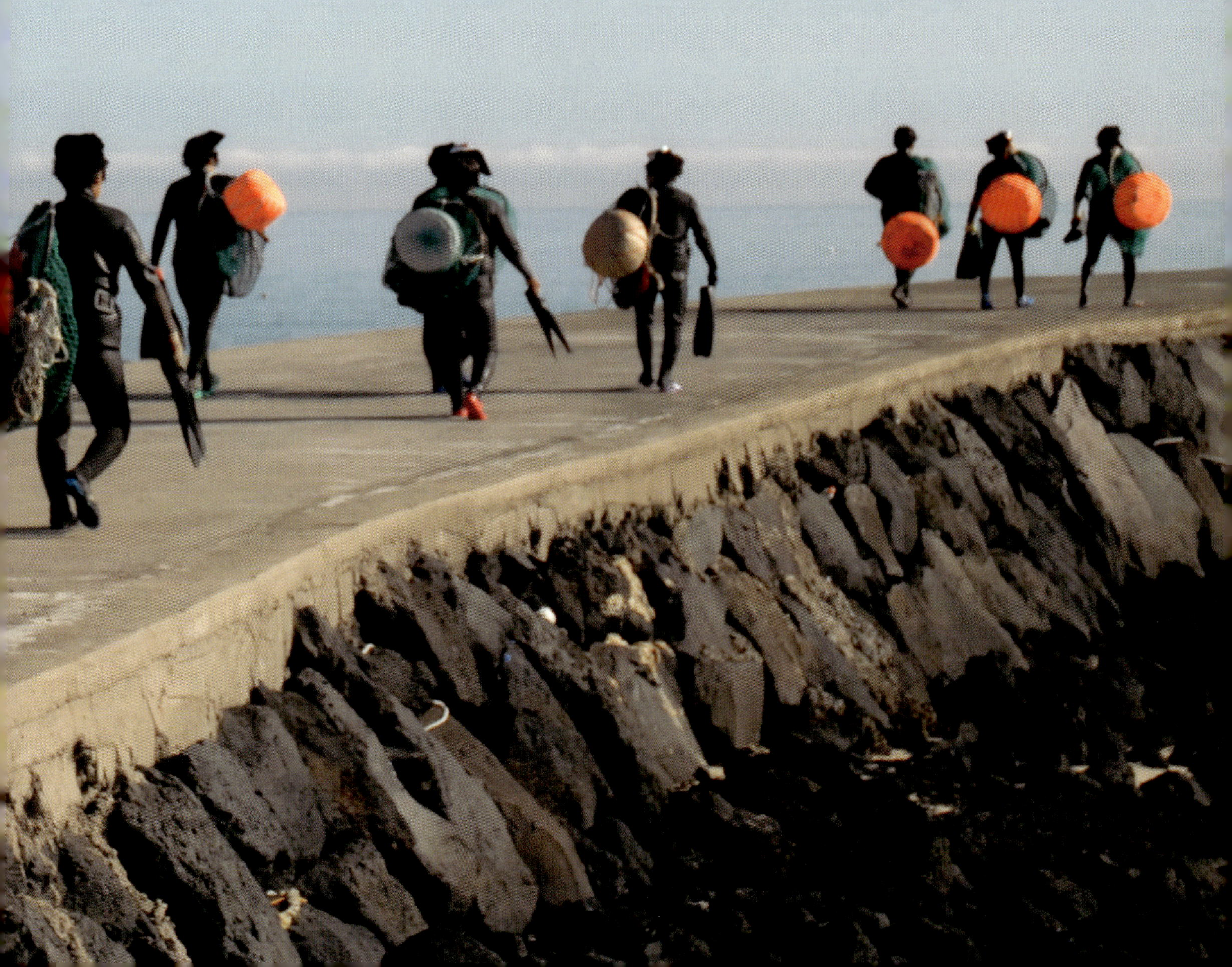

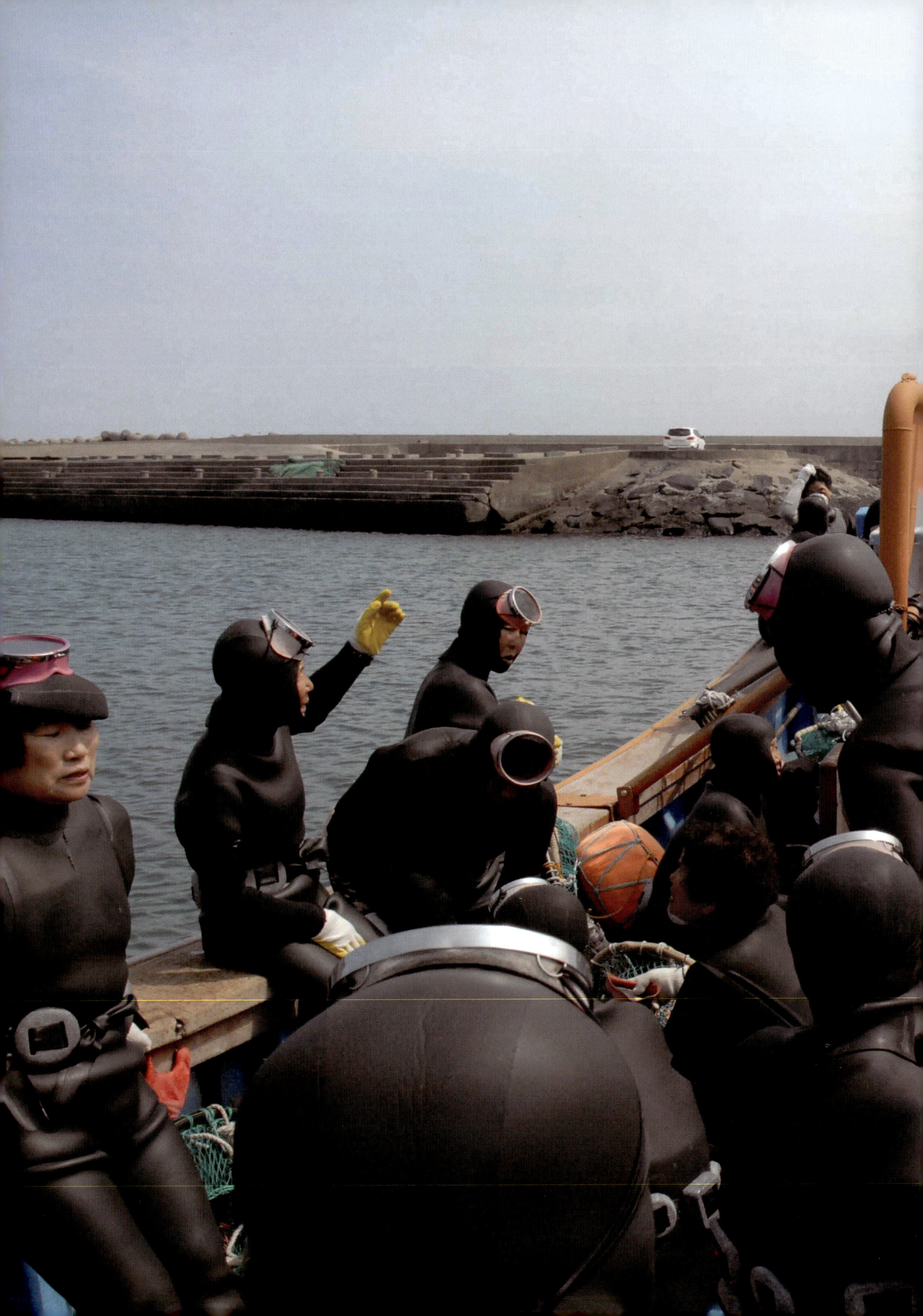

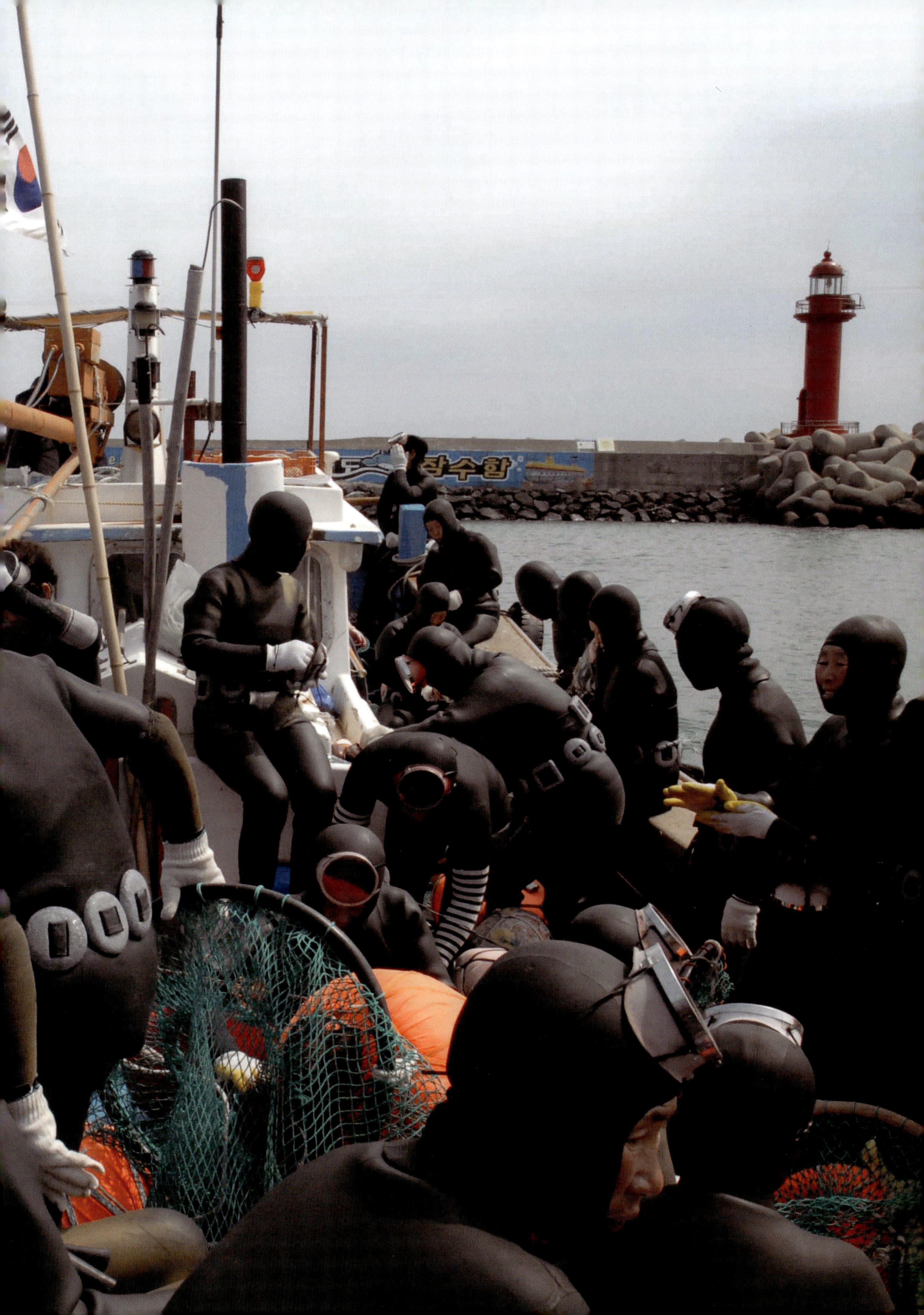

" I've never been afraid of the water, even in the deep ocean. I can dive 10 to 15 meters deep, and remain underwater for about two minutes. My skills have improved over the years because I know all about the sea—where and what I can catch."

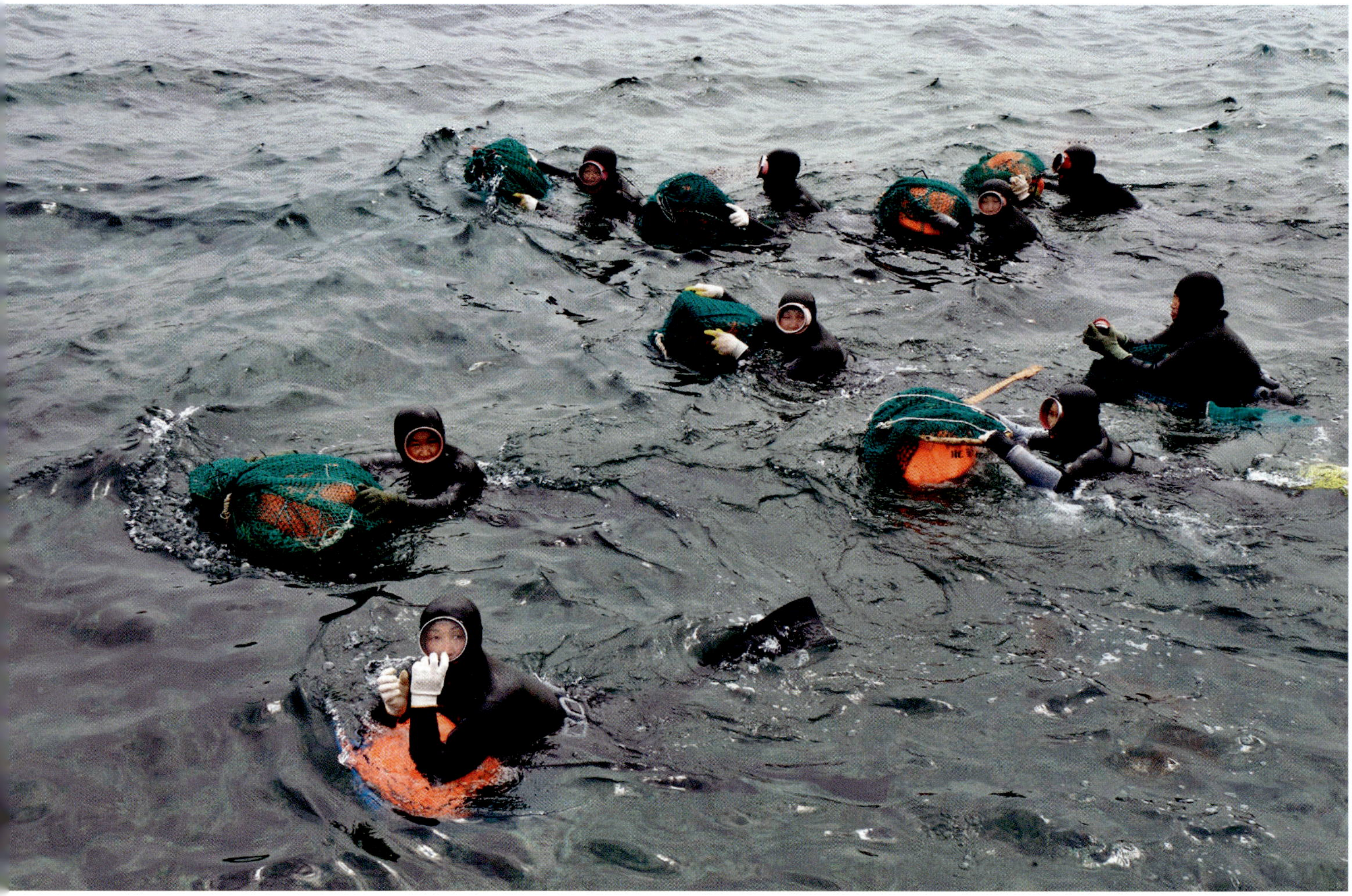

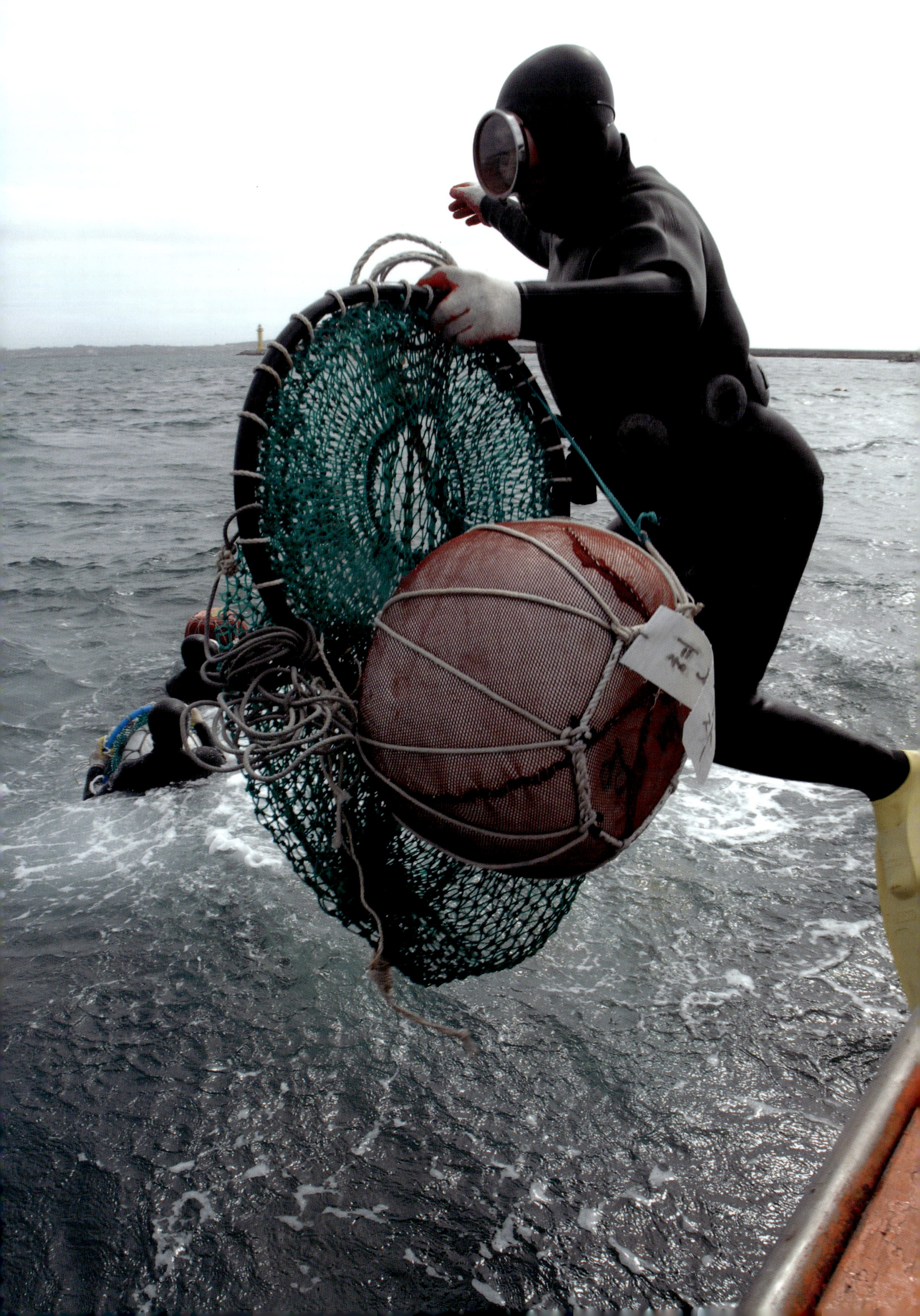

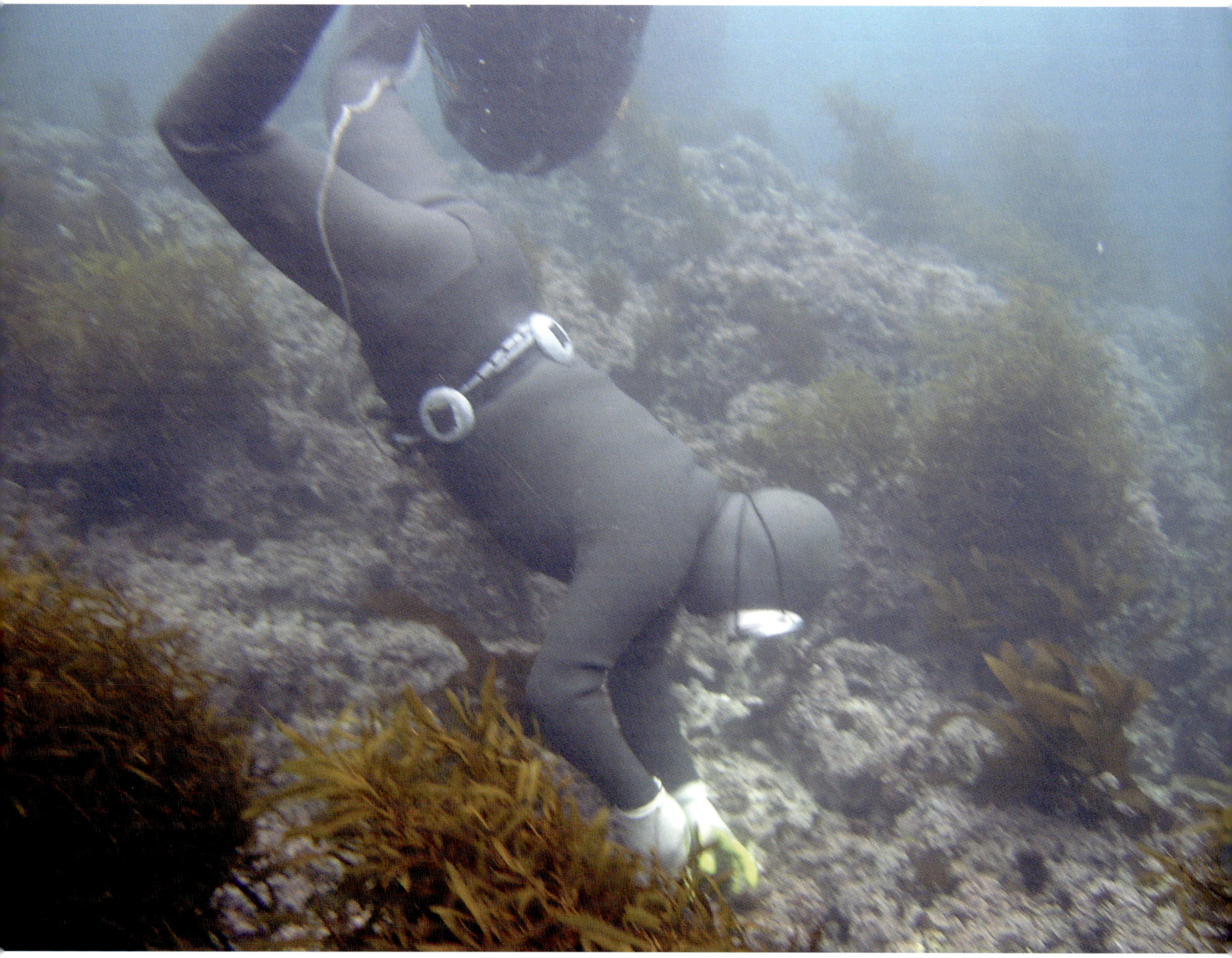

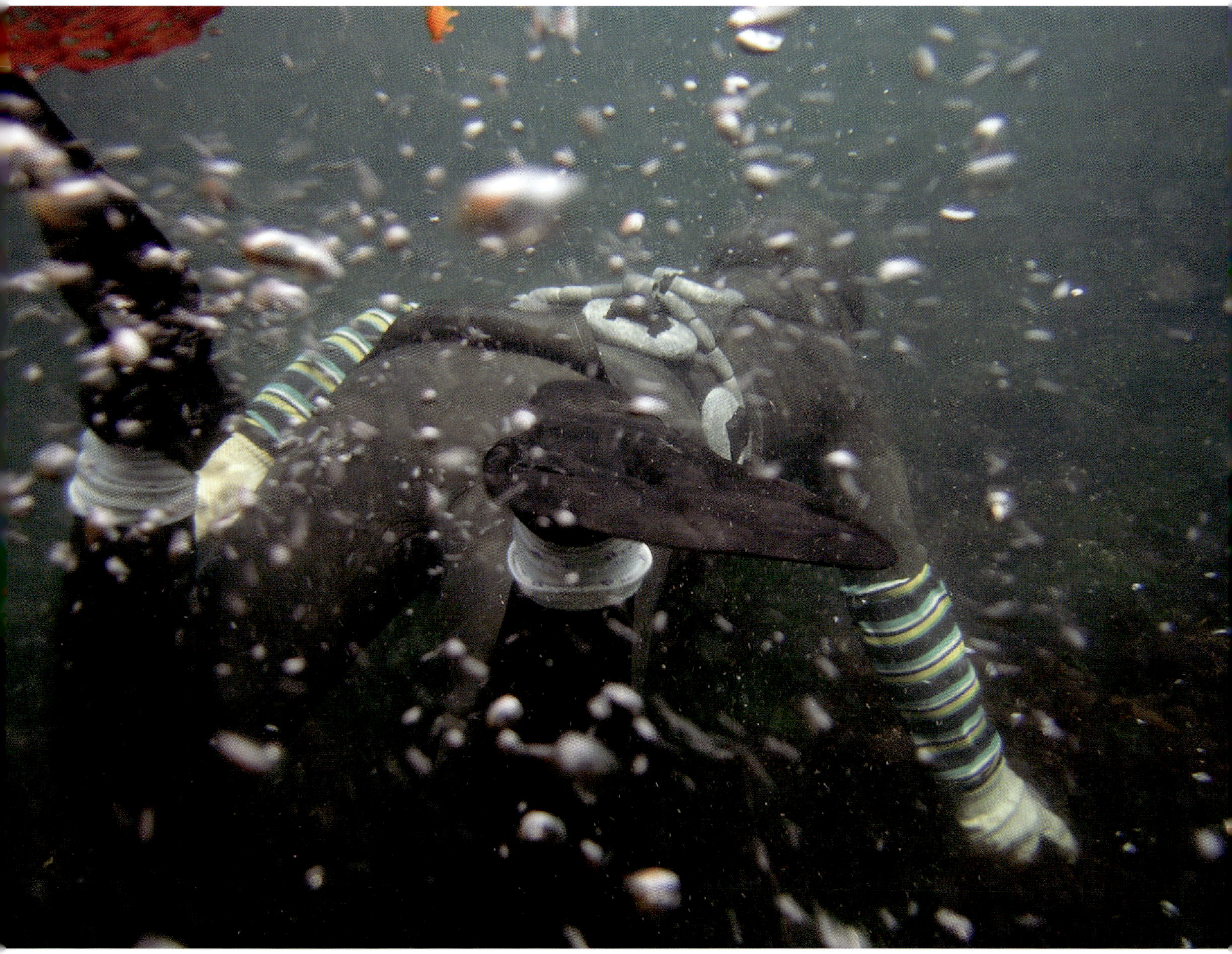

" One of the diving risks is spotting abalone or turban shells right before coming up to the surface. Rather than lose a catch, a *haenyeo* may risk her life by staying down longer than she is capable of holding her breath for."

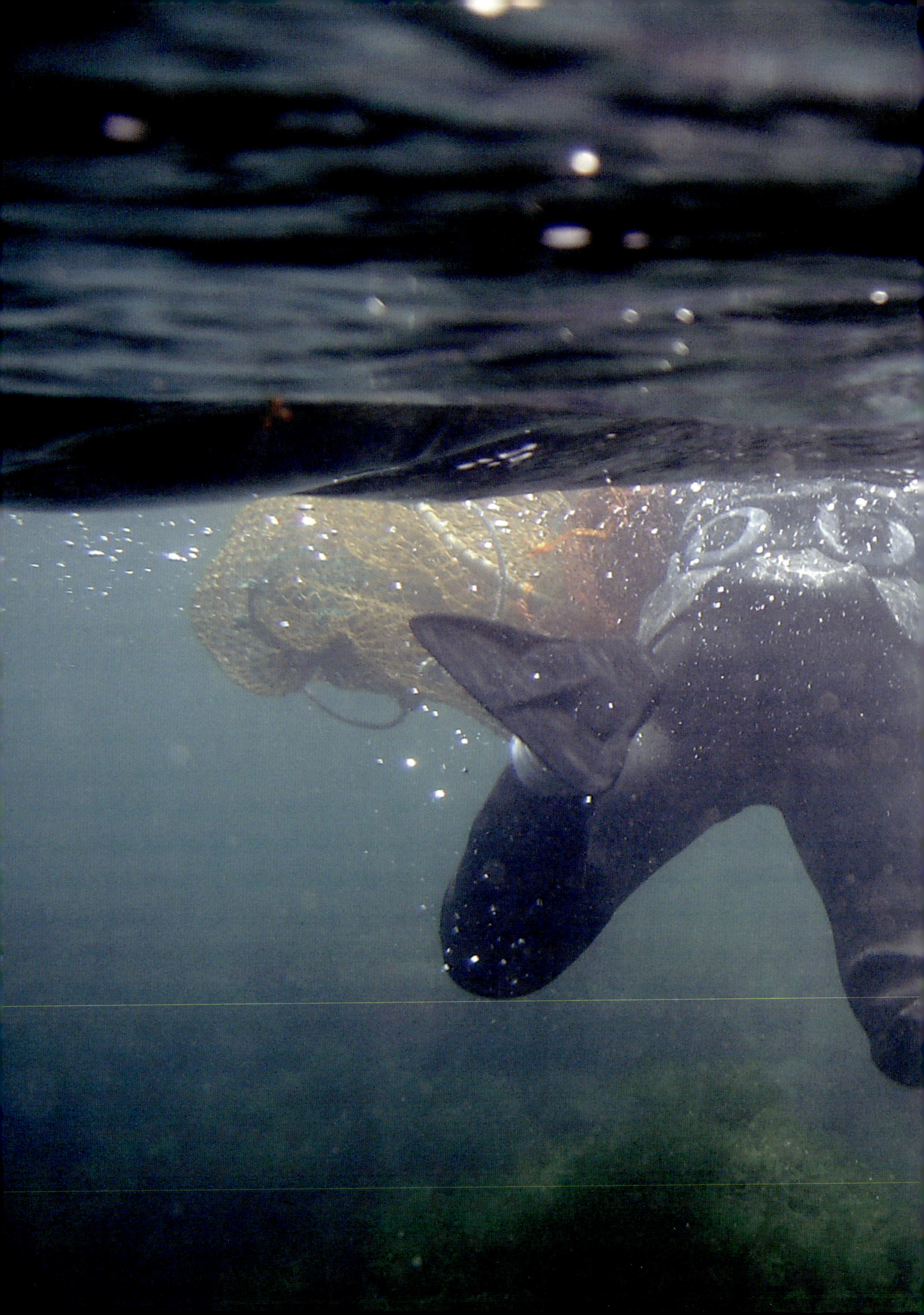

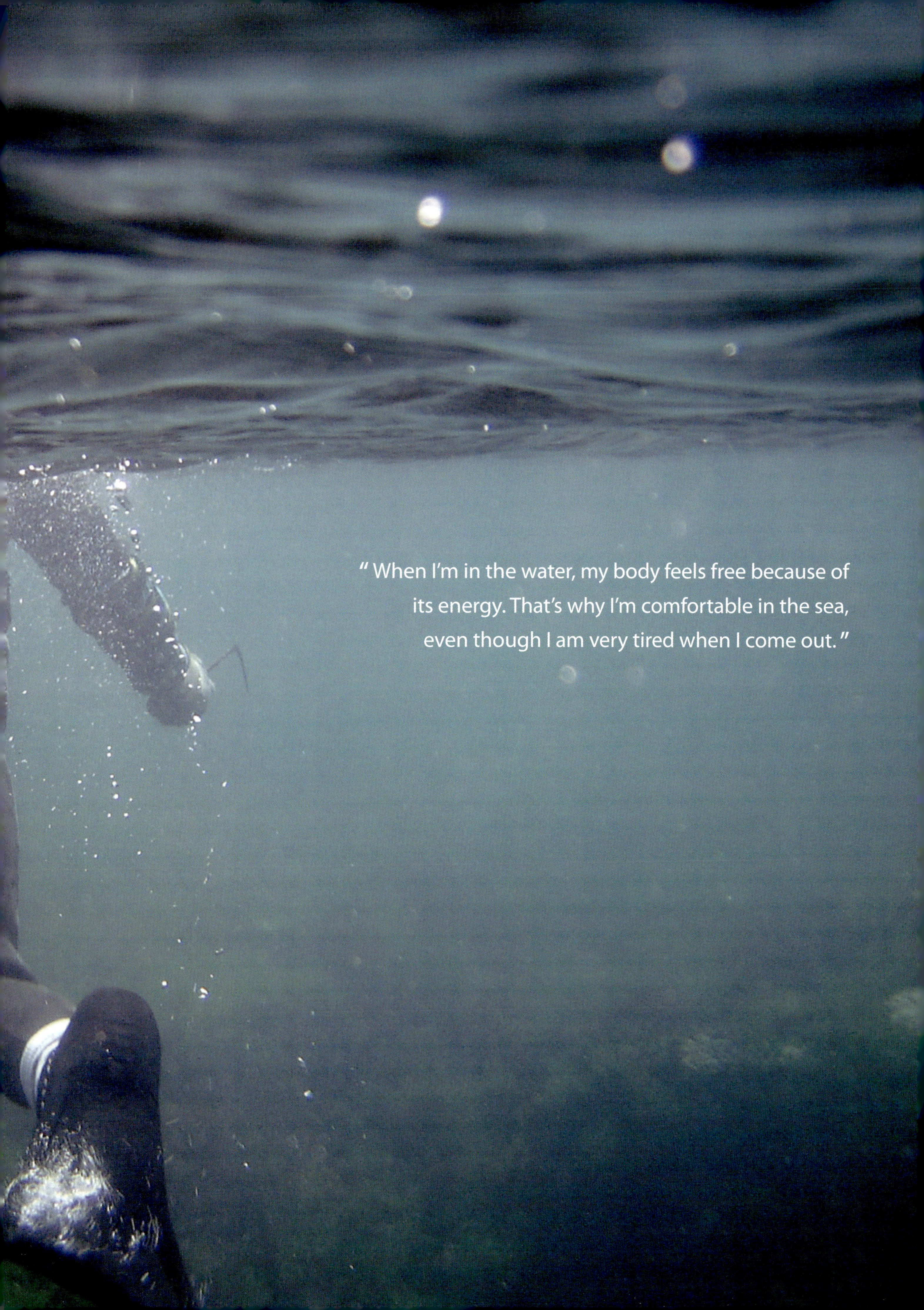

" When I'm in the water, my body feels free because of its energy. That's why I'm comfortable in the sea, even though I am very tired when I come out. "

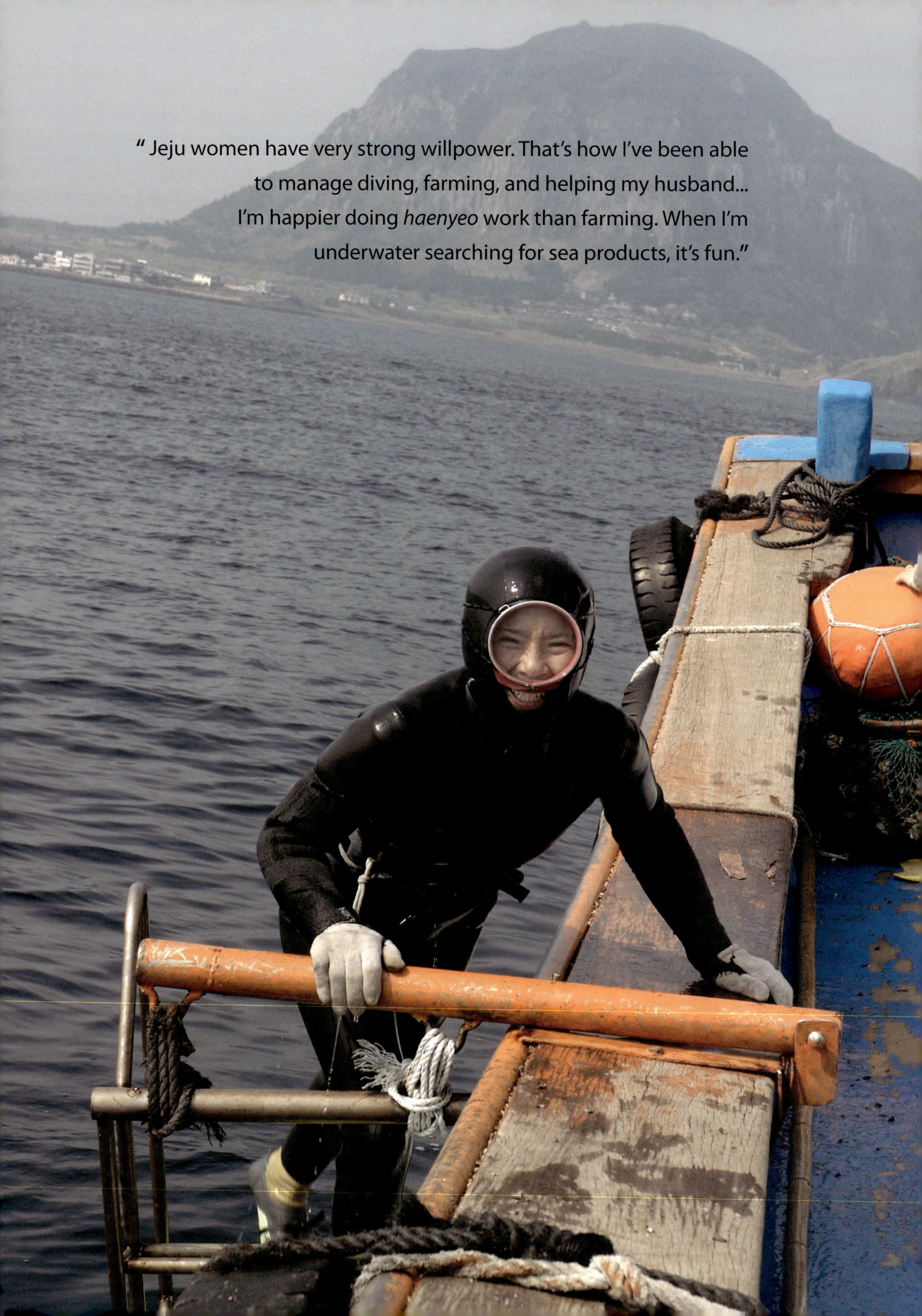

" Jeju women have very strong willpower. That's how I've been able to manage diving, farming, and helping my husband... I'm happier doing *haenyeo* work than farming. When I'm underwater searching for sea products, it's fun."

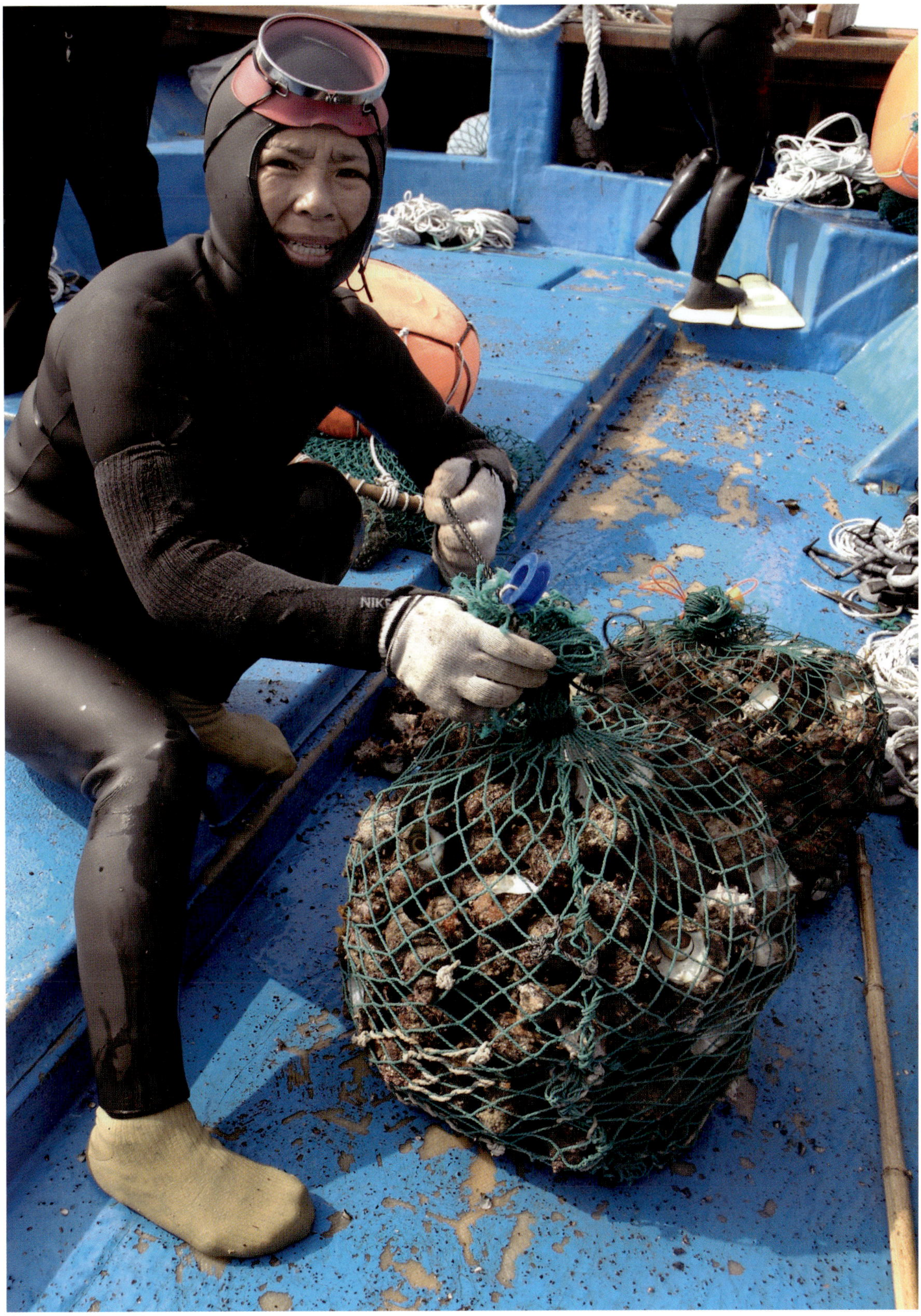
NIKE

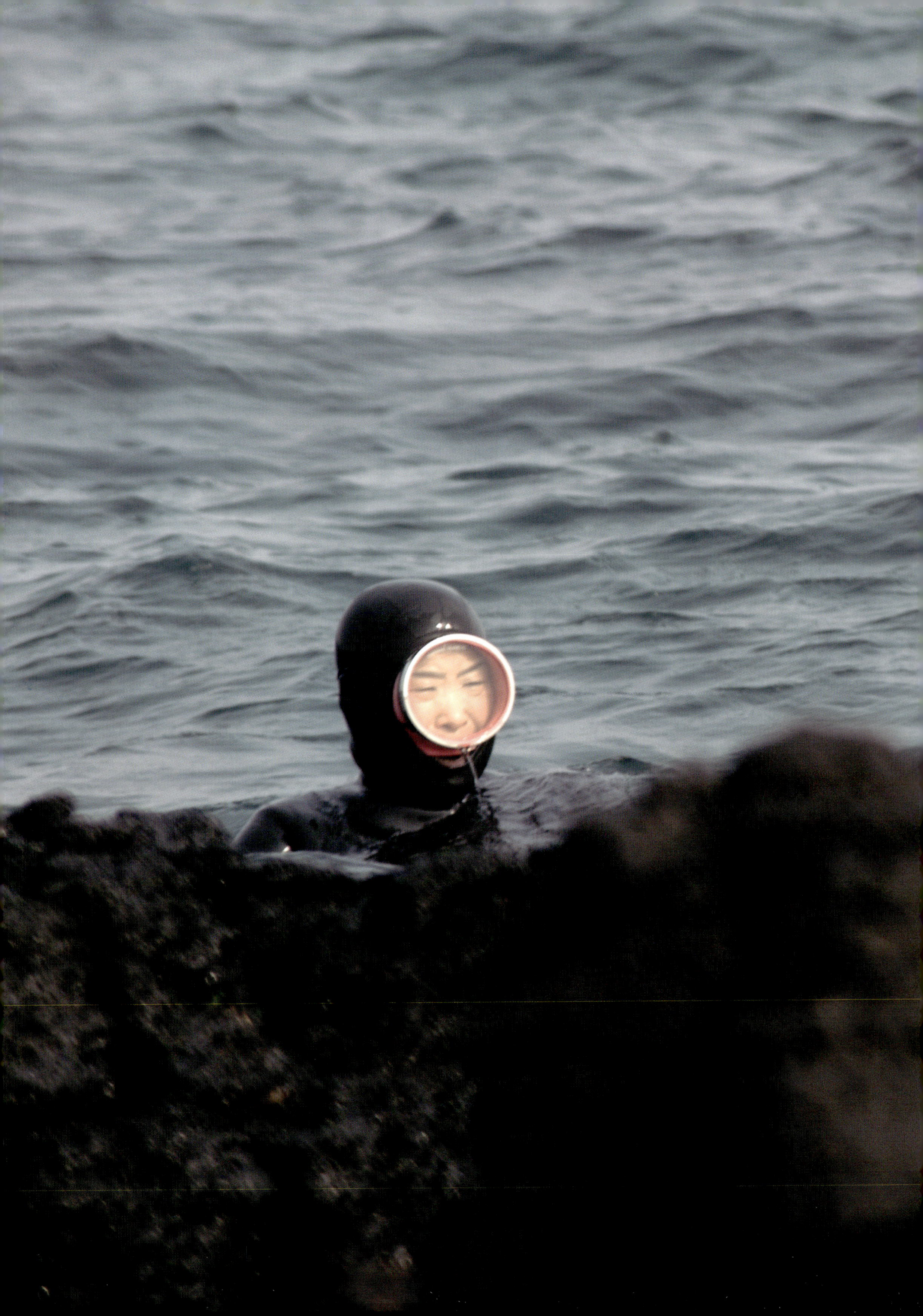

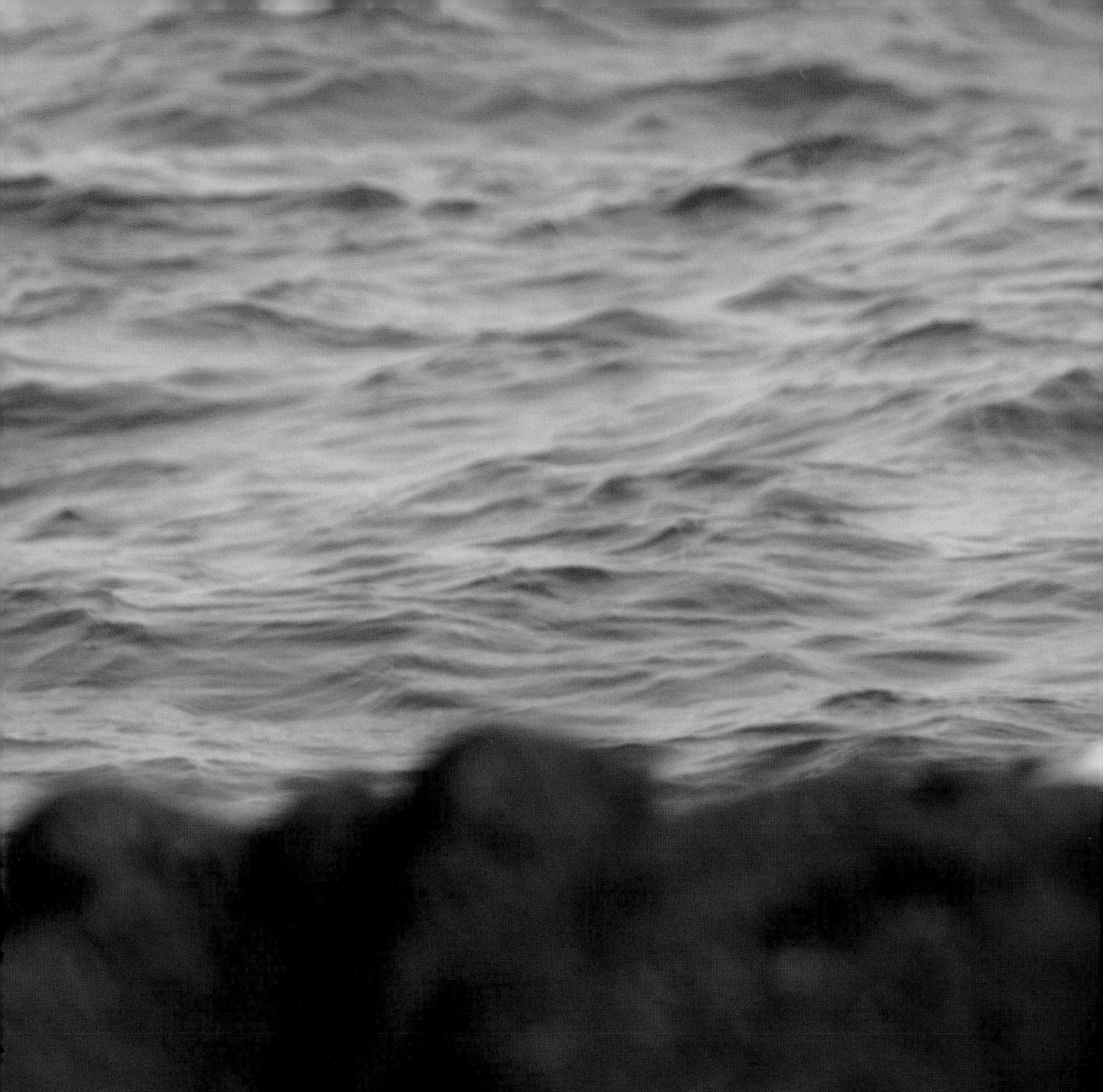

" When you hear a whistling sound, we call that *sumbisori*.
That sound is made when a *haenyeo* inhales
and exhales the breath after rising to the surface.
By doing so, we can remain under the water longer."

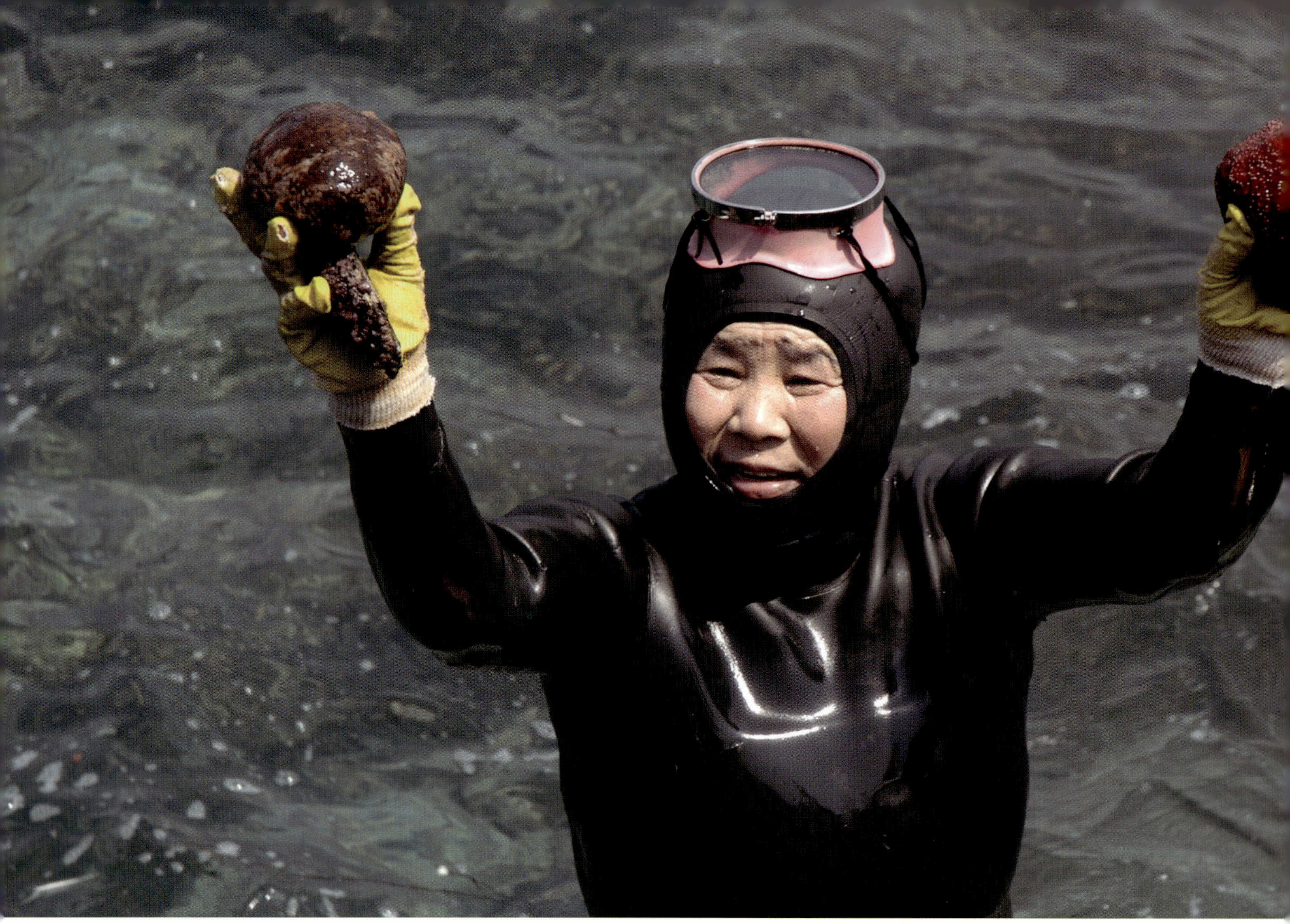

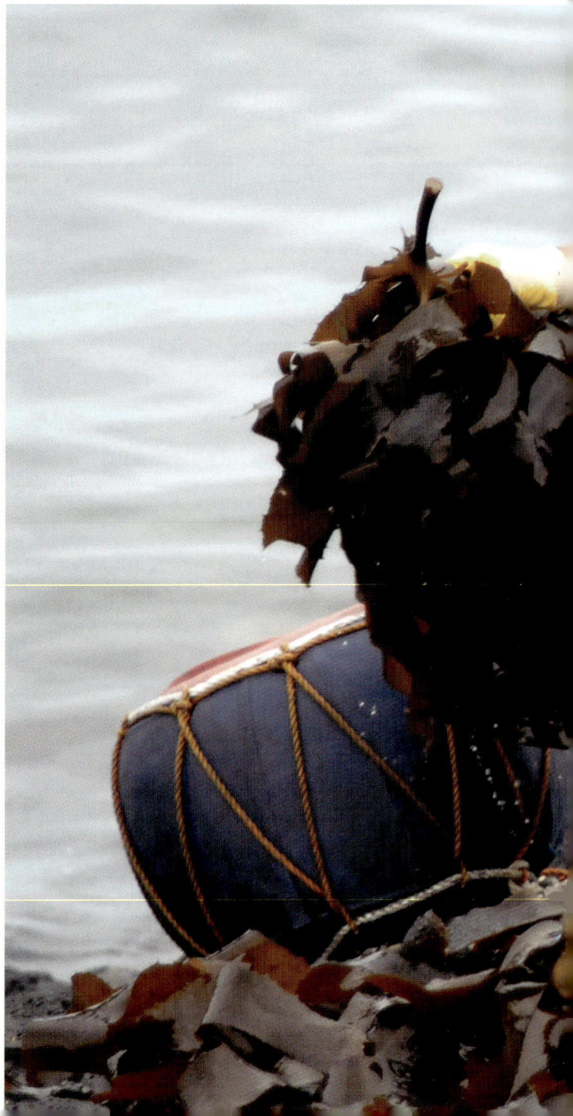

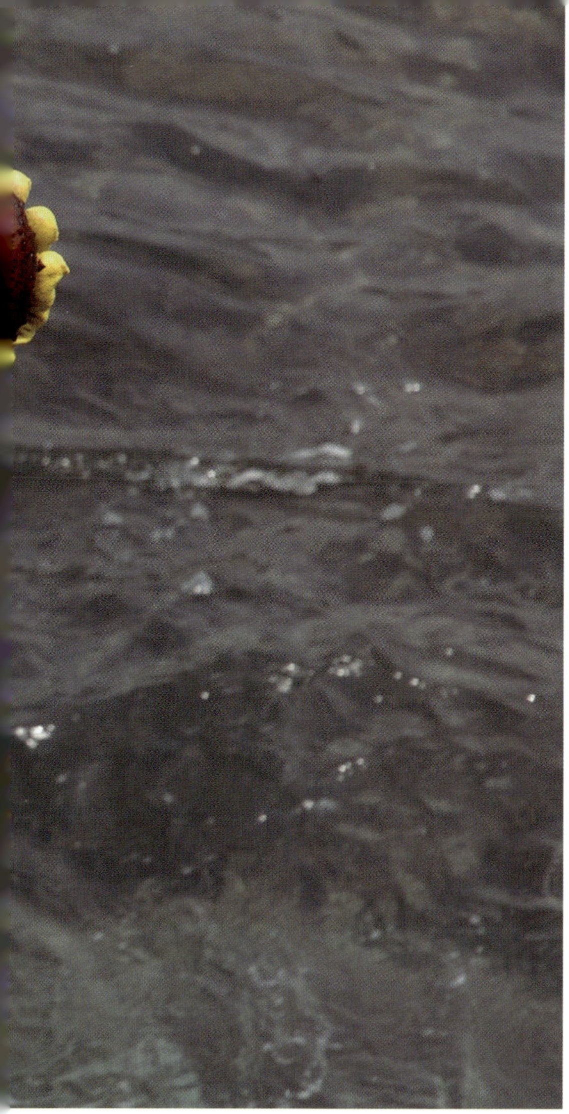

Harvest from the Sea

The *haenyeo* dive for sea products and seaweed, much of which is shipped to the mainland and overseas. Among their harvest: octopuses, sea urchins, small snails, abalone, and squid.

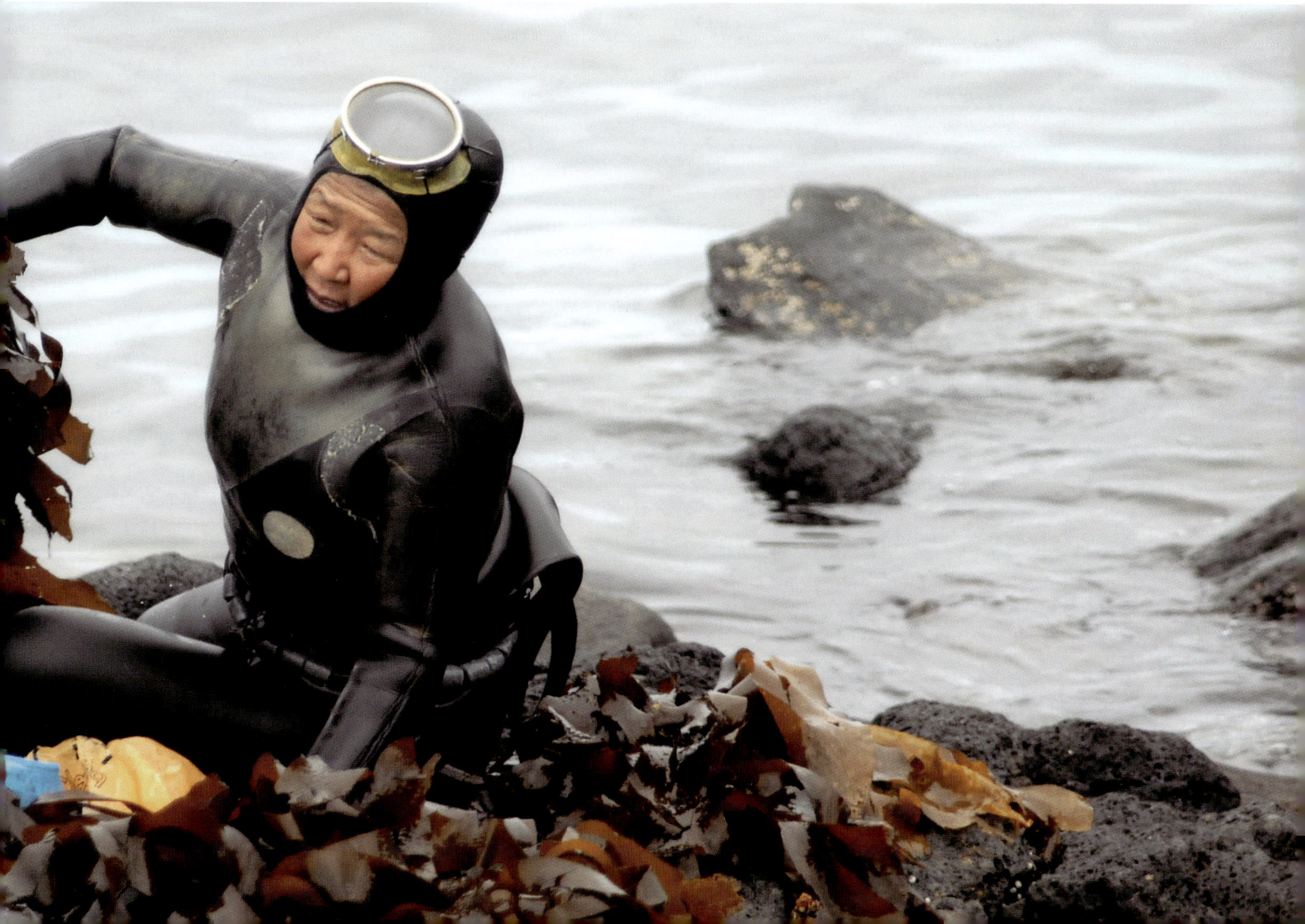

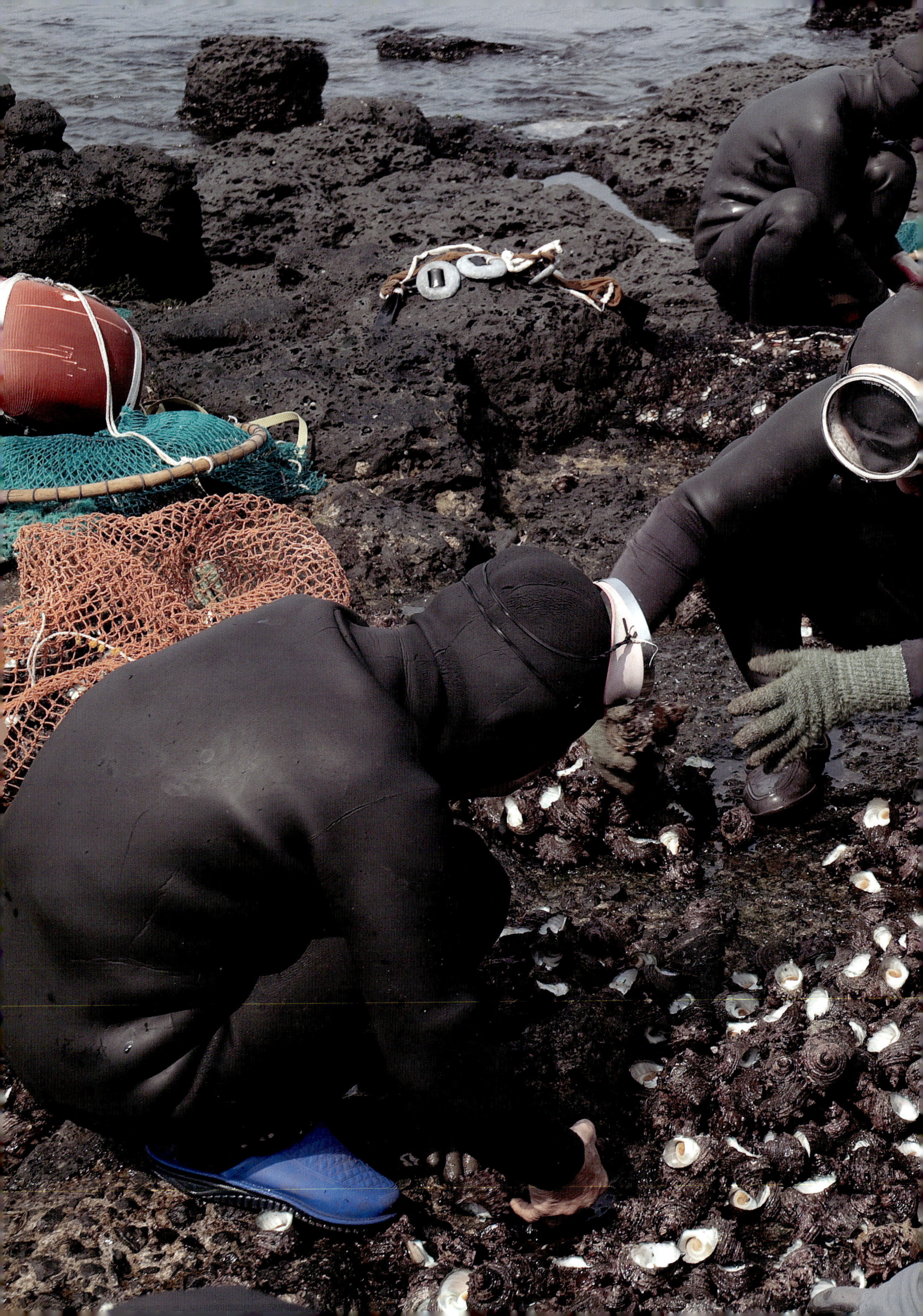

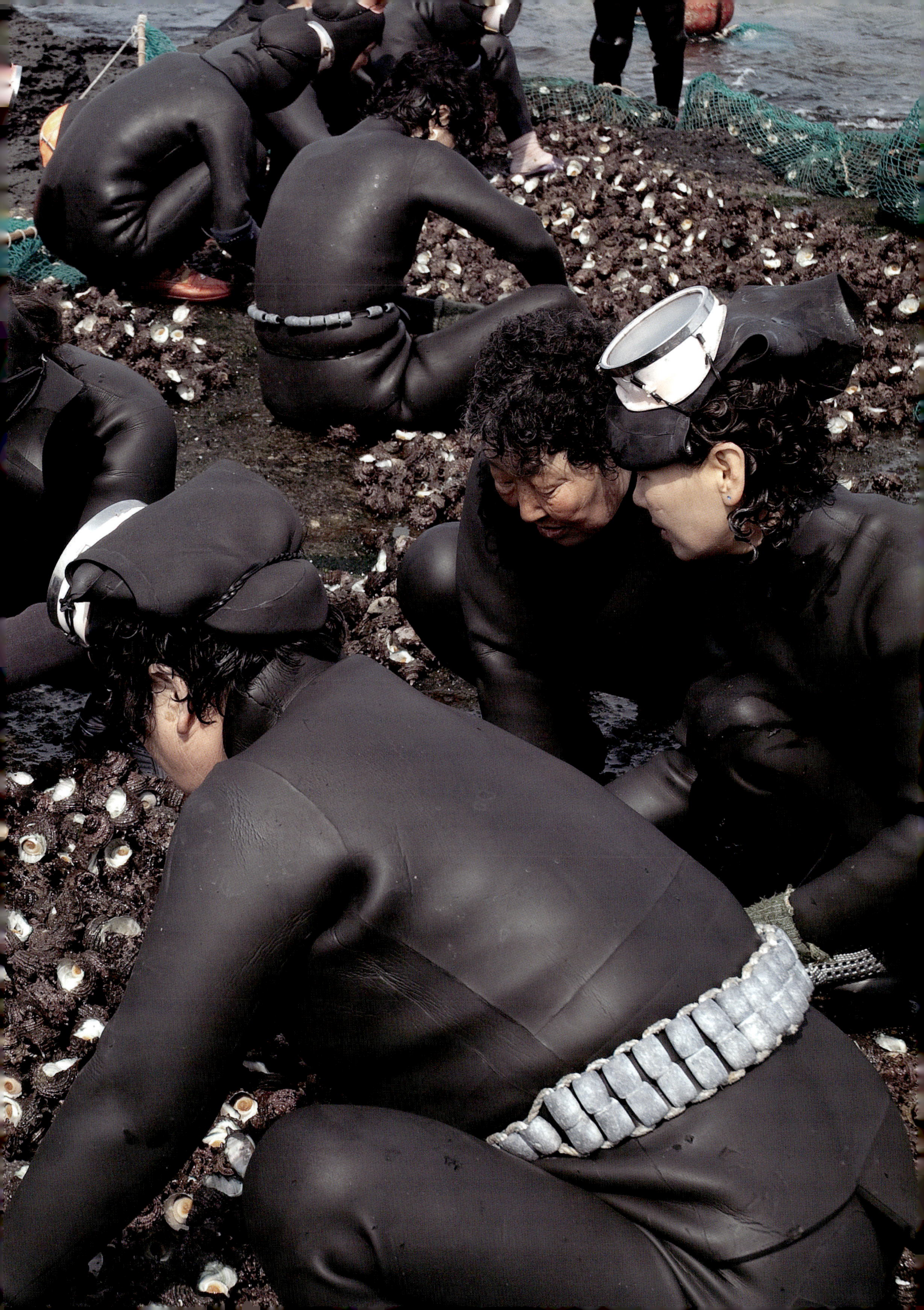

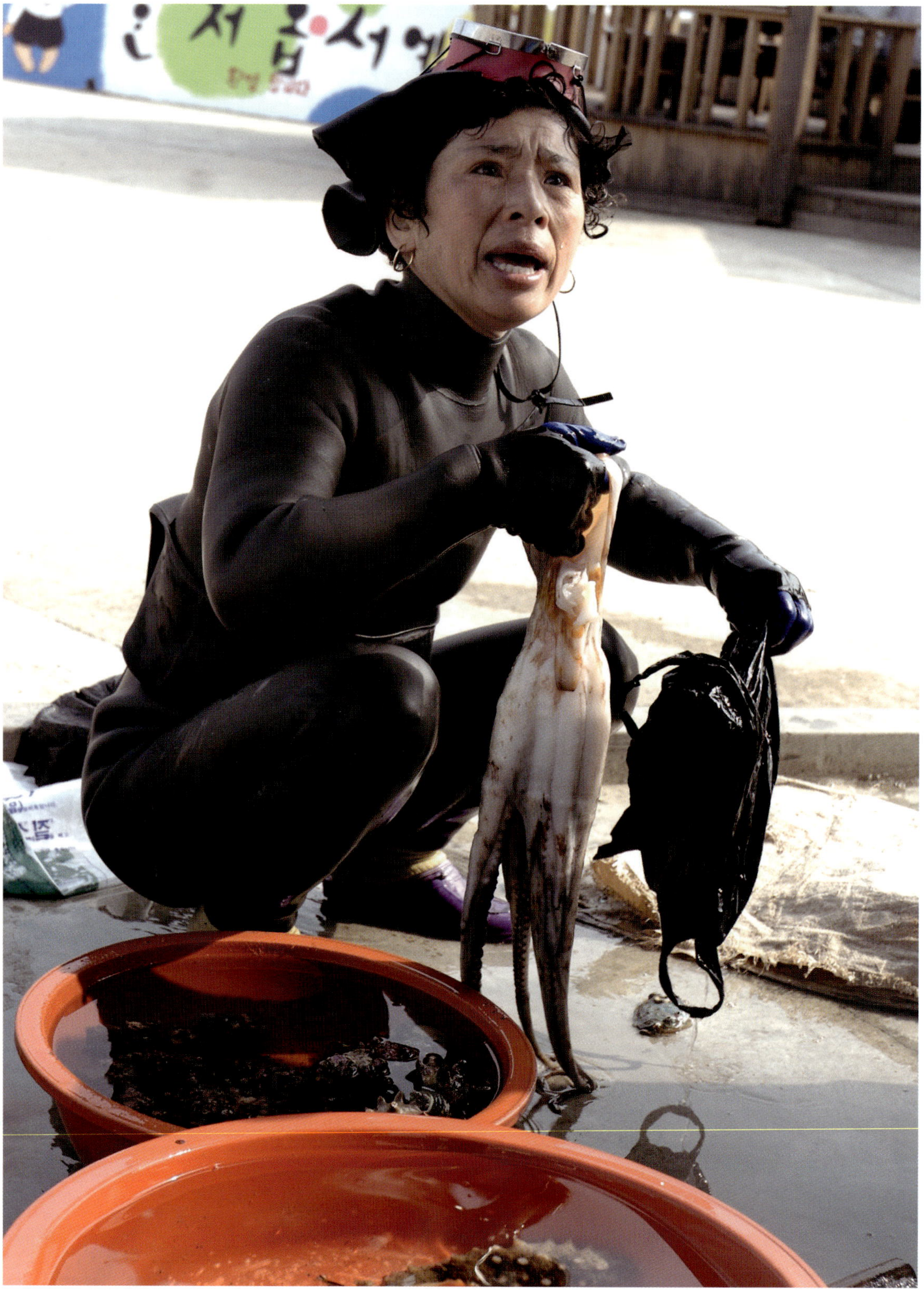

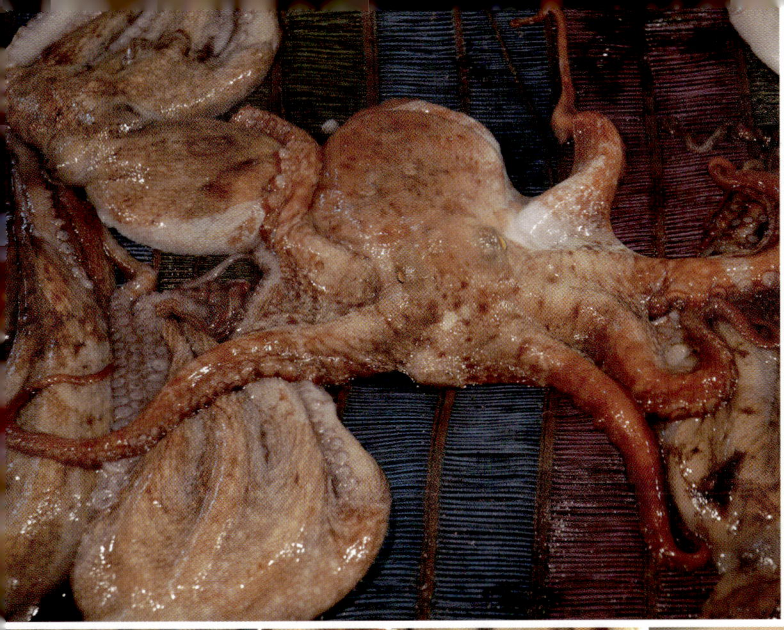
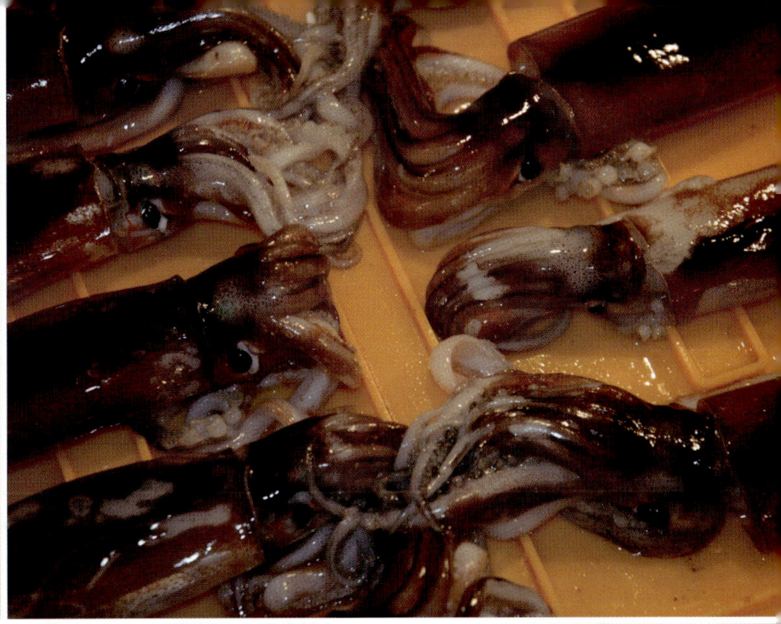
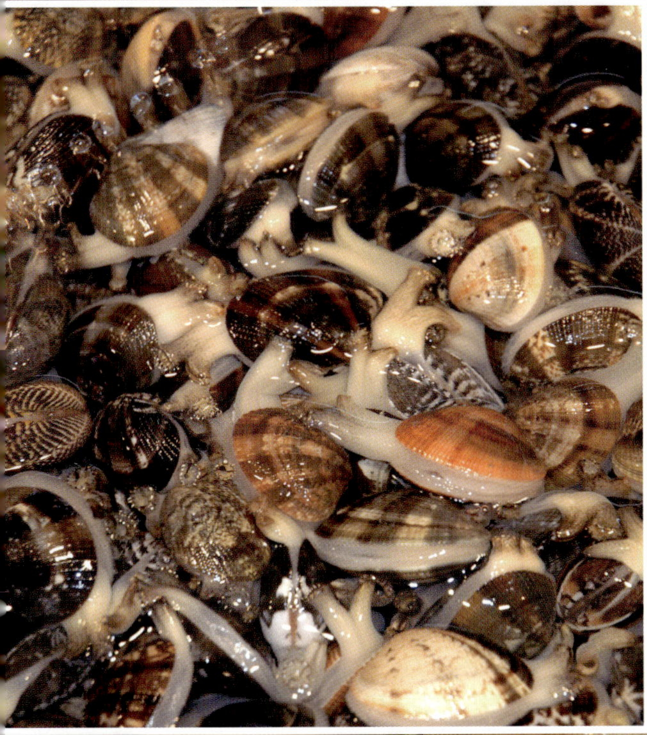
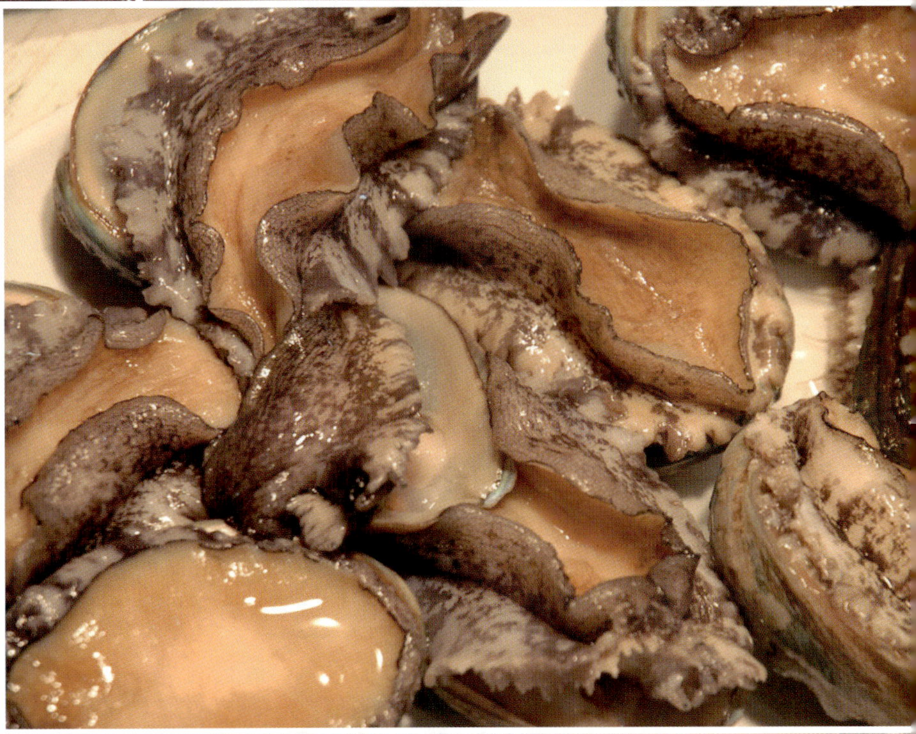
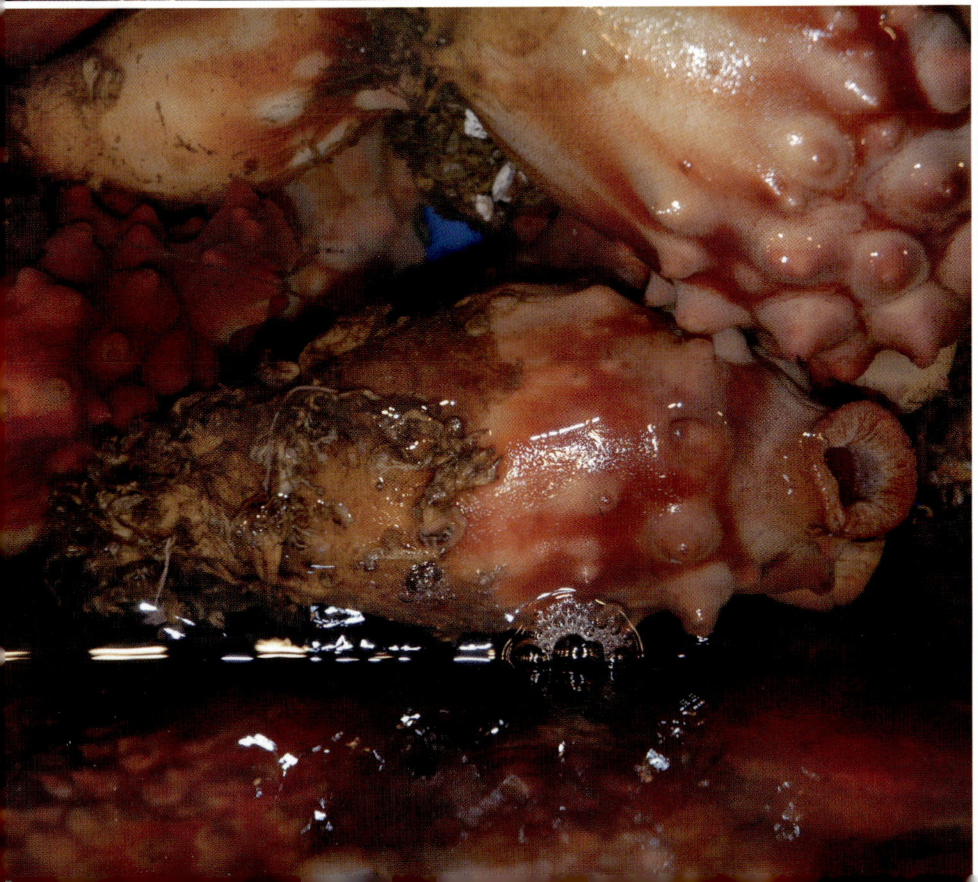
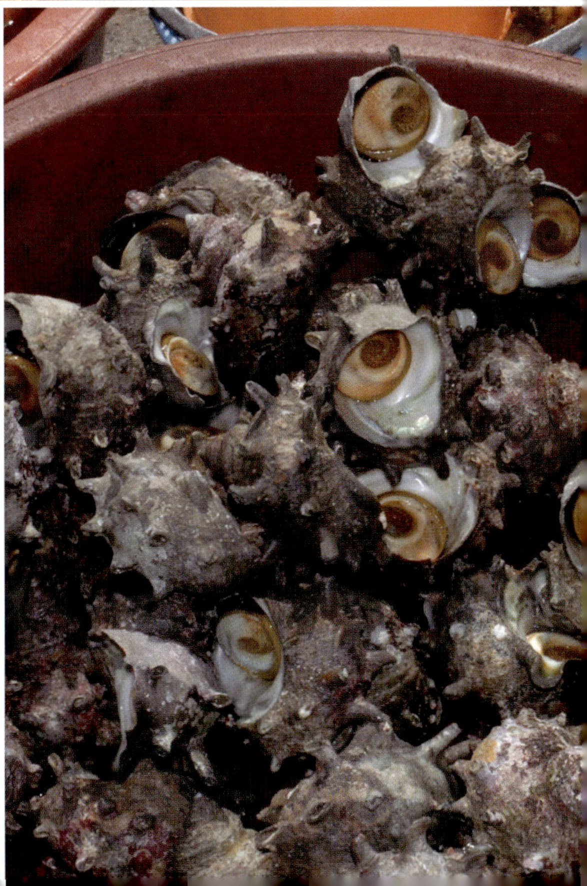

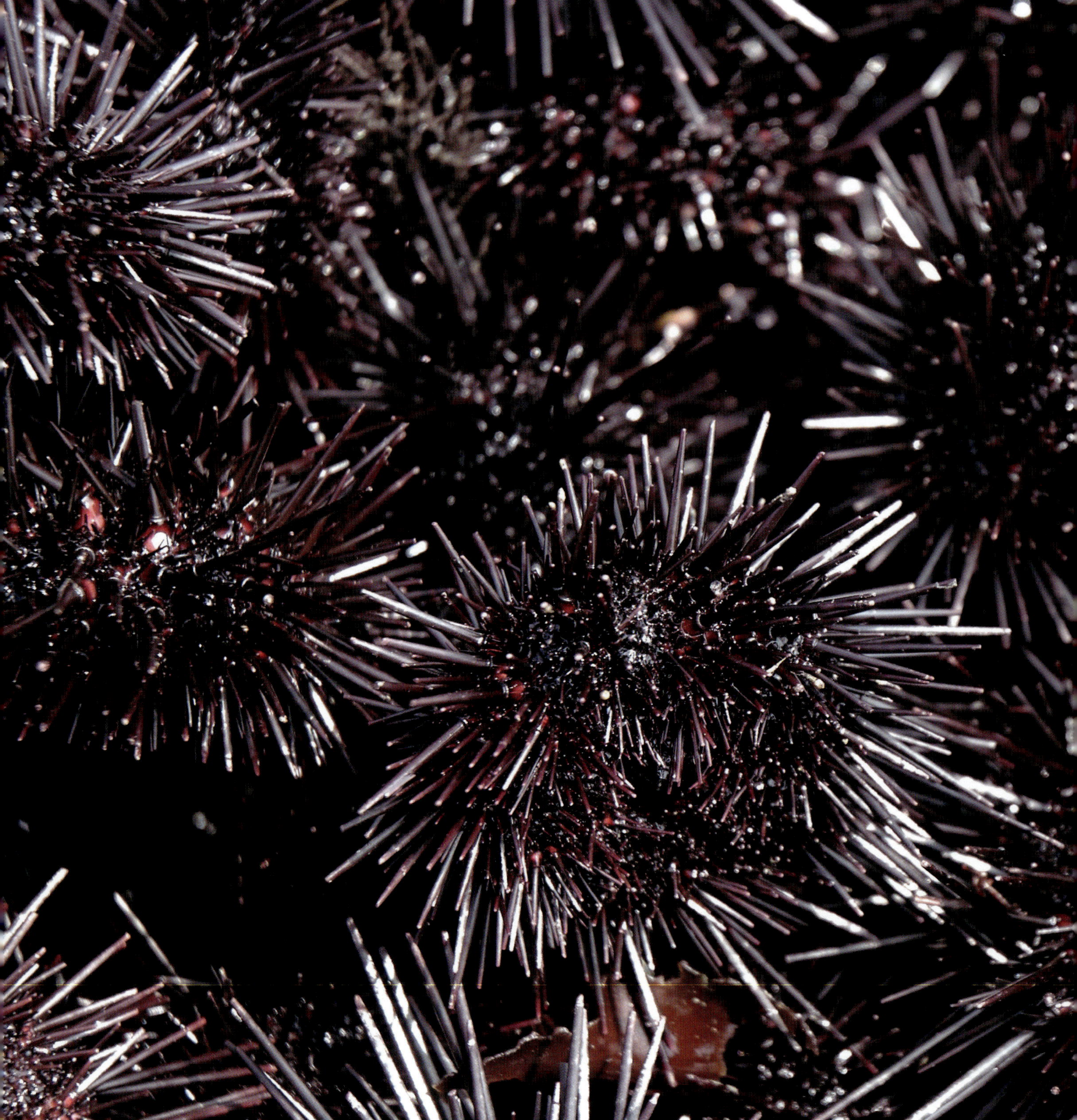

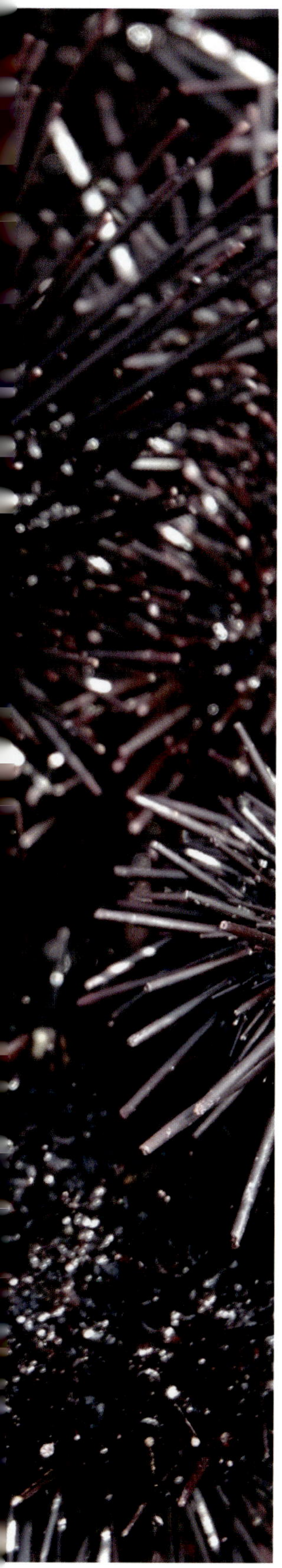

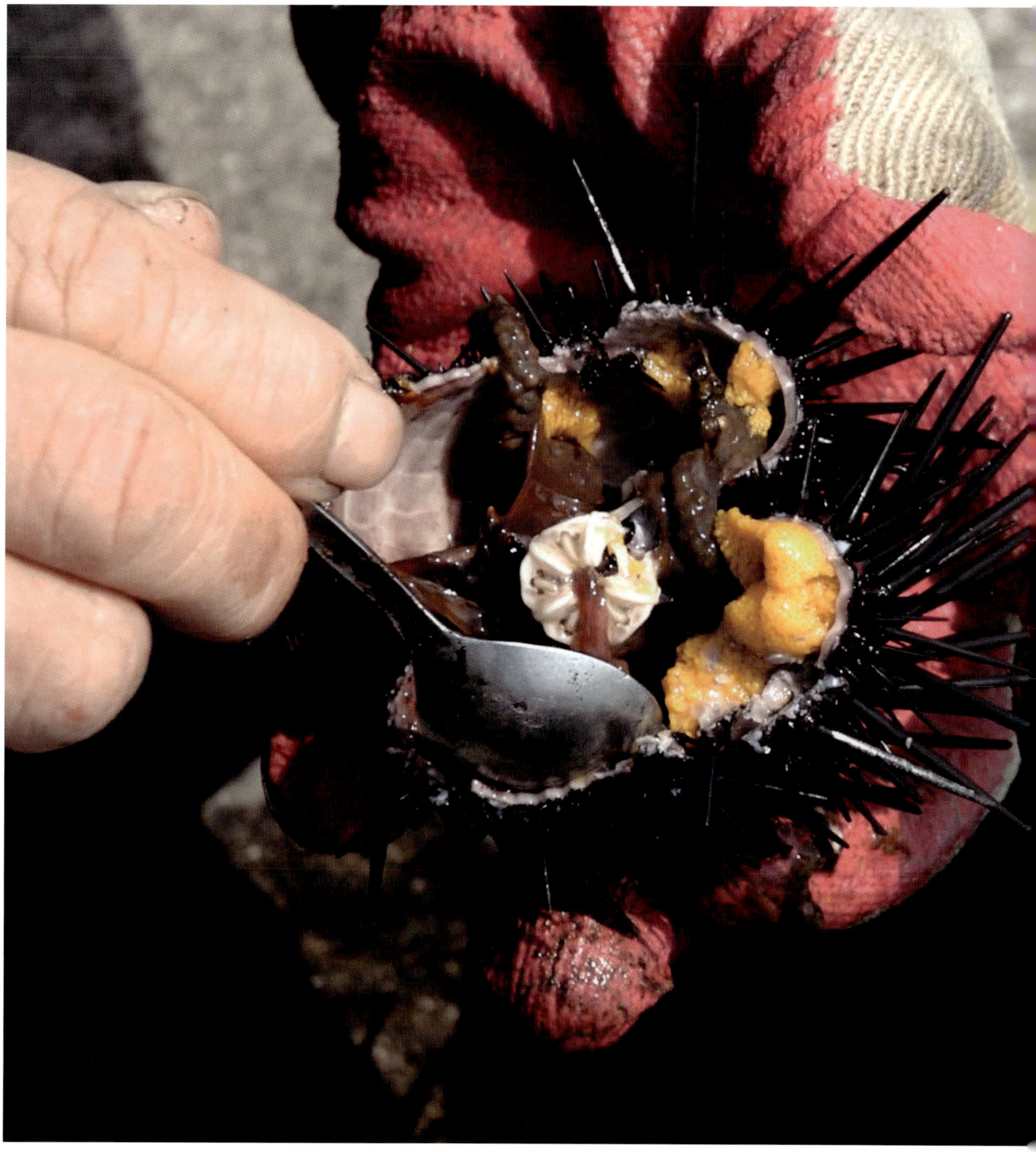

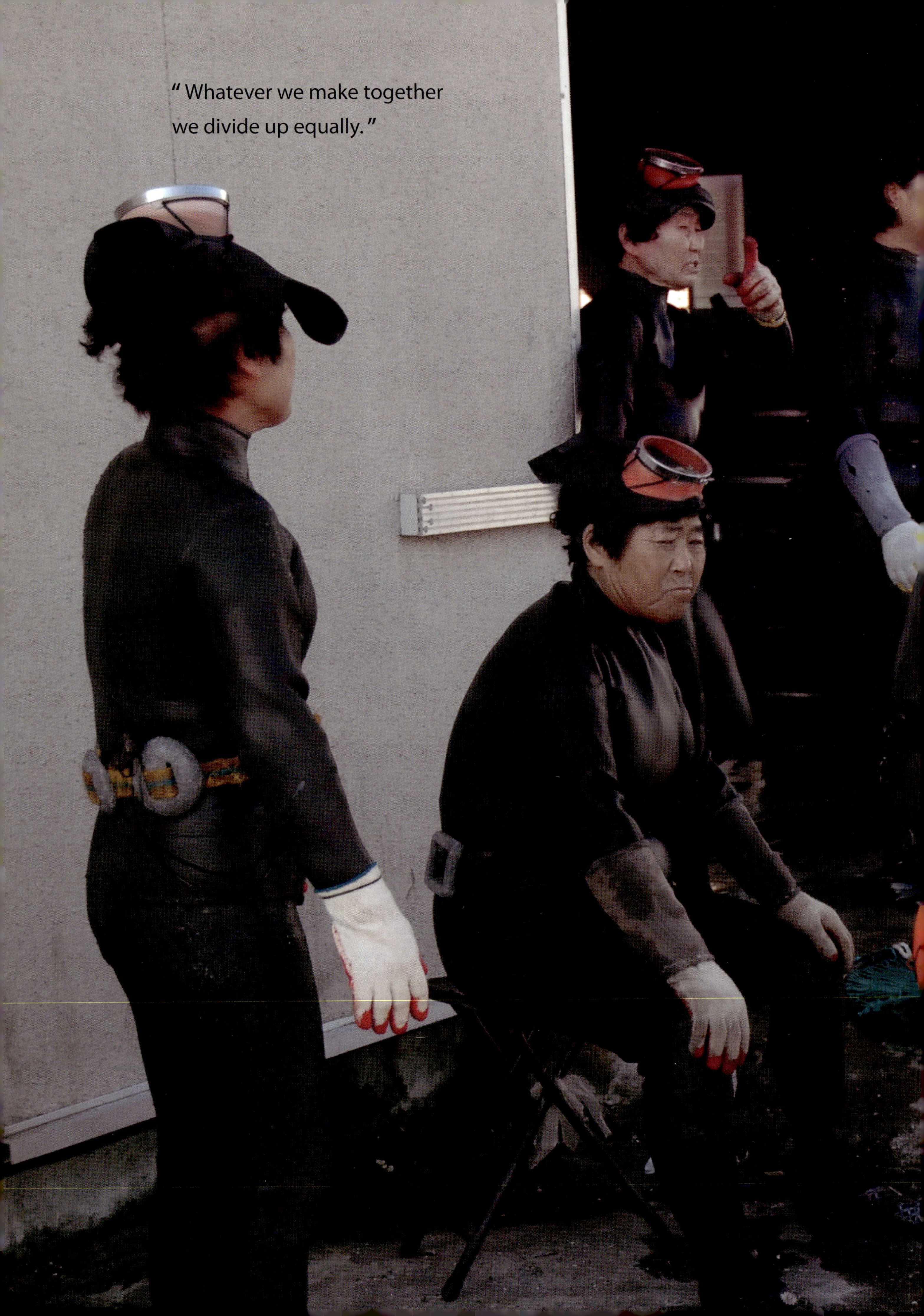
" Whatever we make together
we divide up equally."

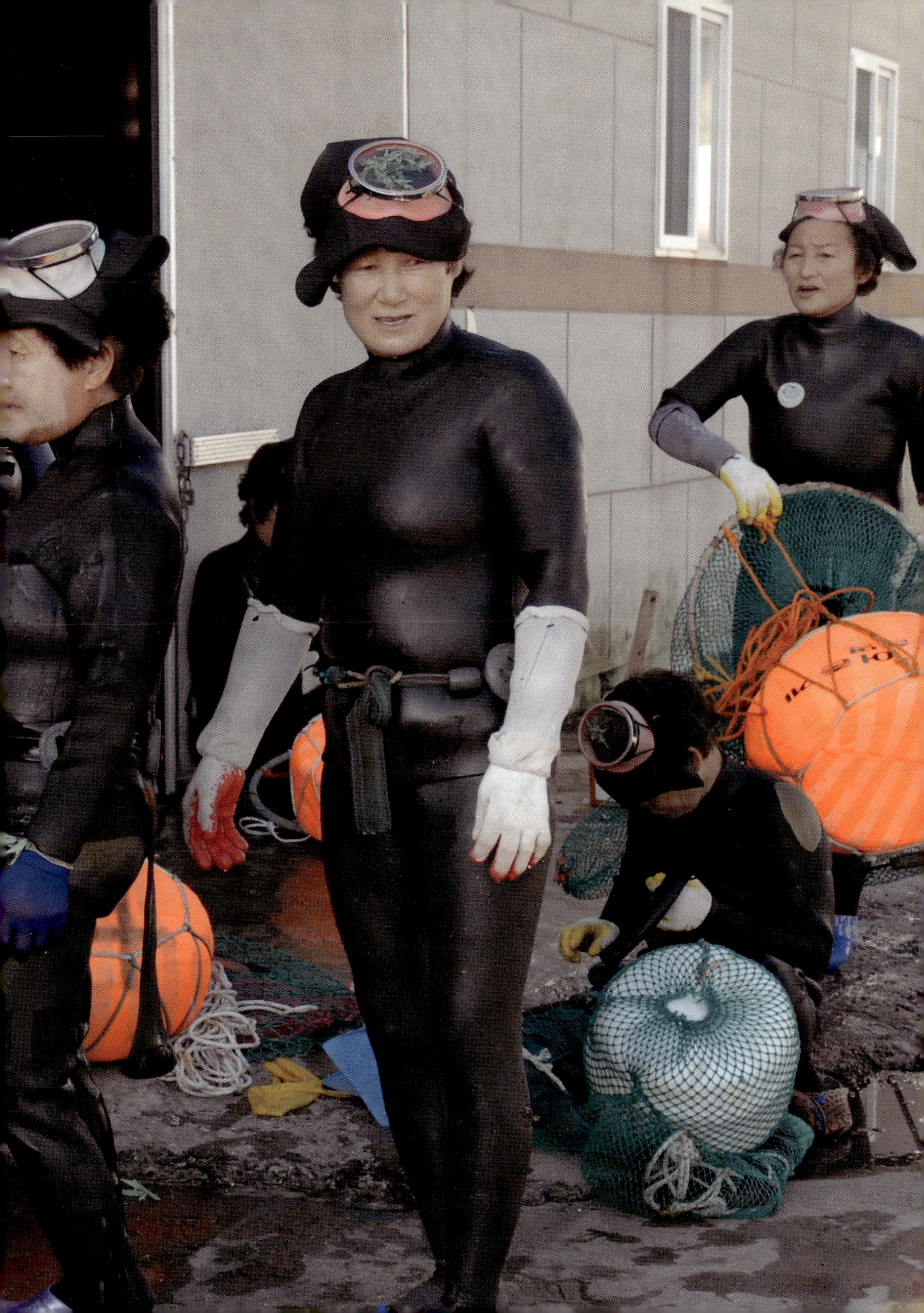

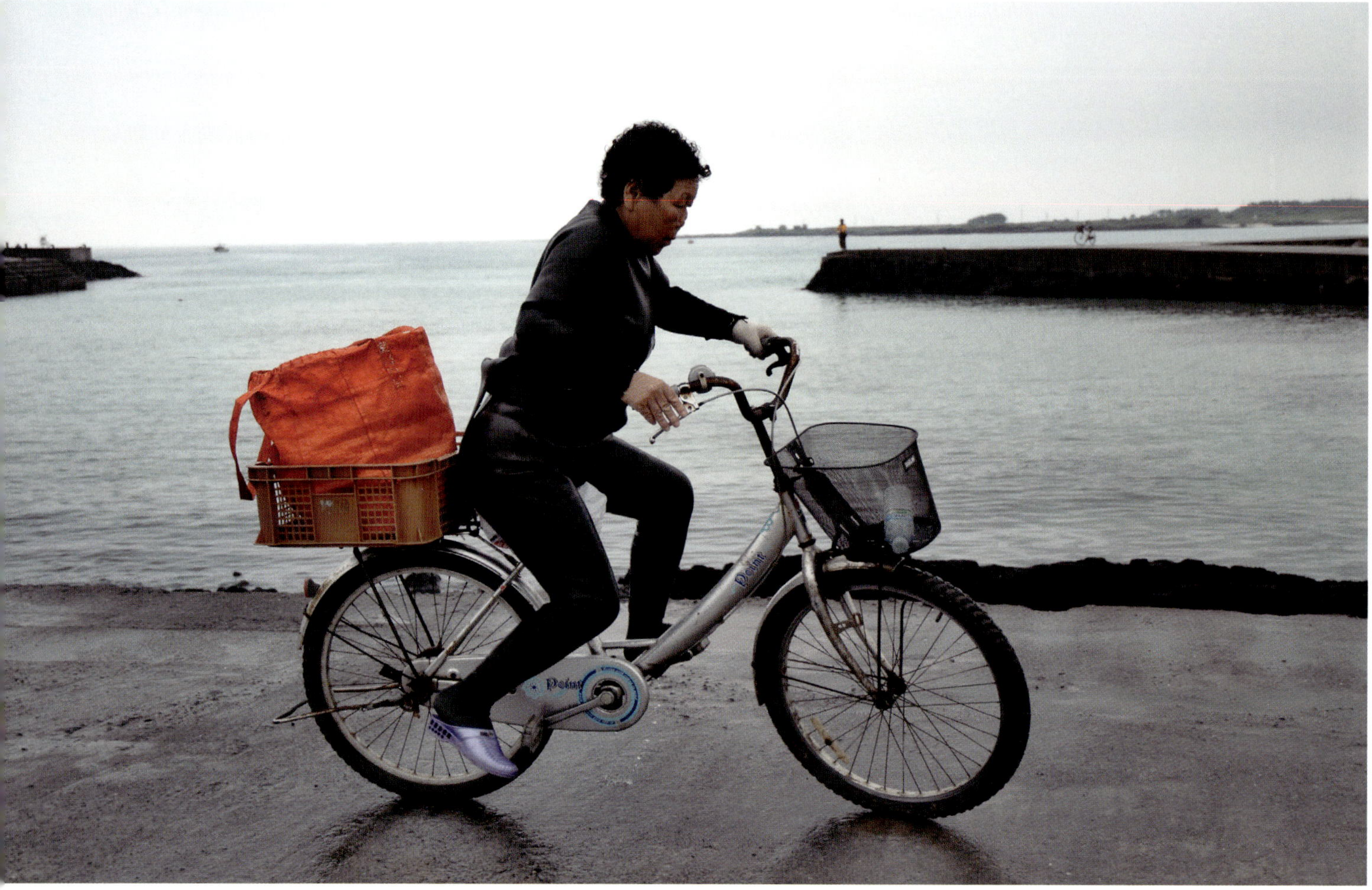

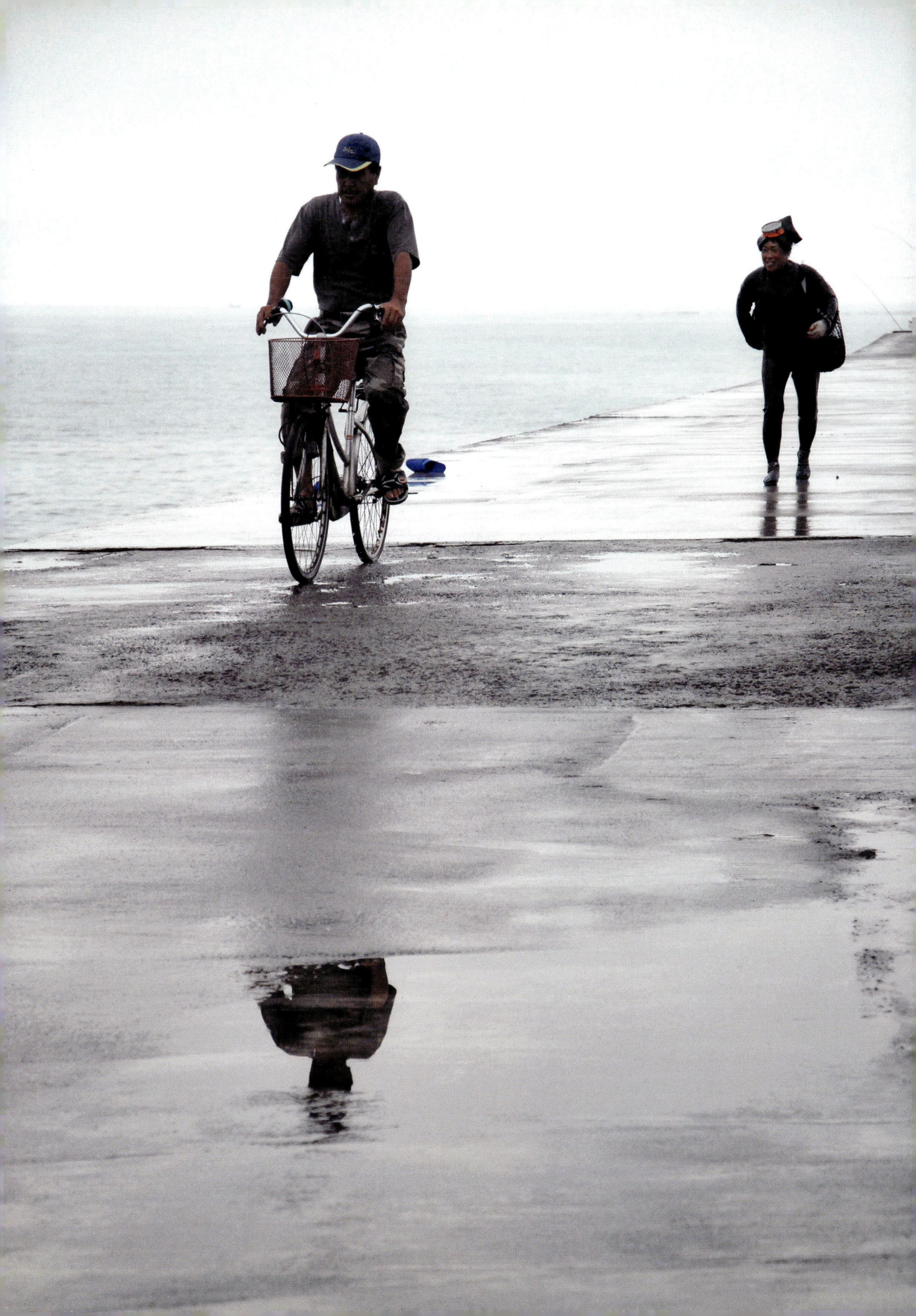

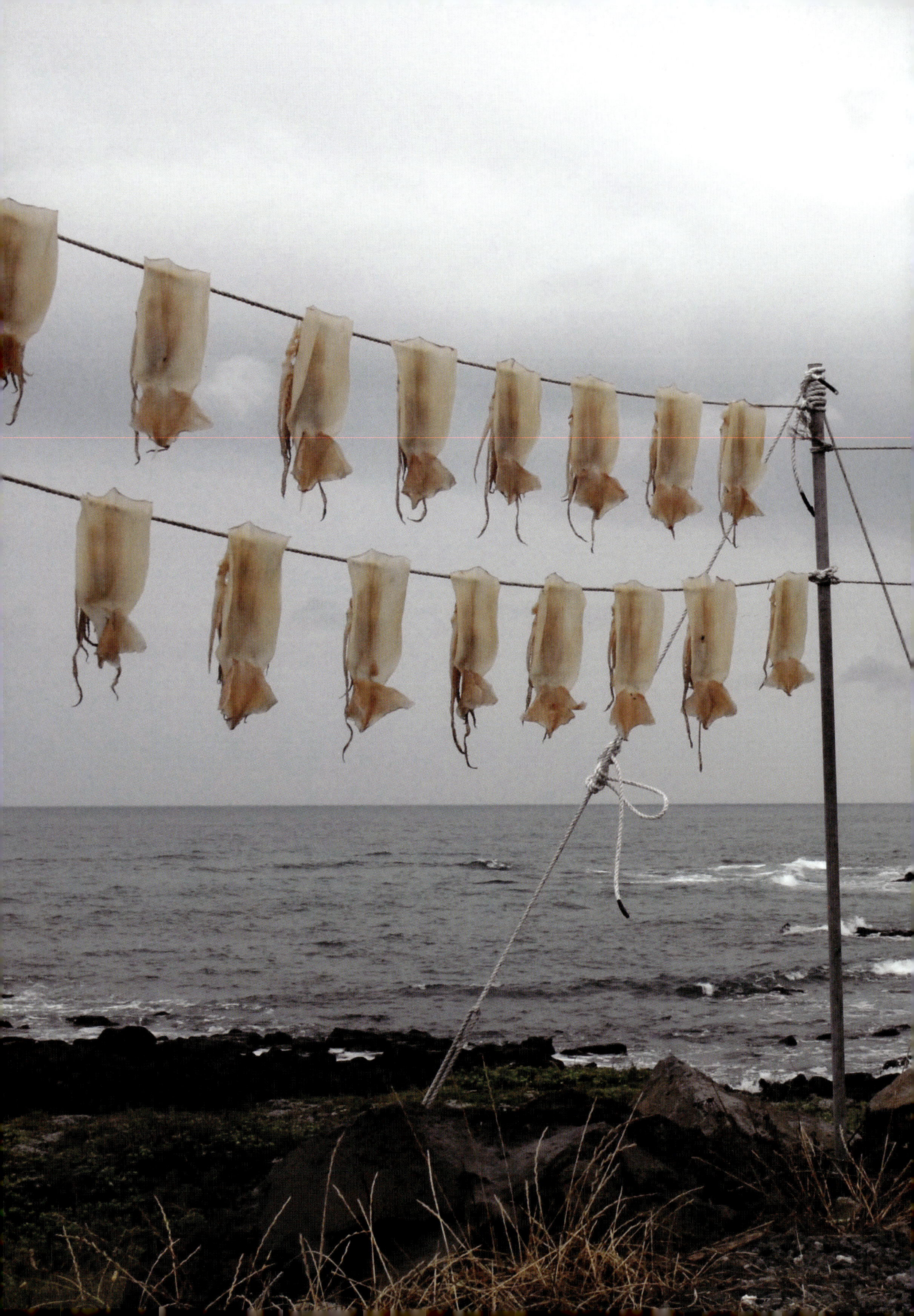

" We can't really predict exactly how much we'll earn.
Our ancestors told us that it all depends on divine will."

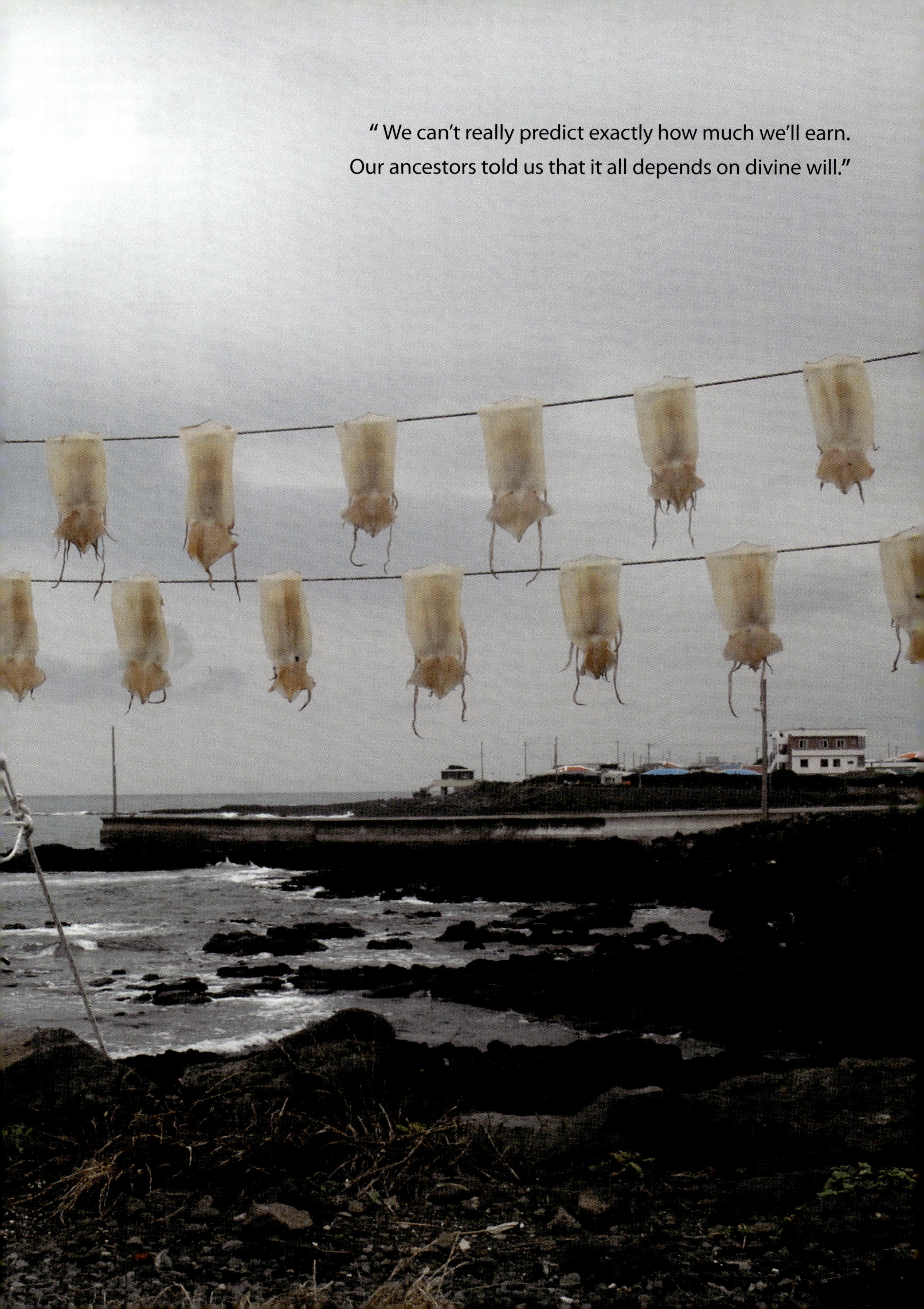

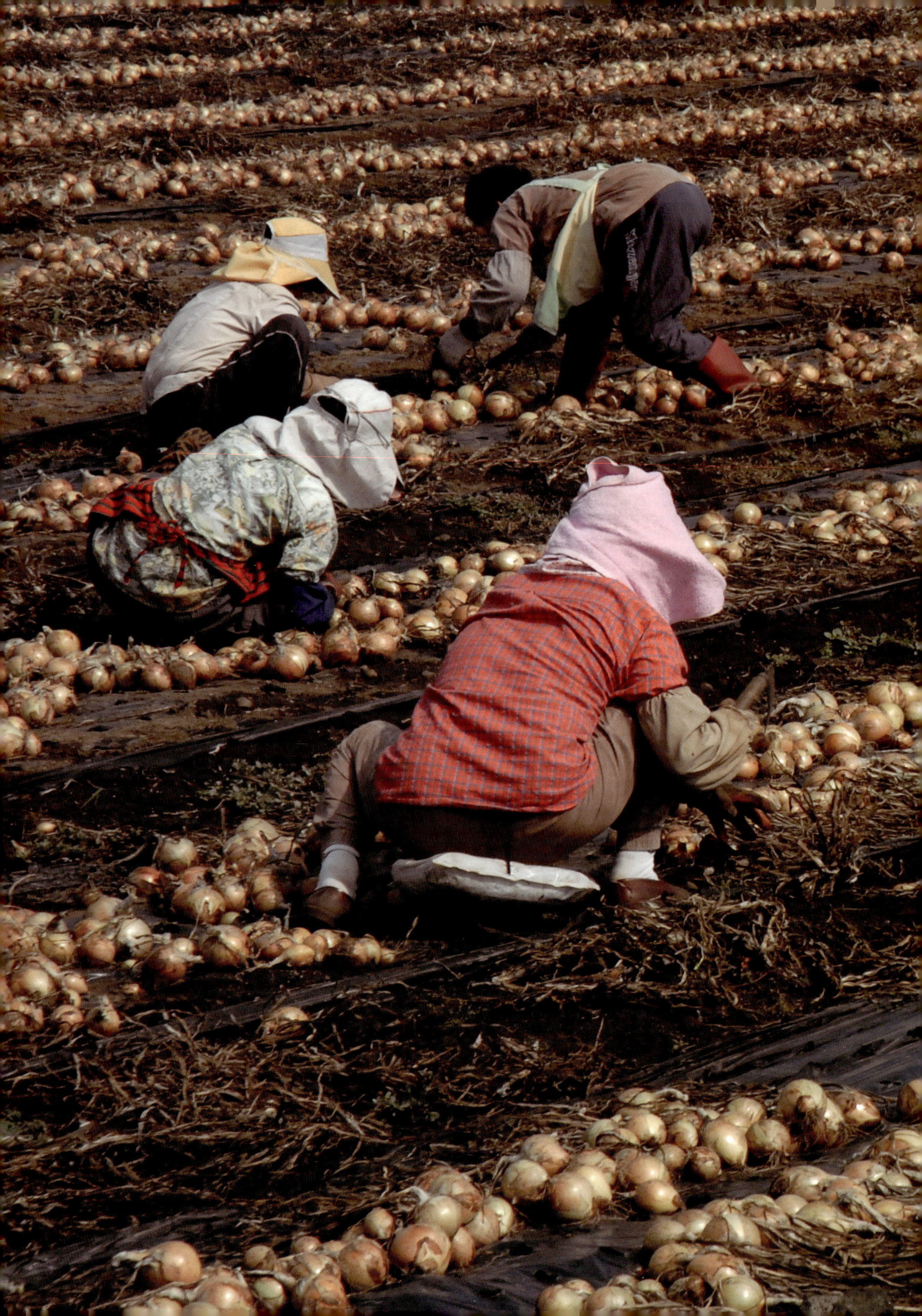

" In between dives, the *haenyeo* work together on the farms and divide the profits equally among us."

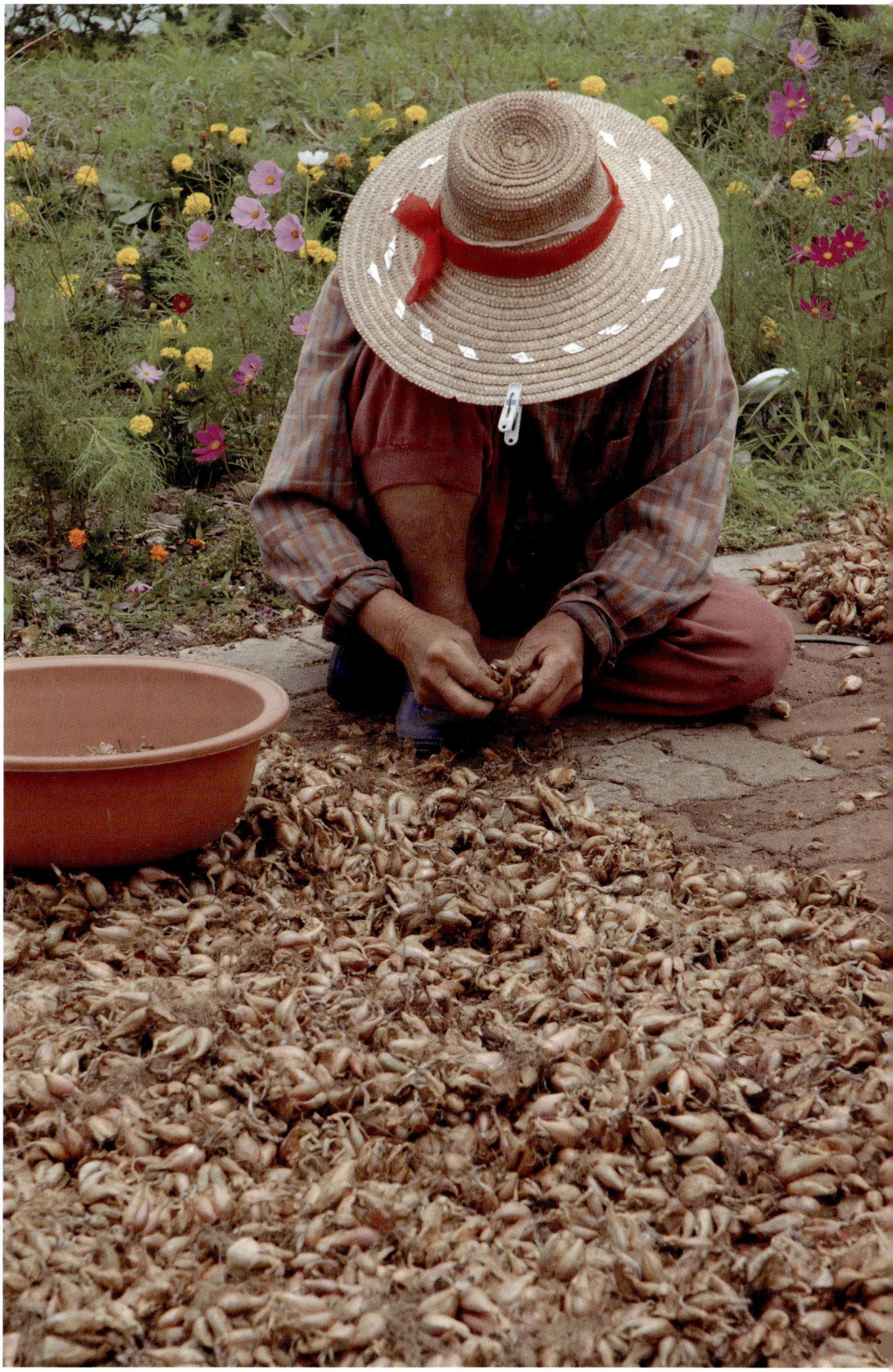

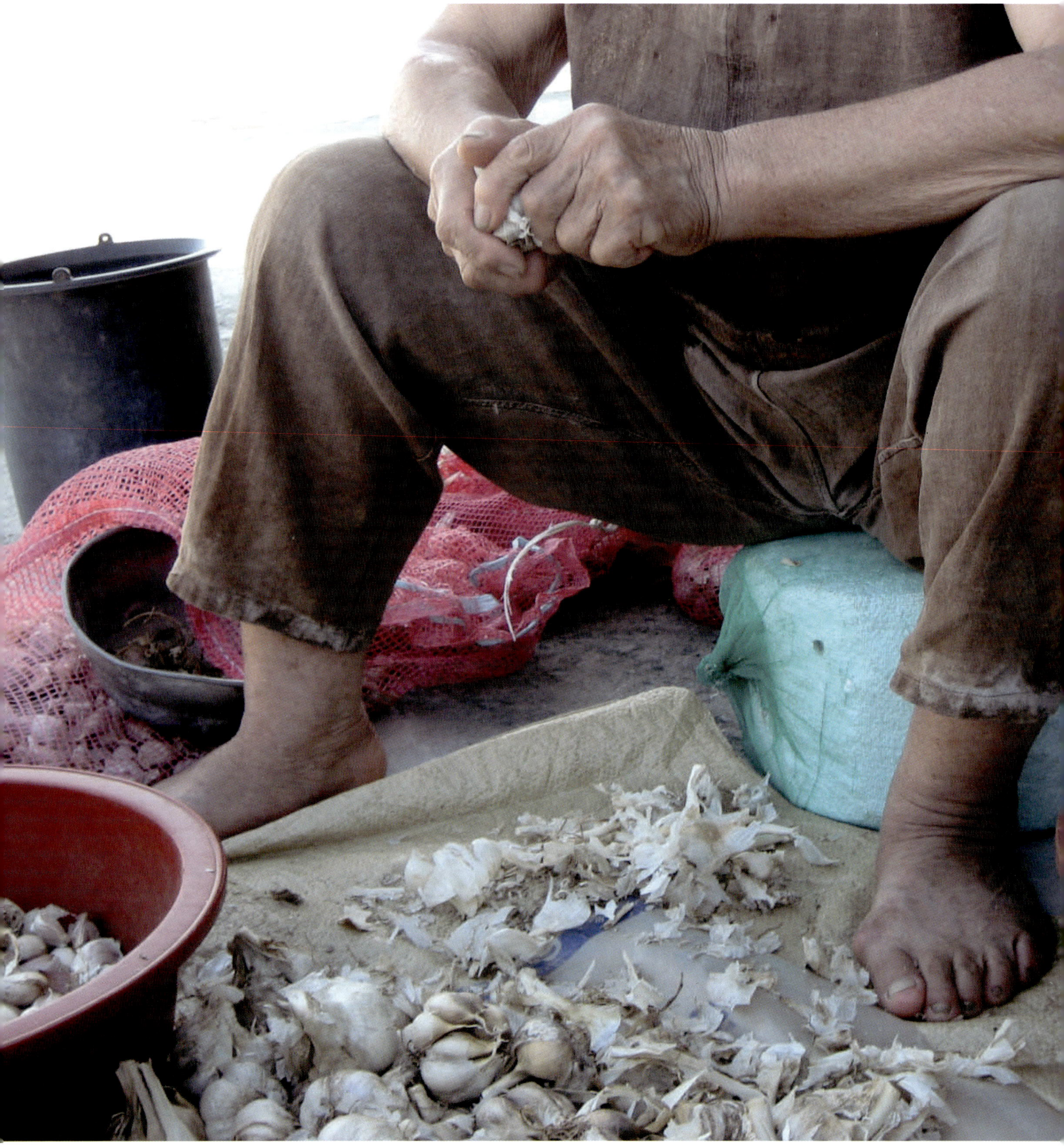

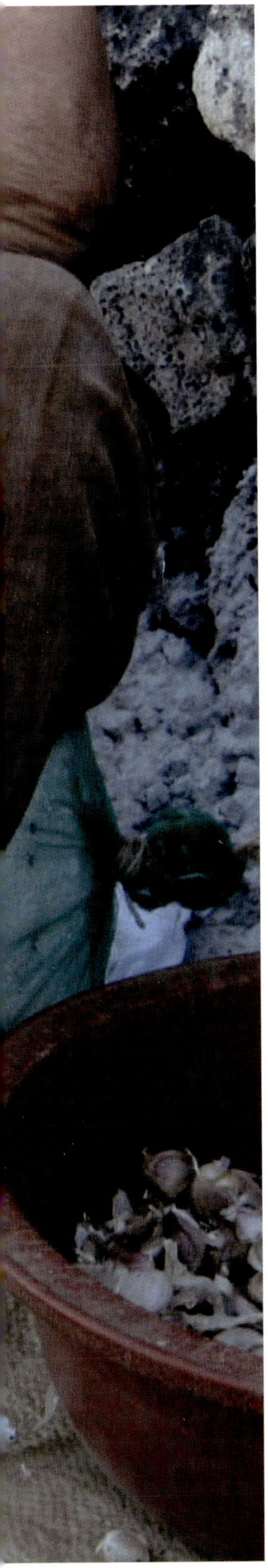

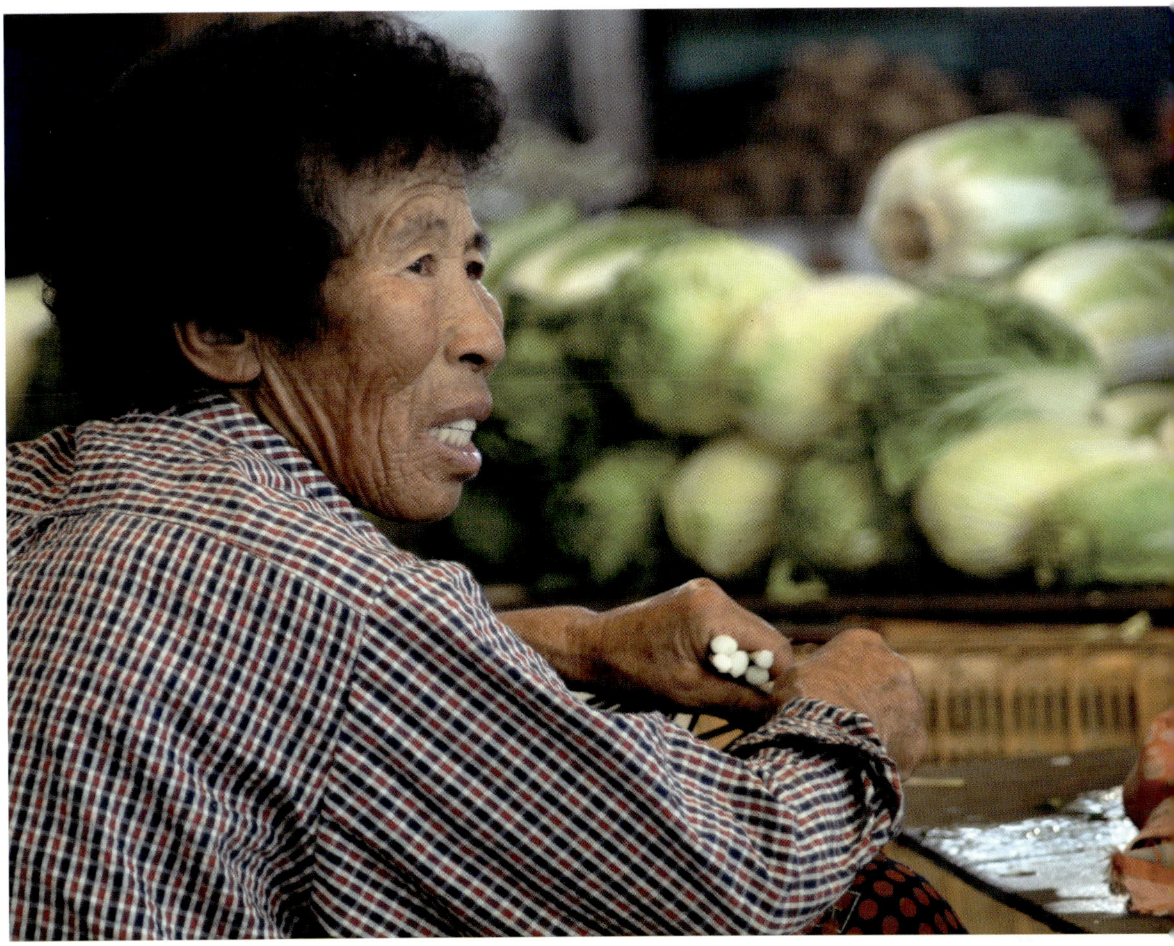

Harvest from the Land

When the *haenyeo* are not diving, many earn a living by farming. Throughout the island, one can see fields of yellow and green onions, garlic, cabbage, and other products, which are often sold at the local markets.

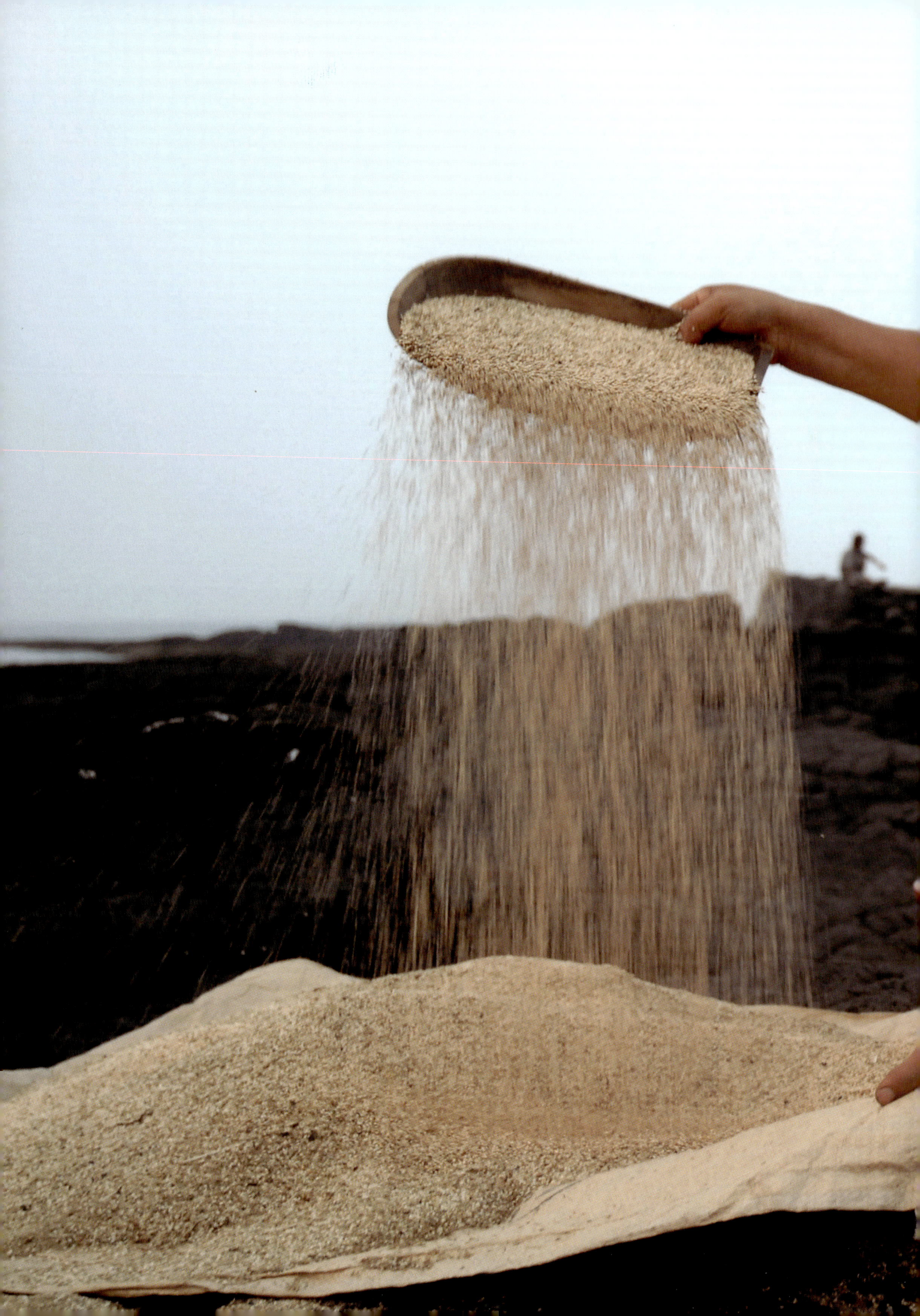

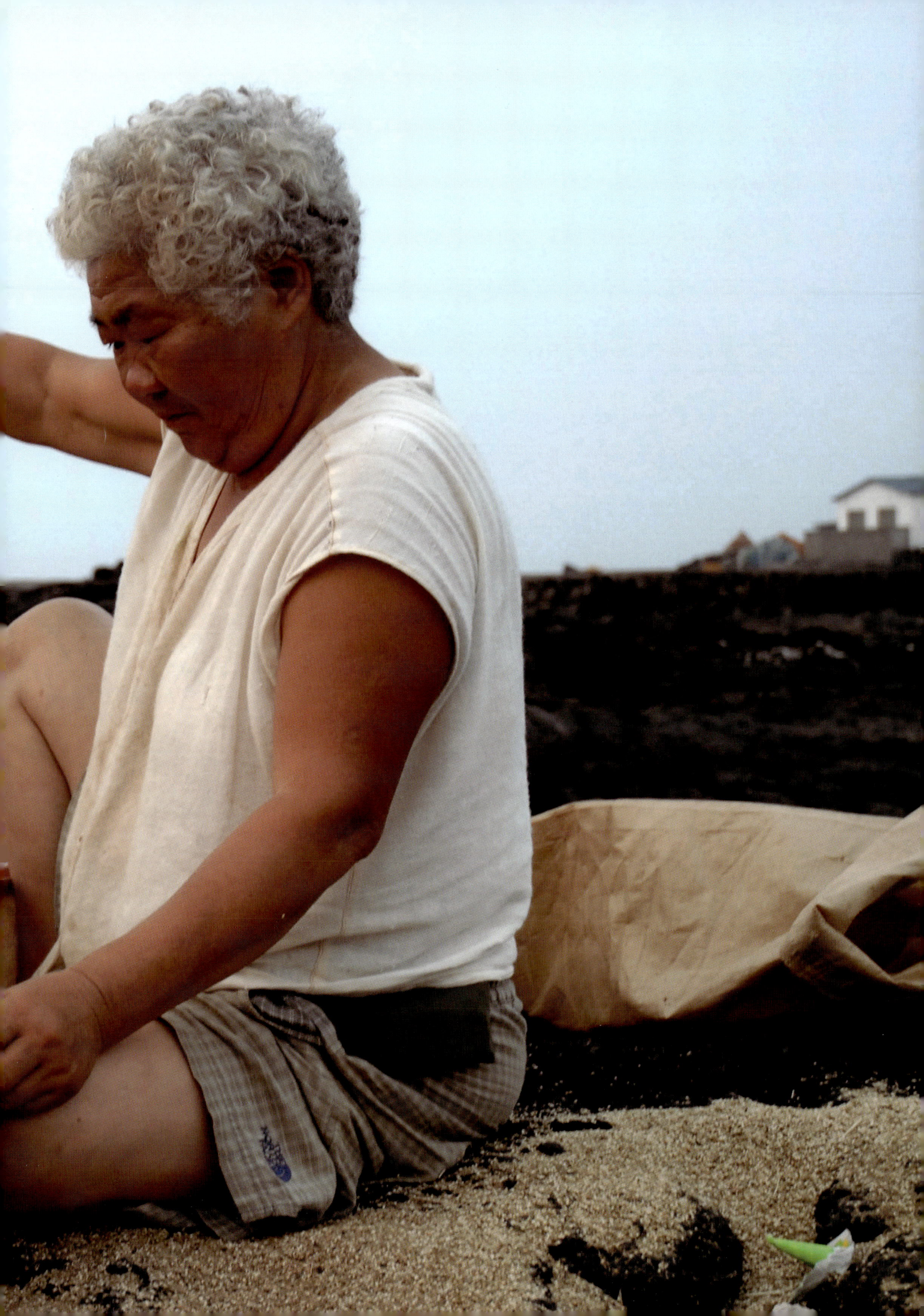

During my second visit, I met Cho Jeong-sun of Sagye-ri at Jeju Medical Center. Her husband would drive over an hour into the city so she could receive oxygen treatments. She is considered one of her village's best divers. Even so, the treatments give her a boost of energy that sustains her while diving.

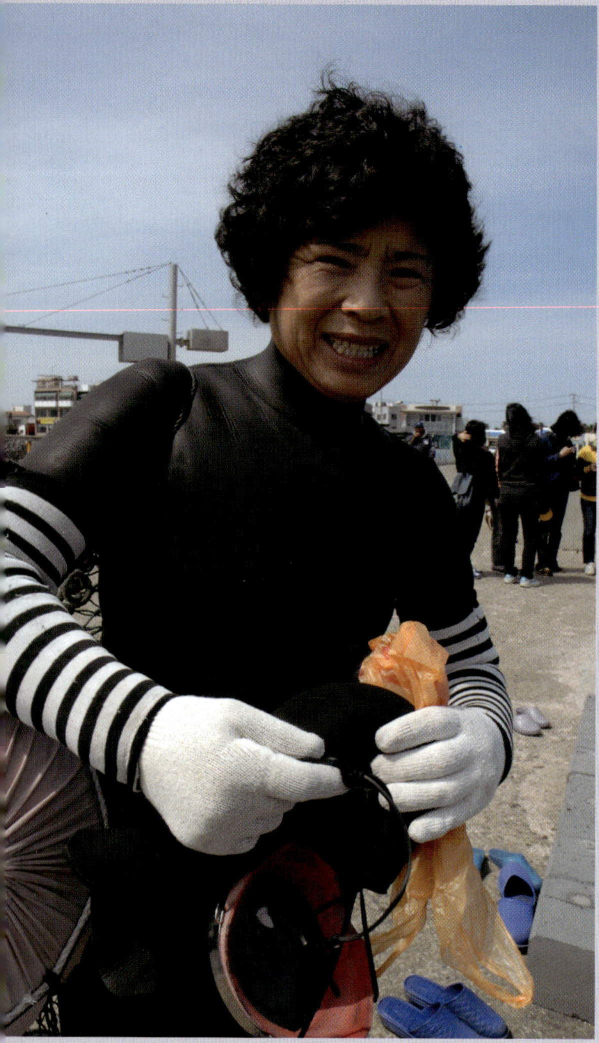

CHO JEONG-SUN

b.1957 | Sagye-ri

Jeju women have very strong willpower. That's how I've been able to manage diving, farming, and helping my husband run a pension for tourists. We both grew up in Sagye, and we have two sons and one daughter.

My mother and mother-in-law were both *haenyeo*. I'm not sure about my grandmother. When I was young, my family was very poor, so I began diving after I finished elementary school. Even though I was young, I went out to the deep sea with the older and higher-level *haenyeo*. I could stay underwater for a long time, much longer than most.

We don't use diving equipment because we don't feel safe with it. We don't trust that kind of thing. One overseas Korean *haenyeo* died in Japan while using such equipment. We think she had the underwater accident because of greed.

This year is a good year, but we can't really predict exactly how much we'll earn. Our ancestors told us that it all depends on divine will. Earlier this year, we caught many turban shells, but we stopped during the spawning season. We'll begin again in the fall. In between dives, the *haenyeo* work together on the farms and divide the profits equally among us. But when we work in individually designated areas, we each earn according to our ability.

Every six days during the diving season, we have six groups of *haenyeo* who sell fresh sea products to the tourists visiting Sanbang-san. Whatever we make together we

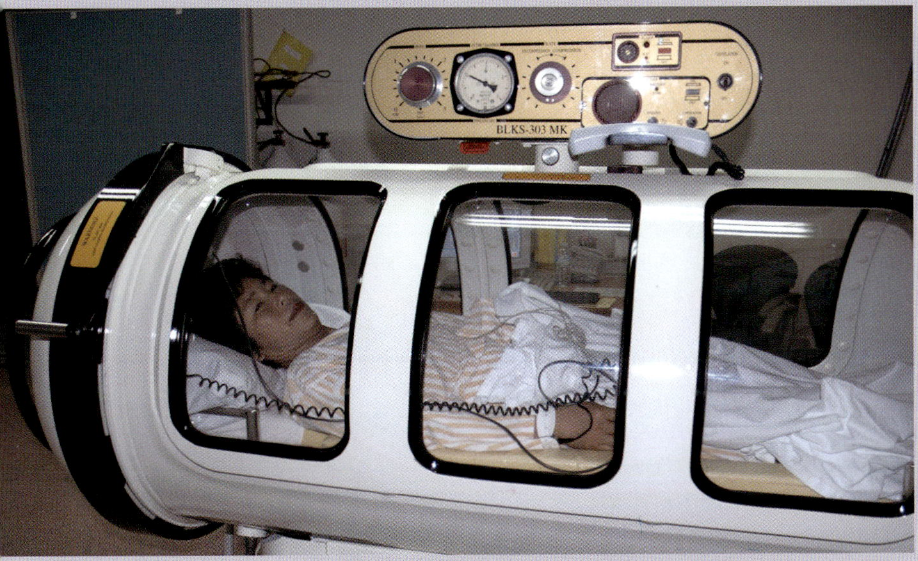

Cho Jeong-sun in oxygen treatment chamber. Most *haenyeo* suffer from headaches, arthralgia, pain in their hands and feet. By being treated with this equipment, their symptoms can gradually be relieved. The treatment is free for the *haenyeo* if they have a registration card to verify their line of work. It's supported by the local government.

divide up equally.

My health is good. Sometimes my body is stiff. But after being in the water a bit, I feel more comfortable. Before diving, I take medicine to prevent headaches. I've taken *noeseon* (an acetaminophen-based pain reliever) for eight years.

I used to go into Jeju-si to receive oxygen chamber treatments. Whenever I had a cold, I felt dizzy and my ears would start ringing. In 2006, a warship came to the harbor near my village. It had one of these machines, so I got to experience the benefits. Later I found out that Jeju Medical Center had one. My husband would drive me whenever I needed the treatment. I've had more than 15 sessions. After the treatments, the symptoms disappeared. I haven't gone for a while.

I'm happier doing *haenyeo* work than farming. When I'm underwater searching for sea products, it's fun. I've actually never swum in a pool. If I were allowed to, I'd want to go into the sea everyday.

(April, 2008)

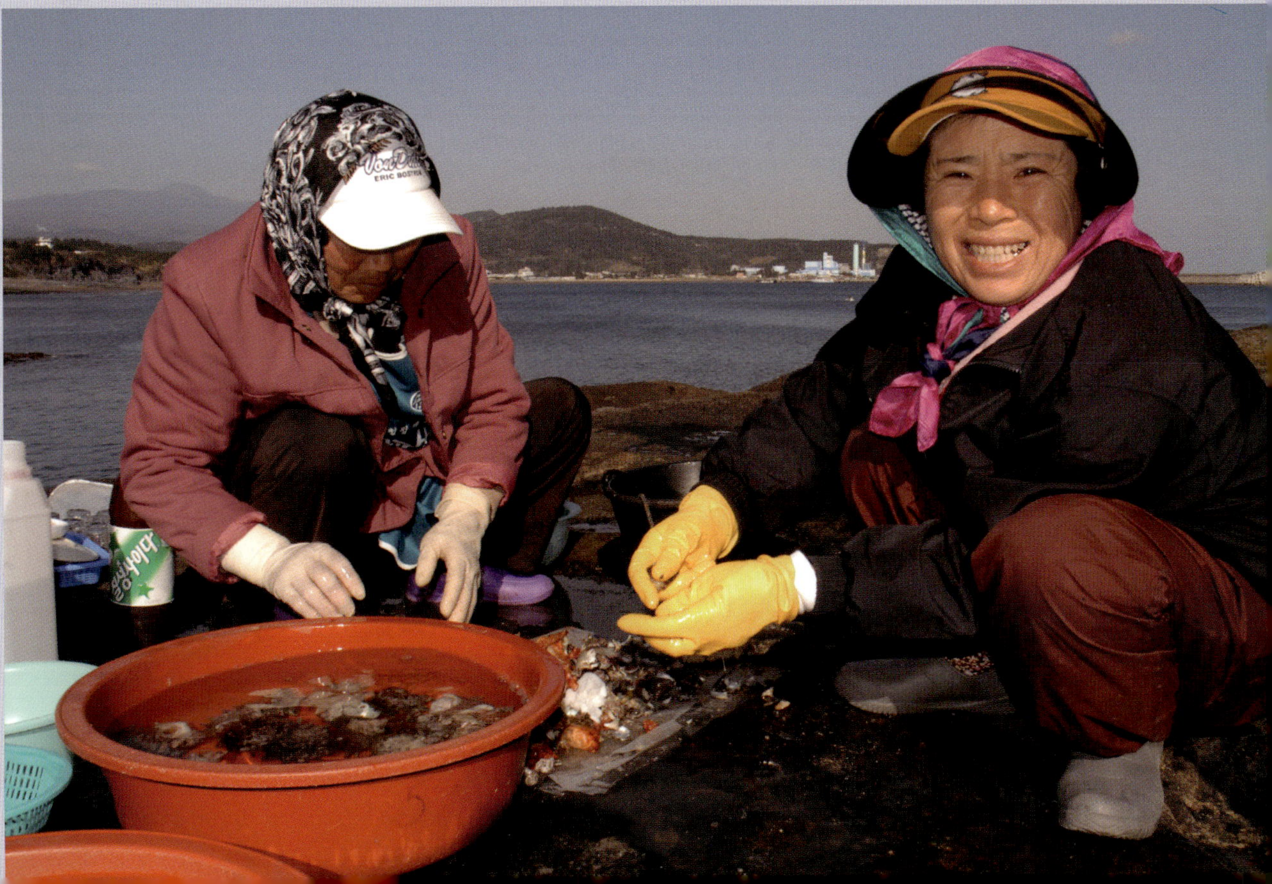

Oh Jeong-ja has never feared for her life. She enjoys her work as a *haenyeo* in spite of its risks. If her daughter, Kim Myeong-sin, decided to follow in her mother's tradition, that would be fine, she says. But Myeong-sin, a white-collar professional, still remembers the pain of separation when her mother would go diving in Japan for months at a time up until 2001. "Whenever she left, I missed her. I wouldn't wear anything she made for me because it made me sad and worried. I heard about a neighborhood *haenyeo* who died after getting tangled in a fishnet. So when my mom went off to the sea, I always prayed."

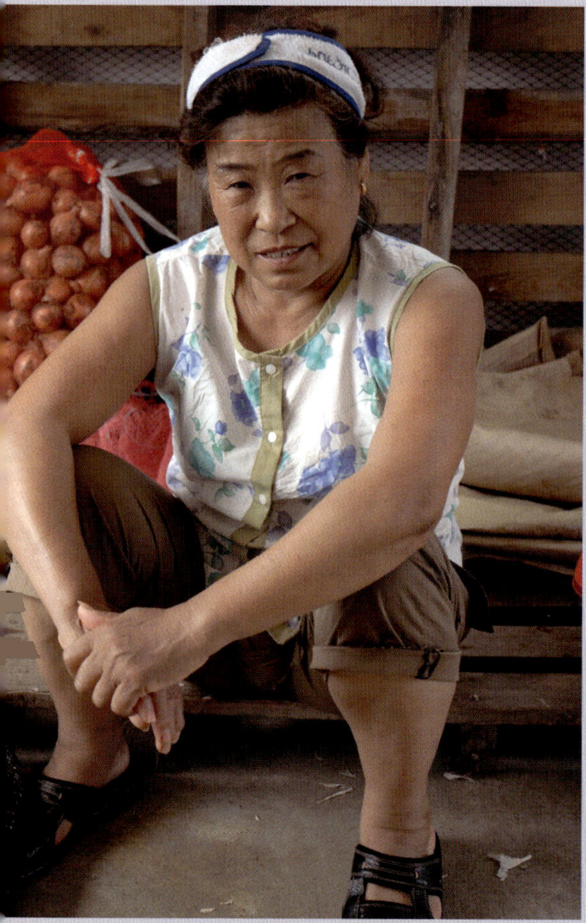

OH JEONG-JA

b.1940 I Woljeong-ri

A long time ago, I followed my grandmother while carrying a bamboo basket with wood to make a fire at the *bulteok* (fire pit). My *taewak* (flotation device) was made from a small gourd with a net sack attached. After we collected the *miyeok* (brown seaweed), we put it in the basket and carried it back home on our backs. We usually collected the seaweed in the spring and continued until the fall. We didn't dive after that because it was too cold. But now *haenyeo* can dive all year round thanks to our rubber suits.

I've never been afraid of the water, even in the deep ocean. I couldn't have become a *haenyeo* had I been afraid as a child. Even then, I could dive around eight meters (26 feet) deep and catch seaweed and turban shells. These days, I can dive 10 to 15 meters (33 to 50 feet) deep, and remain underwater for about two minutes, I guess. My skills have improved over the years because I know all about the sea—where and what I can catch.

I've been diving for 52 years. It's not dangerous to me. It's enjoyable to look around. Sometimes we'll encounter a school of dolphins. They don't menace us, but sometimes we feel alarmed. I also enjoy diving for the good physical exercise. Even when I feel sick or fatigued, I feel relief from going into the water. It has healing effects that make you feel comfortable. I love the sea and working in the water for the catch.

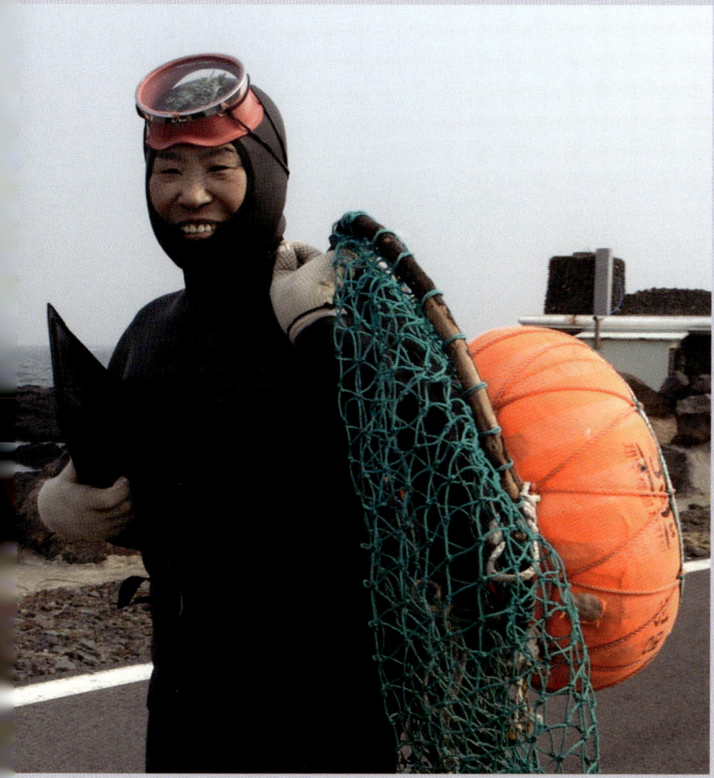

When we come up for air, we rest on a *taewak*. A long time ago, we used to use a gourd, but now we use styrofoam to make our *taewak*. The styrofoam pieces float ashore from mainland fish farms. That's where we get it. When you hear a whistling sound, we call that *sumbisori*. That sound is made when a *haenyeo* inhales and exhales the breath after rising to the surface. By doing so, we can remain under the water longer.

Up until 2001, I used to go to Japan every year for 10 years. I'd stay there for three months at a time. I had an operation on my leg and eventually quit going overseas. I made $1,000 during the three months I dived for a total of 25 days or so. We weren't allowed to dive more than 10 days per month.

My relatives were the ones who invited me to Japan. There were three *haenyeo* in the village where I stayed. Each of us worked for a different boss. Each man had his own boat and worked separately. I worked four hours a day—two hours in the morning, and two hours in the afternoon. From 8 to 10 am and from 1 to 3 pm. I had a boss who arranged the work. We split the earnings in half. I stayed on the second floor of his house, and shared meals together with his family.

(July, 2007)

Jeong-ja and her daughter, Myeong-shin

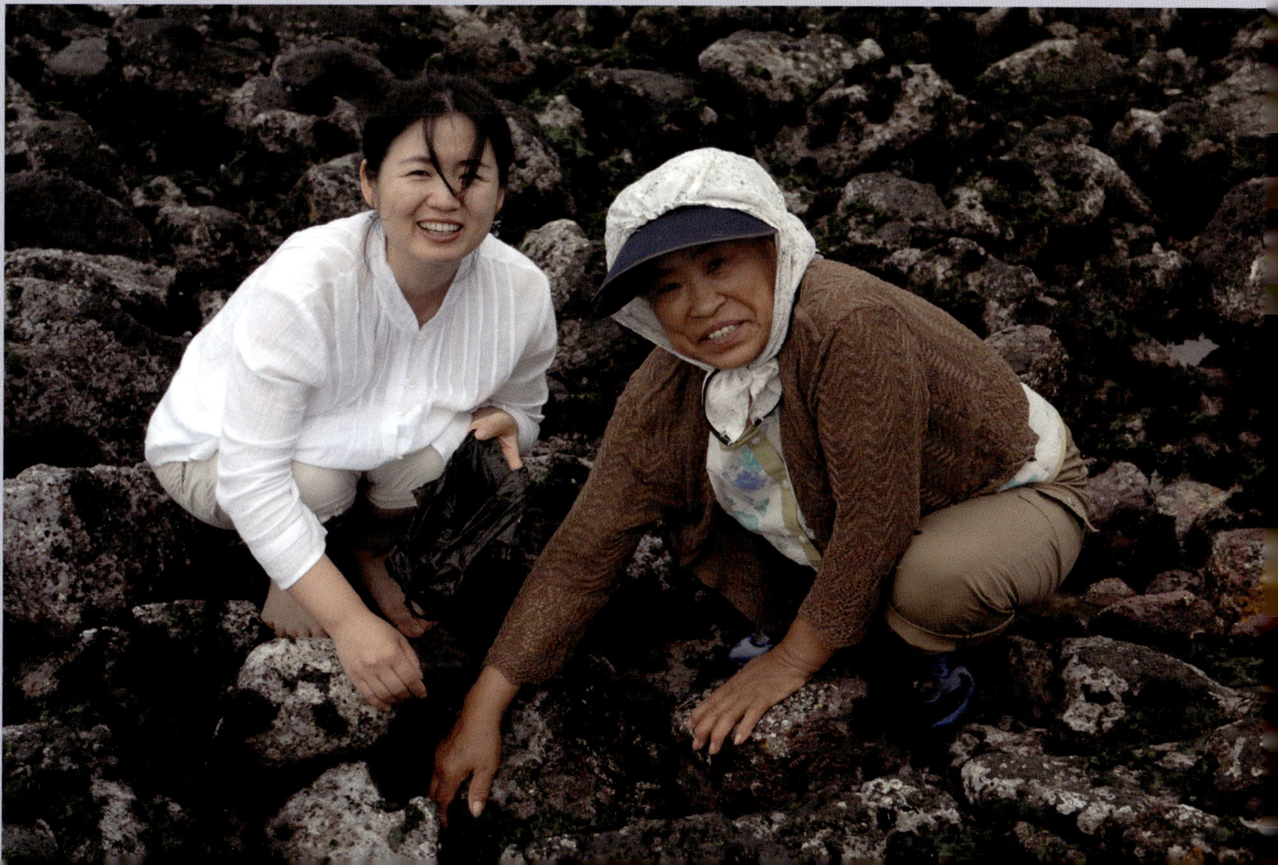

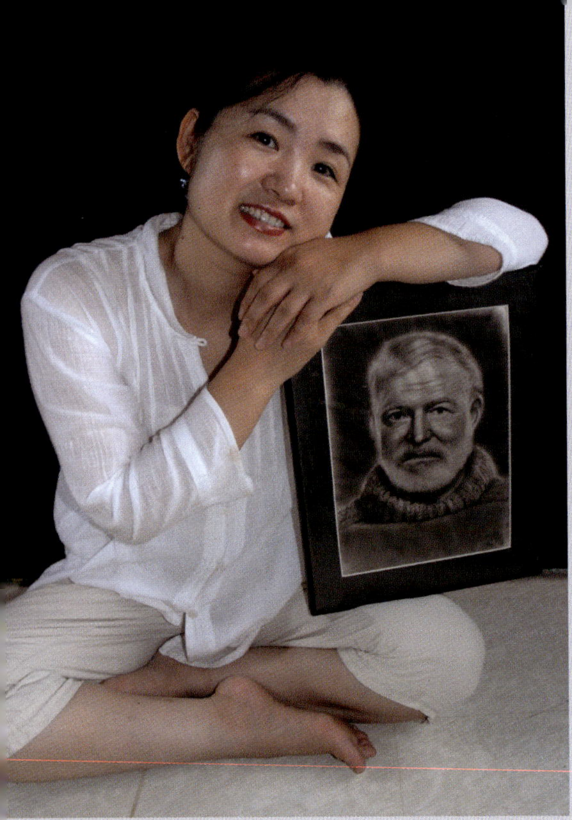

Kim Myeong-sin poses with her sketch of Ernest Hemingway.

KIM MYEONG-SIN

b.1976 | Gimnyeong-ri

Most of my dreams are related to water—a lake near the mountain, not the sea. I believe my dreams of water are premonitions of some sort. For example, one time I dreamt I went to the mountain and saw many small rocks and clean water. It was very strange to me at the time. But later I received a promotion when my company was not expected to give any. That dream, I felt, was like a sign.

Water has played a big role in my life since birth. I was delivered by a *haenyeo* midwife in my family home in Woljeong-ri. She was my mother's neighbor—a *halmeoni* who lived next door.

I learned how to swim at age seven from my mother, brother, and neighborhood friends. During my elementary school days, my mom made me a *taewak*. I just wore short pants and T-shirts. By age 12, I was going to the beach and catching agar-agar and *miyeok*. As I grew taller, my mother gave me a rubber suit for *haenyeo*. It was fun wearing it for the first time. But my mom forgot to make my lead belt, so I just wore the suit into the sea—and I couldn't dive underwater because I kept floating. I was so frustrated. After getting my own lead belt, I could dive deeper into the sea. A swimming pool isn't as much fun. I always preferred swimming near the rocky coast where I could spot the seaweed, turban shells, and octopuses. If I caught an octopus, that was a lucky day.

When underwater, I used to feel like I was princess of the sea. I enjoyed talking to the fish and the seaweed. I always felt as though the seaweed was dancing; I would synchronize with moves that I had seen on TV. My eldest son is learning how to swim well because I would like to show him the underwater sea life in the future. There are many big rocks, but underneath are so many living creatures.

When I caught *sora* (turban shells), I could make 100 won, which was a lot at that time. That was my pocket money. I could buy notebooks, pencils, and some cookies. I remember taking some of the money I earned and buying cigarettes and sugar for my father. In those days, sugar was considered very delicious because we didn't have

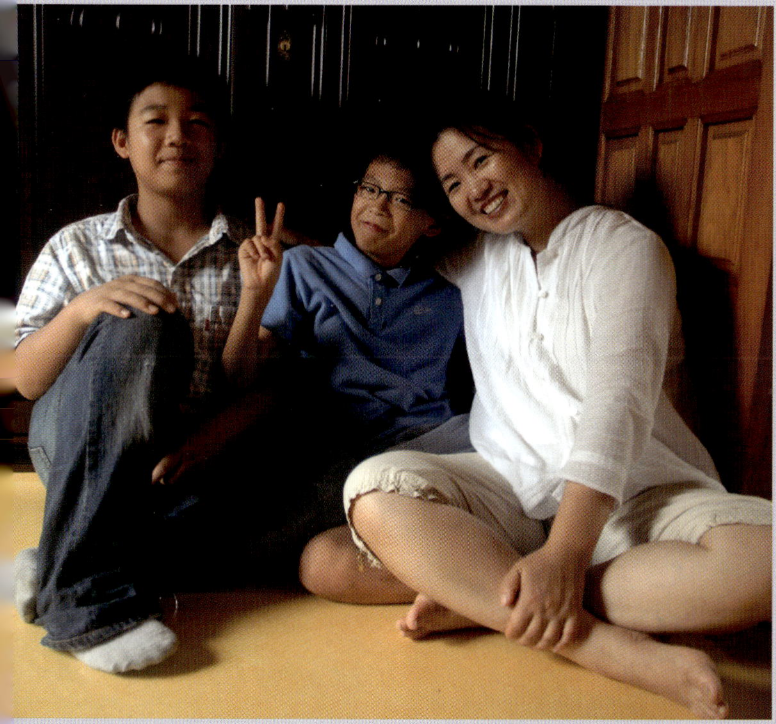

any cookies. There wasn't enough food to eat. I remember eating the sugar with a spoon.

My mom dived in Japan for many years. Whenever she left, I missed her. I wouldn't wear anything she made for me because it made me sad and worried. I didn't study well in school either because I missed her. That was not a lucky time for me. In those days, there wasn't phone access or email. Nothing. I just had to wait. One time, I wrote her a letter and sent it off to Japan. When I heard she cried after reading my letter, I didn't write to her any more.

You see, as a young girl, I thought *haenyeo* work was very dangerous. I had heard about a neighborhood *haenyeo* who died after getting tangled in a fishnet. So when my mom went off to the sea, I always prayed. I prayed while cooking, boiling water, and washing the floor.

My father took care of me and my two brothers. He did everything. In order to feed us good food, he'd prepare fried egg sandwiches and milk. There was nothing like that in my hometown. I don't even know where he got it. My friends envied me, so I was very proud of my father. Sometimes I would wake up in the middle of the night. The moon was shining, and my father was still awake making fishing equipment to catch the octopuses. He would often go to the beach in the middle of the night. I would be nervous, worried, and afraid for me and my brothers. Dad would come home by sunrise with lots of fish and cook a pot of soup.

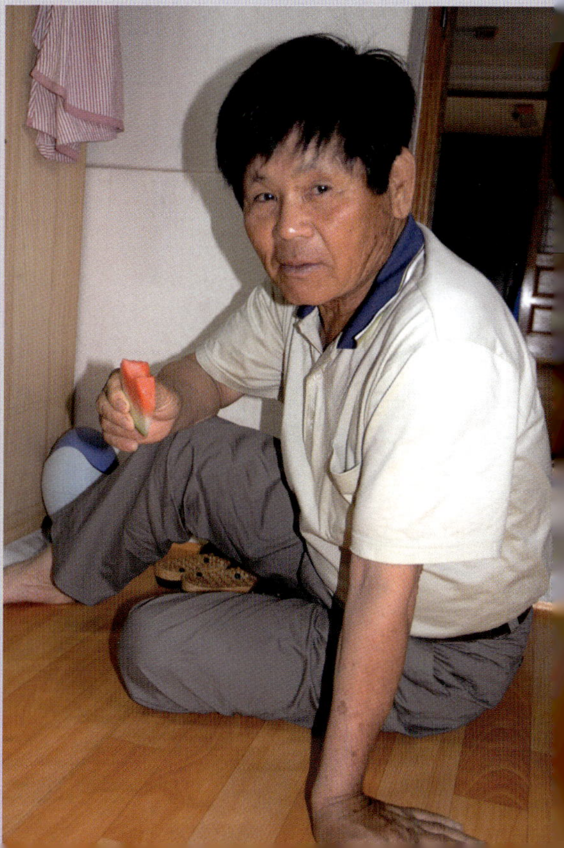

Do you know that in the countryside and villages, some men drink too much and beat their wives? My father never did that. He loves my mom very much. I witnessed his behavior since childhood. He would boil water and wait for my mom to return from the sea. He'd give her the boiling water and towels to warm up and dry herself. He also prepared dinner for her. I always remember this.

(August, 2007)

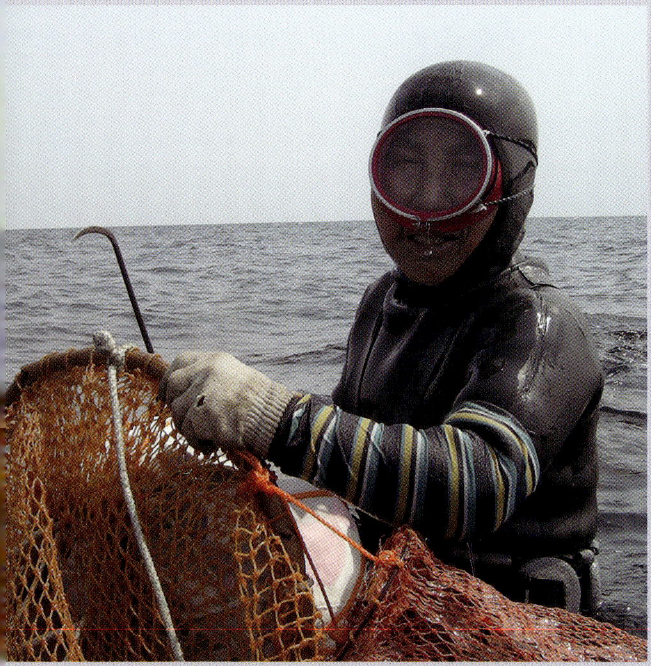

KO MYEONG-HO

b.1931 I Gwideok-ri

Years ago, in my native village, we used to go out by boat. We sang and chanted songs while we rowed at sea.

One of the diving risks is spotting abalone or turban shells right before coming up to the surface. Rather than lose a catch, a *haenyeo* may risk her life by staying down longer than she is capable of holding her breath for. One could lose her life from such poor judgment. Some *haenyeo* have died because of this. We should give up when we have to. I've experienced a few close calls myself when I nearly became unconscious. But finally I let the catch go. When I'm in the water, my body feels free because of its energy. That's why I'm comfortable in the sea, even though I am very tired when I come out.

(March, 2008)

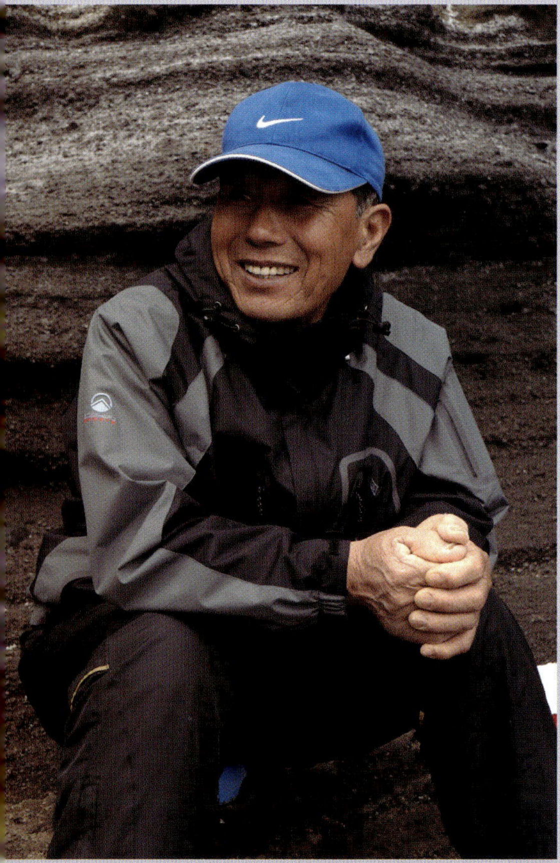

IM SEONG-OK

b.1940 I Sagye-ri

I was born here in Sagye, so I grew up playing in the water and swimming quite well. In fact, I won a prize in a swimming competition when I was young.

I learned how to dive by myself. As I mentioned before, I've loved the water ever since I was a kid. Whenever I was free, I went to the sea to swim and fish. One day while swimming, I saw some turban shells and began catching them. My older sister used to tell me stories about *haenyeo* work, and boasted of catching sea products. So one day I suggested that I help her. She accepted. And that was the beginning of my becoming a male Jeju diver.

At first, the *haenyeo* kept their distance from

me. Some of them complained about my diving because I did so well. So I would dive alone. They thought I would eventually quit and didn't allow me to join the *eochongye* (diving association). Later, when they accepted me, I became a member. That was in 1983.

Since then, I really haven't faced any difficulties, except that I can't use the same shower facilities and changing room as the *haenyeo*. That's all. It sounds like I'm bragging, but I think I understand women better because I work with them. We get along well. In fact, my wife is a *haenyeo*. She is 70 years old. She likes that I am good friends with other *haenyeo*. Sometimes I even participate in their group activities without her. She doesn't mind that I joke around with them. I like talking and joking with the women.

My wife and I met through a matchmaker, and we have three daughters and one son. All of them are married, and all of them live in Jeju. They visit us quite often. But my wife doesn't want any of our daughters to be *haenyeo*. Besides, young people don't want to do such work. It's too hard. We are doing it because we learned how to dive when we were younger.

I know of two other male divers on Jeju. They are younger than me. One of them often goes to Japan. The other, I think, lives in Jeju-si. As for me, I will continue diving for two or three more years, as long as my health allows. I've never gone to the hospital to see a doctor.

I dive the same depths as the *haenyeo*. I've never measured how long I can stay underwater. Most of the *haenyeo* in this village can hold their breath for more than one minute. Now I'm not doing as well as I used to do. But I've never had any accident. Sometimes, when the *haenyeo* are scared of the dolphins, they come closer to me. They feel safer around me, as though I'm their guard.

(March, 2009)

Im's wife, selling seafood at the seashore.

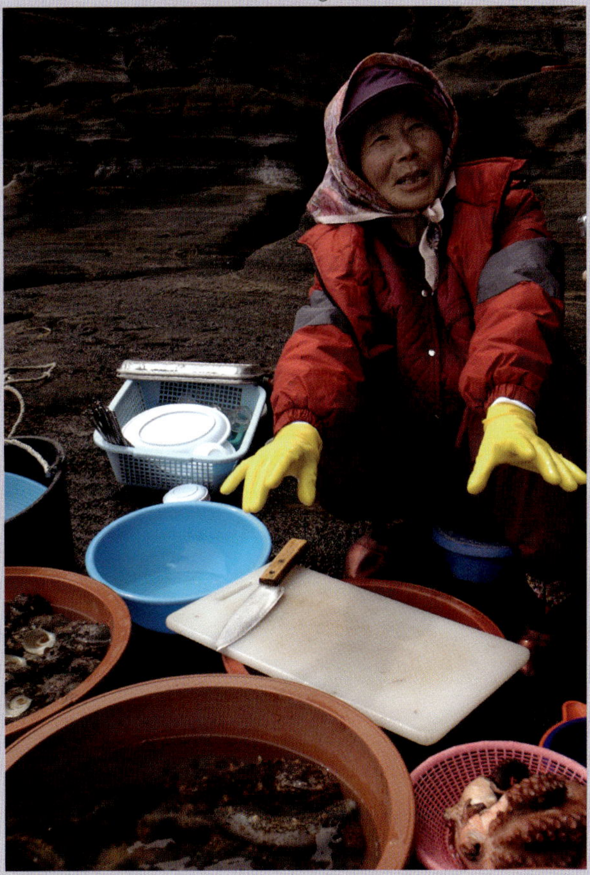

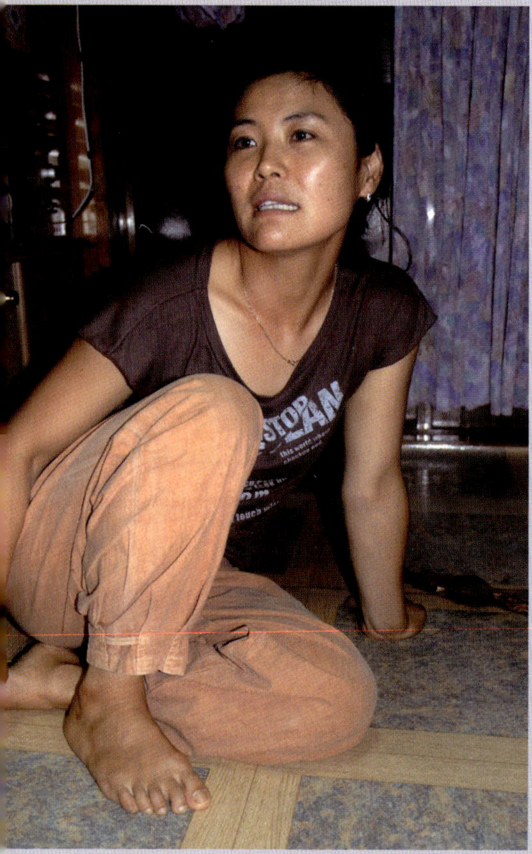

AN MI-SEON

b.1970 | Haengwon-ri

I'm 39—the youngest *haenyeo* in Haengwon, and probably the youngest in Jeju. There are about 150 women in this village who are still diving. My mother and grandmother were both *haenyeo*, and used to work on the mainland at Ulsan, near Busan. I don't know how many generations of women in my family were divers.

From the time I was in middle school, I helped my mother collect agar-agar. Everyone in the village would help clean and dry it, including the elders who could no longer dive.

After high school, I went to the mainland and got married, lived there for a while, and returned to Jeju. I began diving at age 29 because I had to support my family. My children were very young. If I had a full-time job, I wouldn't have been able to take care of them. My husband works on a fish farm. Fortunately, the *haenyeo*'s job is flexible. We usually dive 15 days a month, and only a few hours a day. So I can manage childcare and work.

I don't have any friends my age who are diving. Most young women don't want to be a *haenyeo* because they think the work is too difficult. But I love it. When we all go to the sea, we work together like a family. I'm not as skillful as I'd like, but I'm trying to improve and I'm proud of what I do. Who knows? In a few years, young people may become interested in such work.

On average, a skilled *haenyeo* can earn $1,000 a month. Older *haenyeo*, however, only earn around $200 a month. Not bad, considering that most of the *haenyeo* who are over 60 didn't receive any education. Those in their 50s attended elementary school. And most of those in their 40s and under attended high school.

(August, 2007)

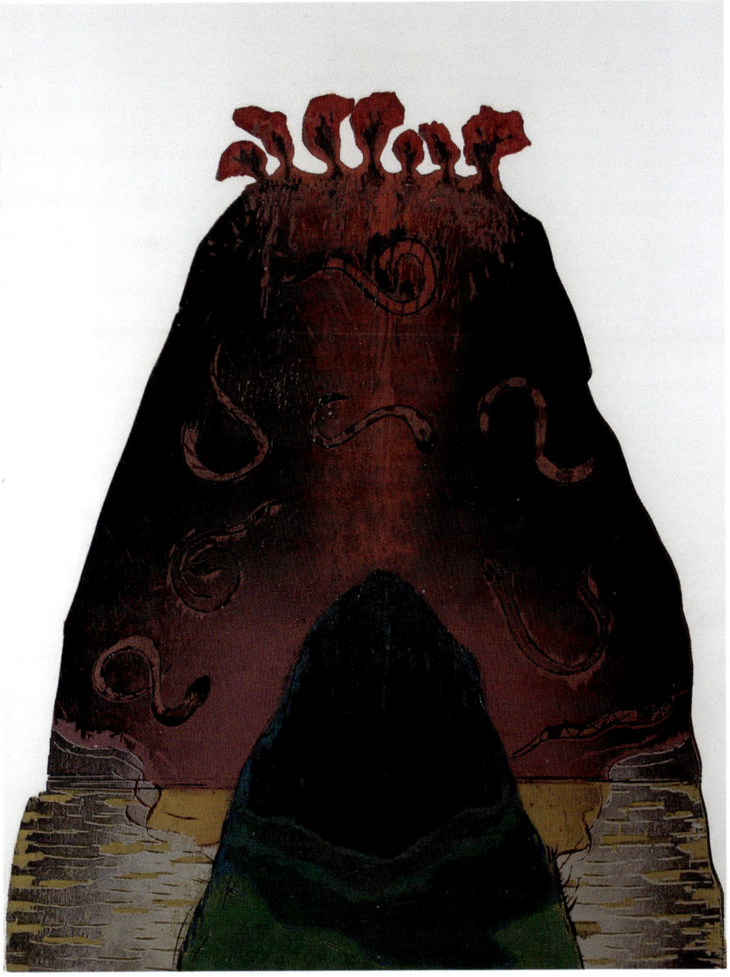

Baeyeomjulyi Illoejungjo 87x70 cm | woodcut | 2008

Yongwang & Yeongdeung

(Dragon King of the Sea & Goddess of the Wind)

The Jeju Chilmeoridang Yeongdeunggut is a ritual held in the second lunar month to pray for gentle seas, an abundant harvest and bountiful catch from the sea. Village shamans perform a series of rituals to the Goddess of the Wind (Yeongdeung), the Dragon King (Yongwang) and the mountain gods. The Yeongdeung Welcome Rite includes a ceremony to summon the gods, prayer for a good catch, and a three-act play to entertain the ancestral gods.

The Yeongdeung Farewell Rite is held two weeks later. It includes offerings of drinks and rice cakes, a ceremony to welcome the Dragon King, and fortune telling with millet seeds. During Yeongdeung's departure on the 15th day, marking the arrival of spring, she sows seeds and calms the turbulent waters. The ceremony ends with the launching of a straw boat into the sea by the village's senior man and woman divers, thereby sending the revered gods back to the sea.

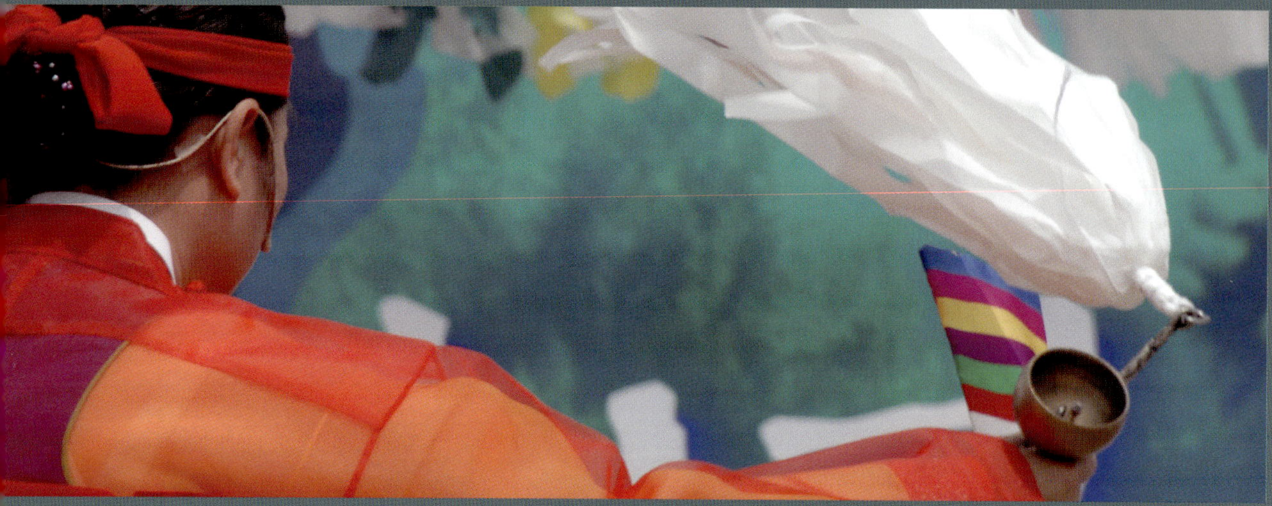

Chapter Two

LIFTED BY THE WIND & TIDES OF SHAMANS

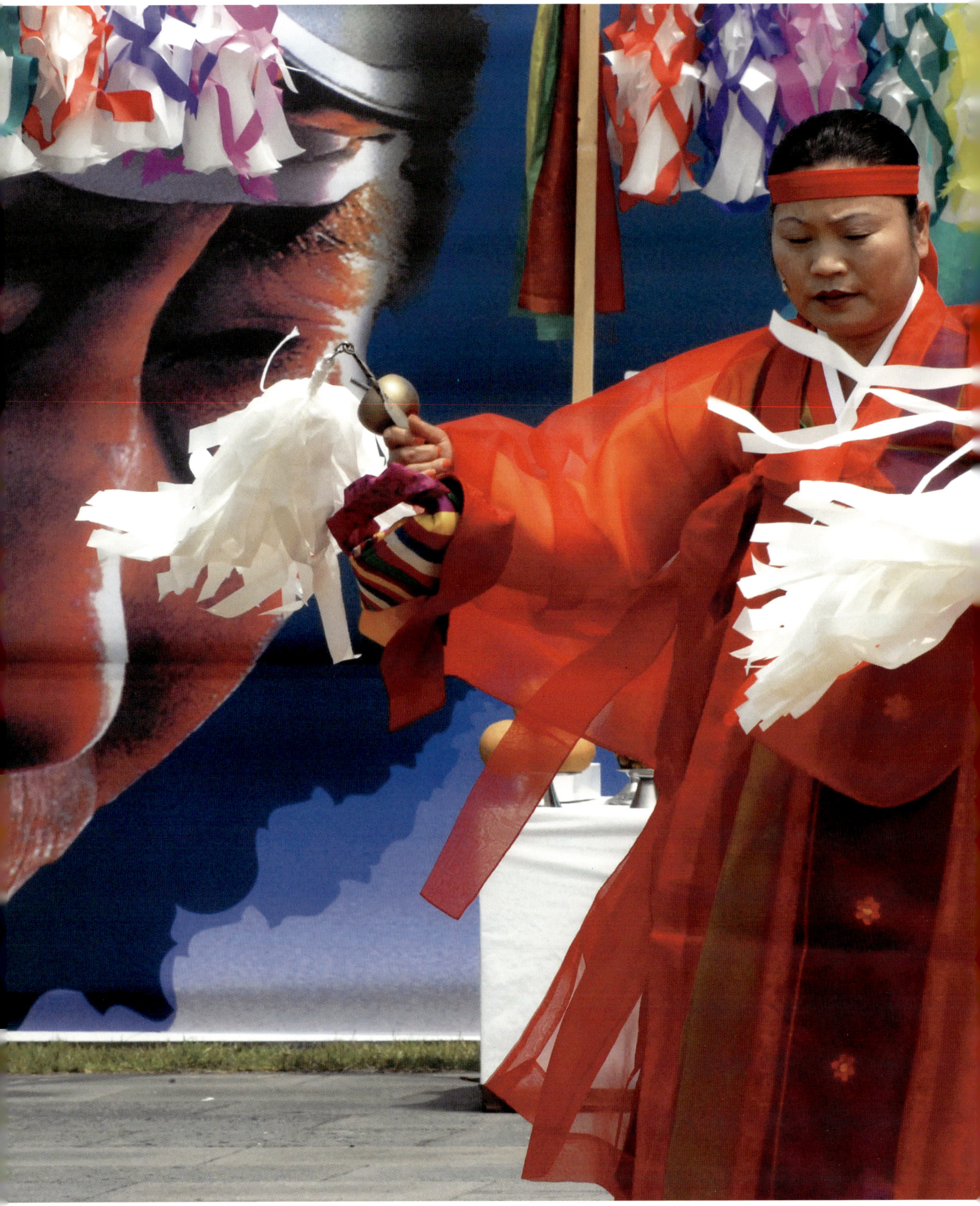

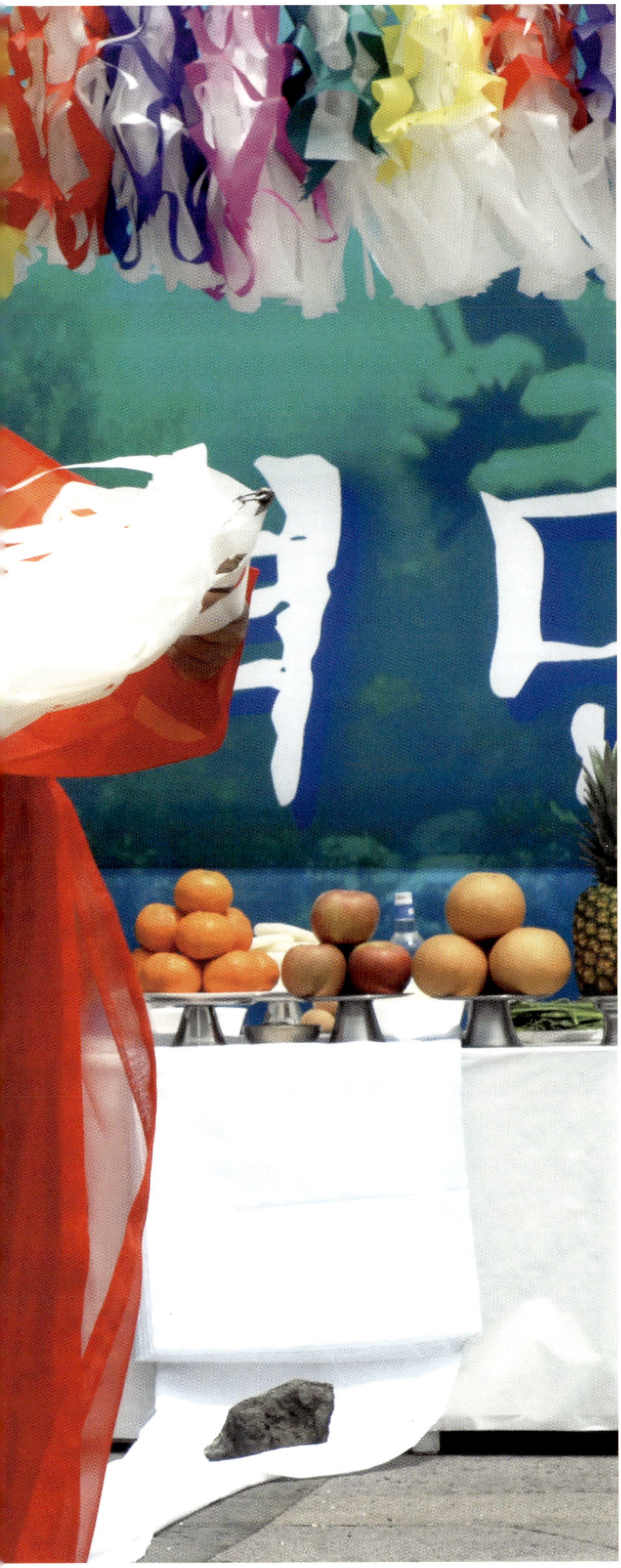

I first encountered a Korean shaman in the 80s, while working for the Korea Times in Los Angeles. Upon entering her house, I was stunned to immediately view the head of a pig, whose orifices were stuffed with dollar bills. Later, the shaman walked outside to her backyard and grabbed a live chicken. While clutching the neck, she began twirling the bird to death in order to secure its blood for ritualistic consumption. I suspended my Christian judgment while witnessing, spellbound, the shaman enter an ecstatic trance. That brief introduction remained unforgettable.

Korean shamans are similar to those in Siberia, Mongolia, Manchuria, and Japan. Most of them are women. They are often reluctant healers. One doesn't aspire to become a shaman. Usually, one's call to shamanism is preceded by a spiritual crisis, often accompanied by a physical or emotional breakdown. They might experience melancholy, solitude, visions, and insomnia. After being treated through a shamanist ritual, the distressed individual is often encouraged to consider the path of a shaman healer, if gifted in this respect.

Shamanism, as a practice, dates back to the Paleolithic era. Originally, the word "shaman" referred to Siberian healers. But in recent times the term has been given to healers in many global traditional cultures—including Jeju Island, Korea—who enter altered states of consciousness in their work.

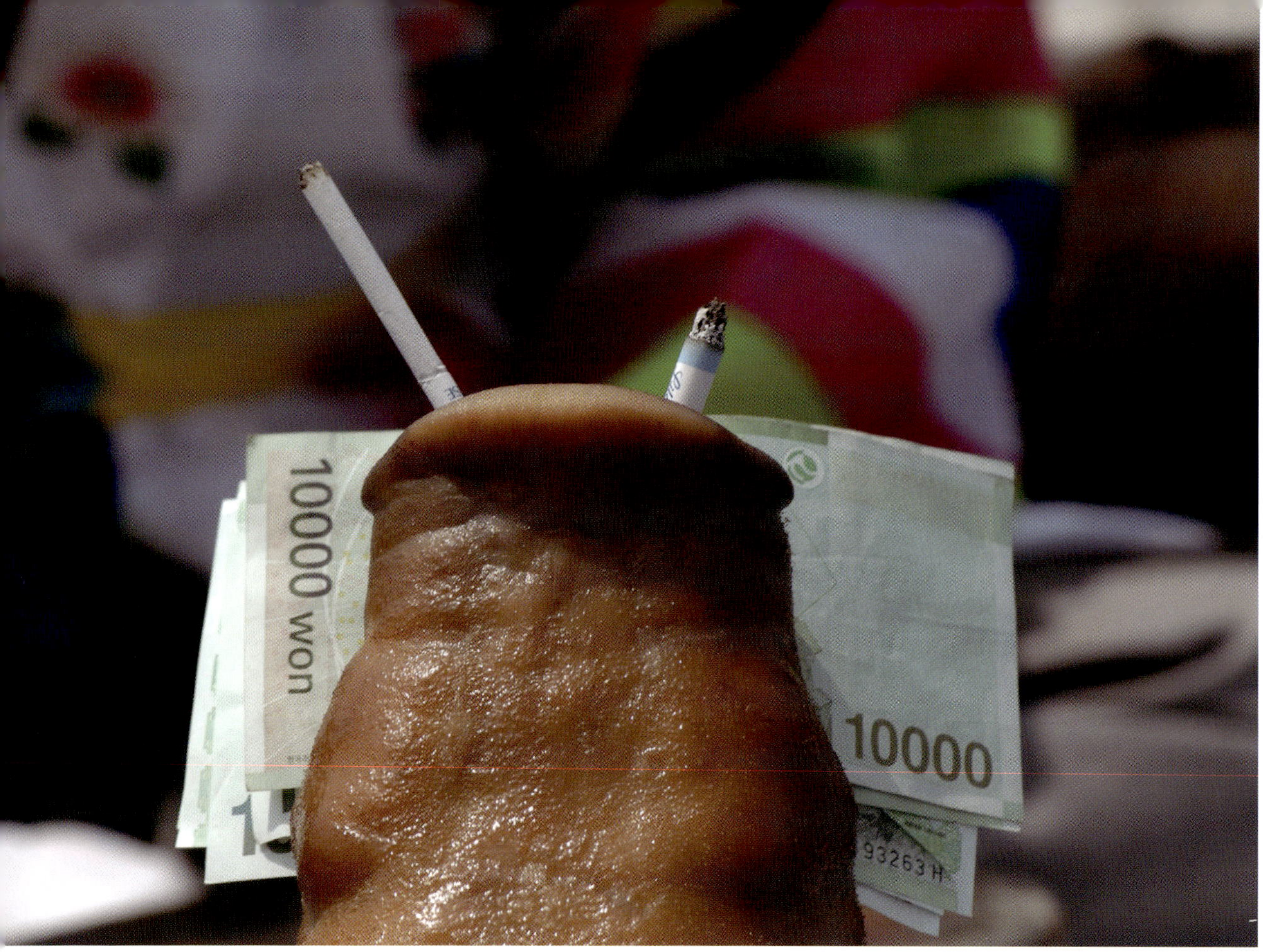

Jeju shamans, like their counterparts, serve as intermediaries or messengers between the physical world and the spirit worlds. They are sought by believers to treat ailments, mend the soul, protect the living, ensure bountiful harvests from the land and sea, bring prosperity to their business, or ask for their children's luck in education and marriage. They also have a specific rite known as *muhongut* to cleanse the spirits of loved ones drowned at sea, and to guide the person peacefully to the land of the dead.

I didn't witness a *muhongut*, but I was able to meet and interview two shamans (*simbang*) in Jeju: Suh Sun-sil and Lee Yong-ok. Both of them performed Yeongdeunggut rituals during the second lunar month in order to protect the fishermen and *haenyeo*, and to ensure a prosperous season. Watching them during the public ceremonies, I was struck by their intensity and charisma. There were moments when they wept for the dead. At other times, they cajoled the spirits, eliciting laughter and clever repartee from the audience. But when I had an opportunity to meet them privately in their homes, I became more aware of their humility and unique circumstances. Both were born into shaman households—their destinies inevitably pre-determined.

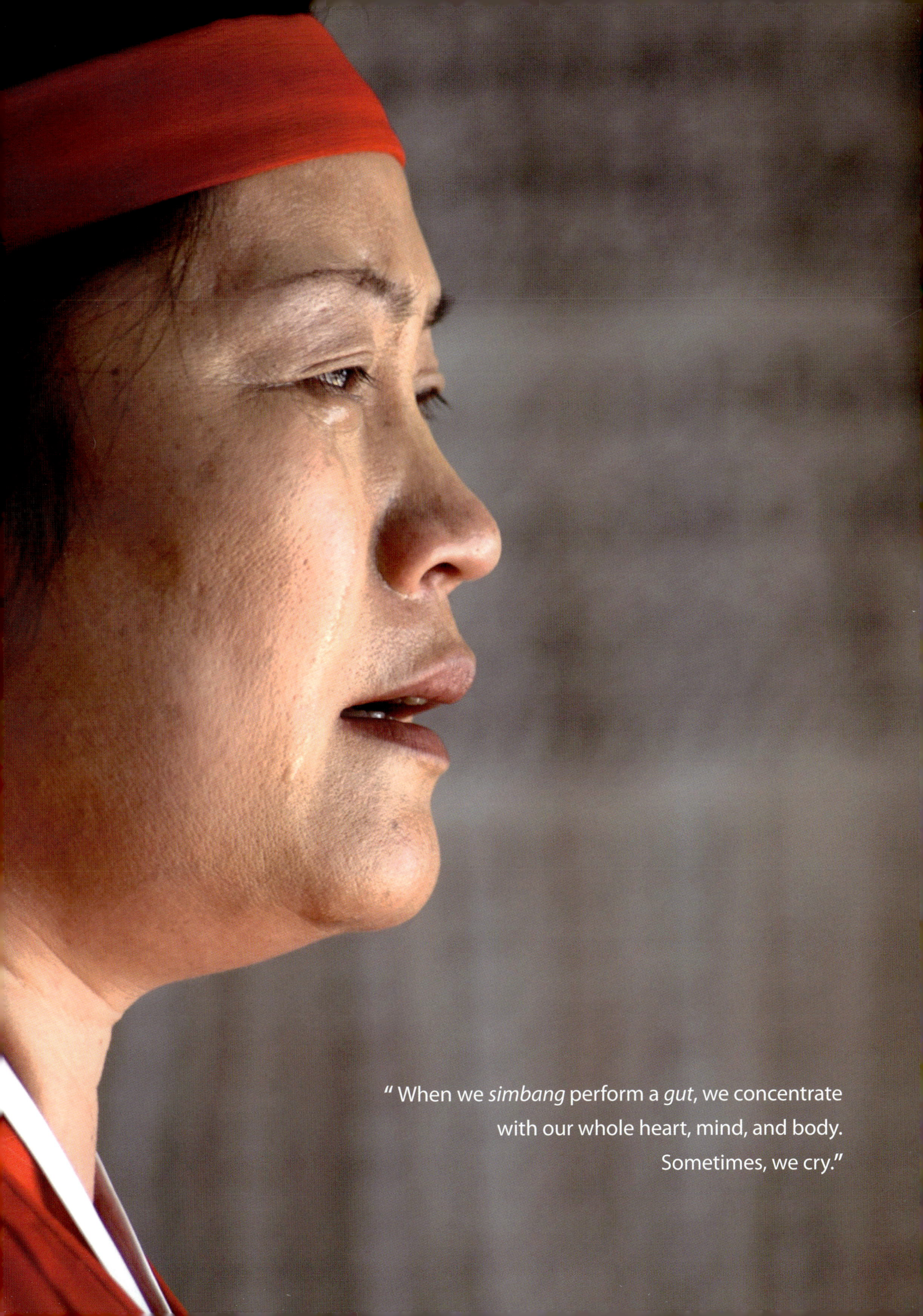
" When we *simbang* perform a *gut*, we concentrate
with our whole heart, mind, and body.
Sometimes, we cry."

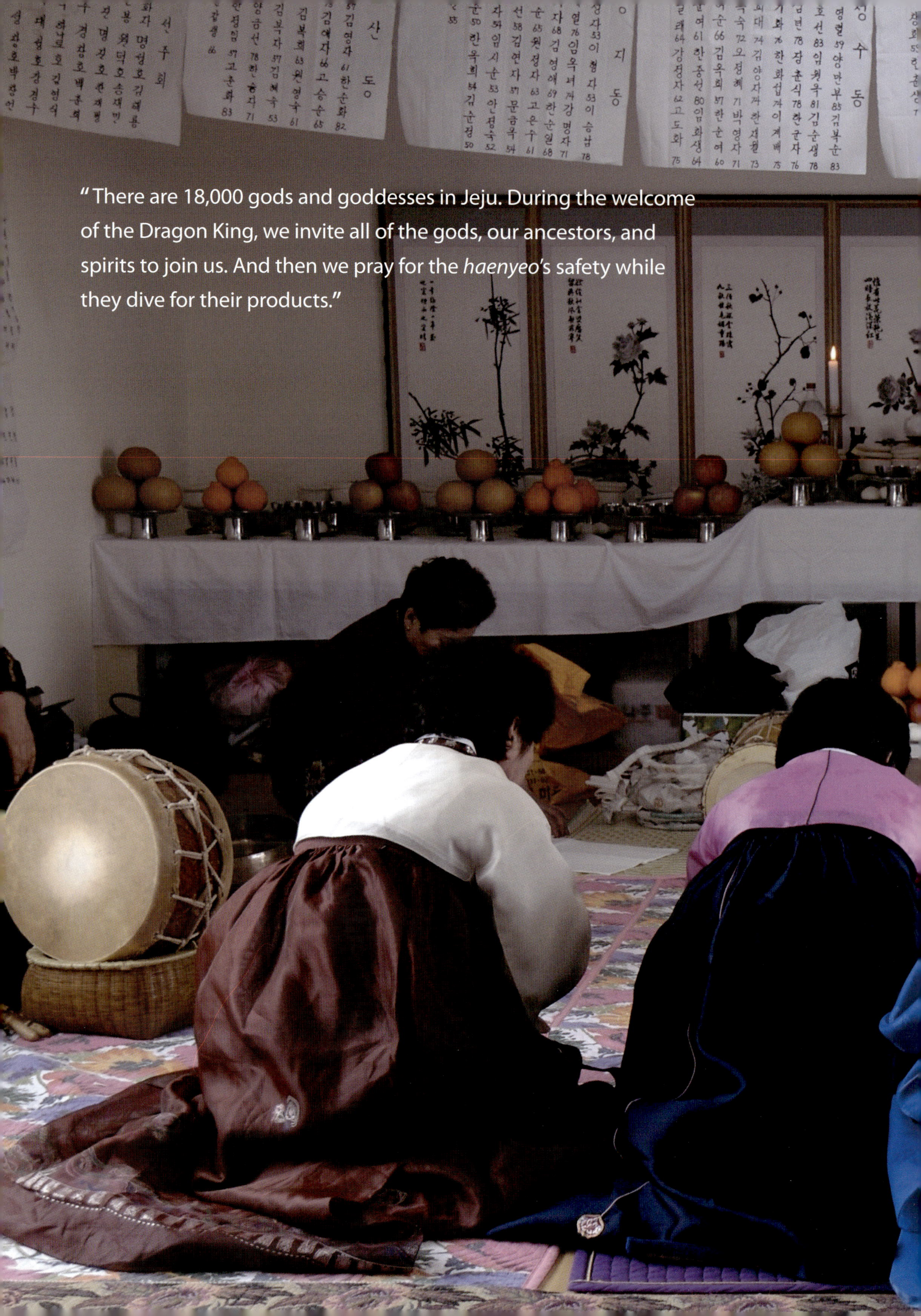

"There are 18,000 gods and goddesses in Jeju. During the welcome of the Dragon King, we invite all of the gods, our ancestors, and spirits to join us. And then we pray for the *haenyeo*'s safety while they dive for their products."

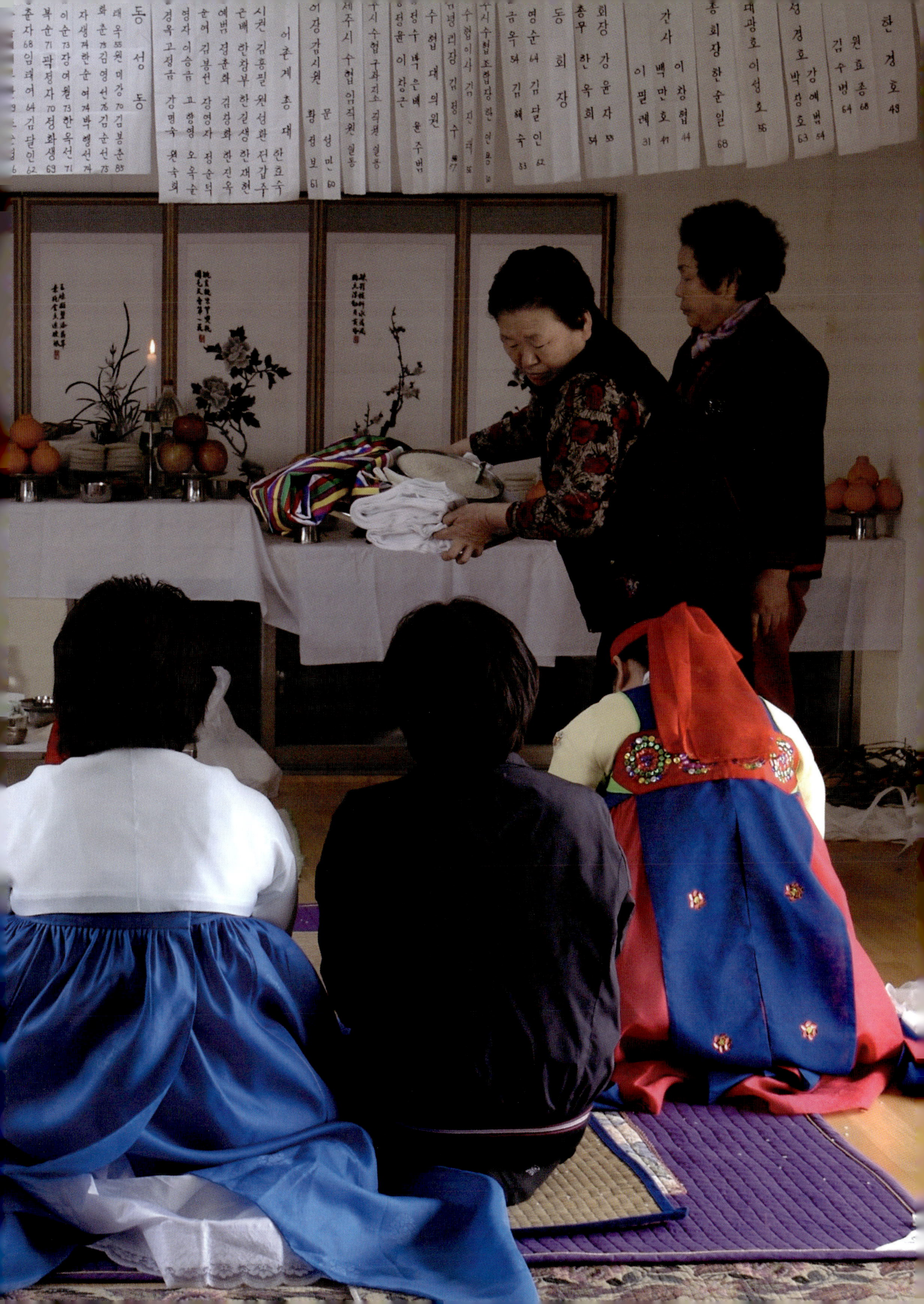
동 회 장
회장 강윤자 53
총무 한옥희 54
한경호 49

"When I am communicating with the spirits—through dancing and talking—I tell the gods what the *haenyeo* want and need for the year, and get the messages from the gods."

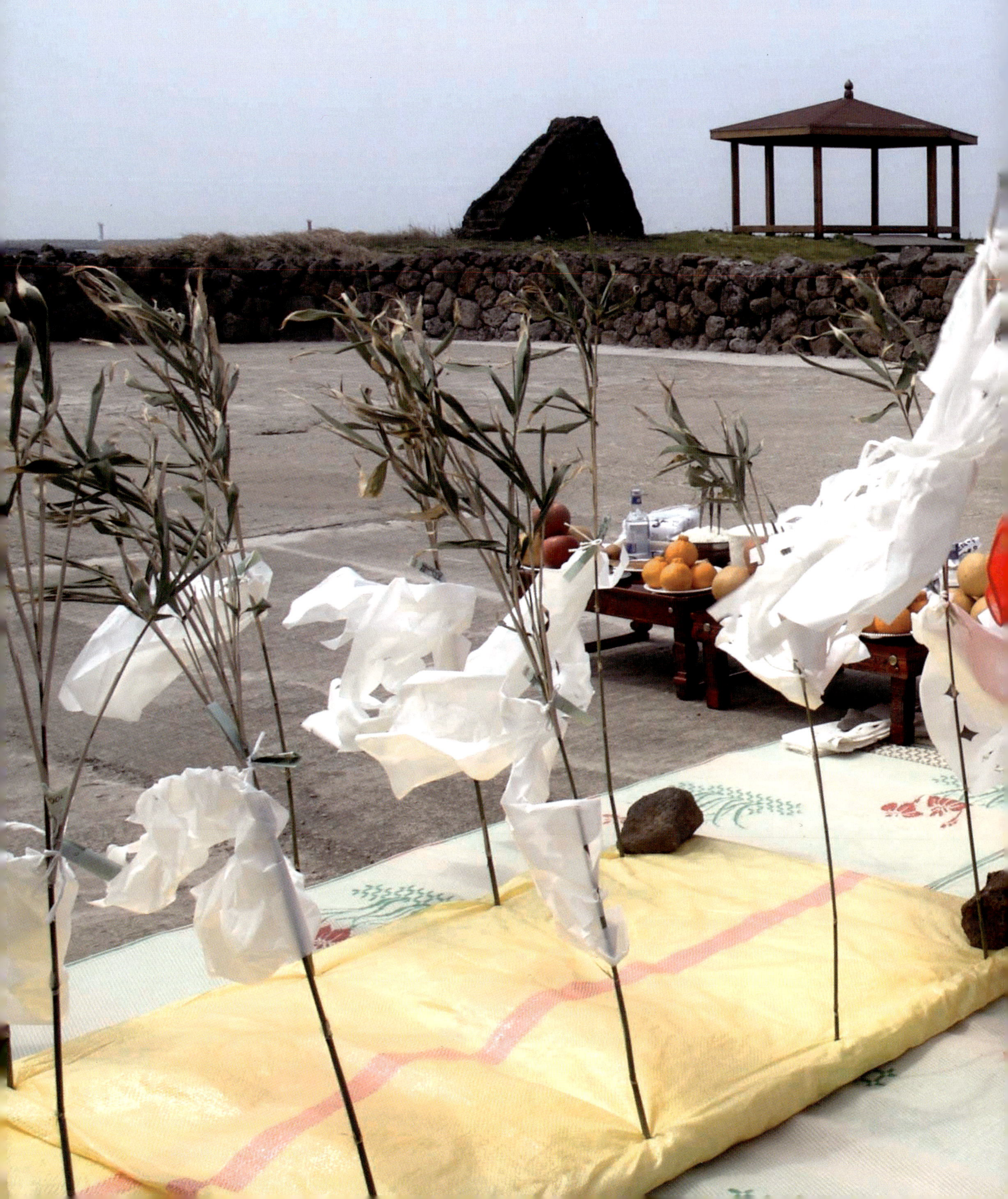

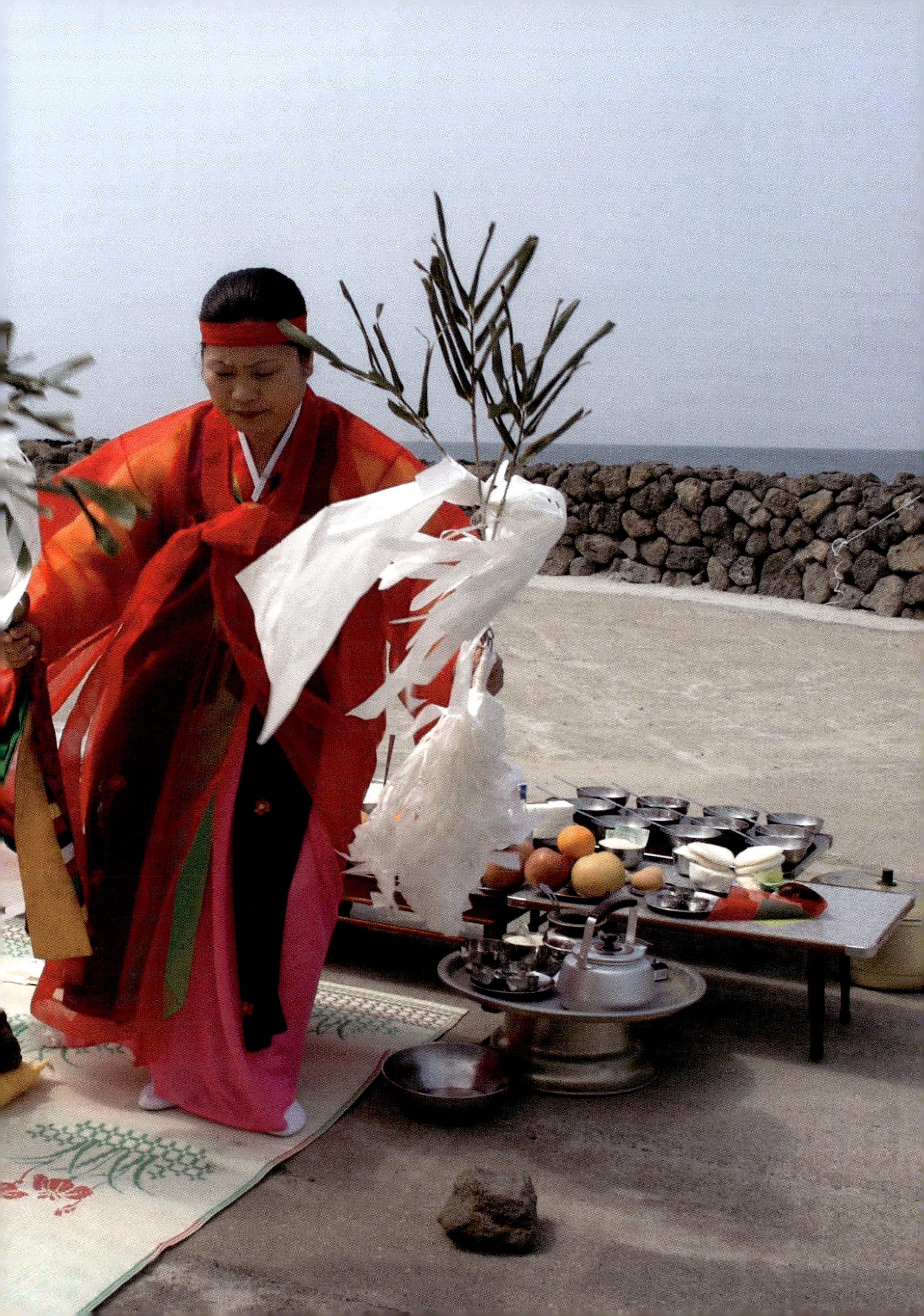

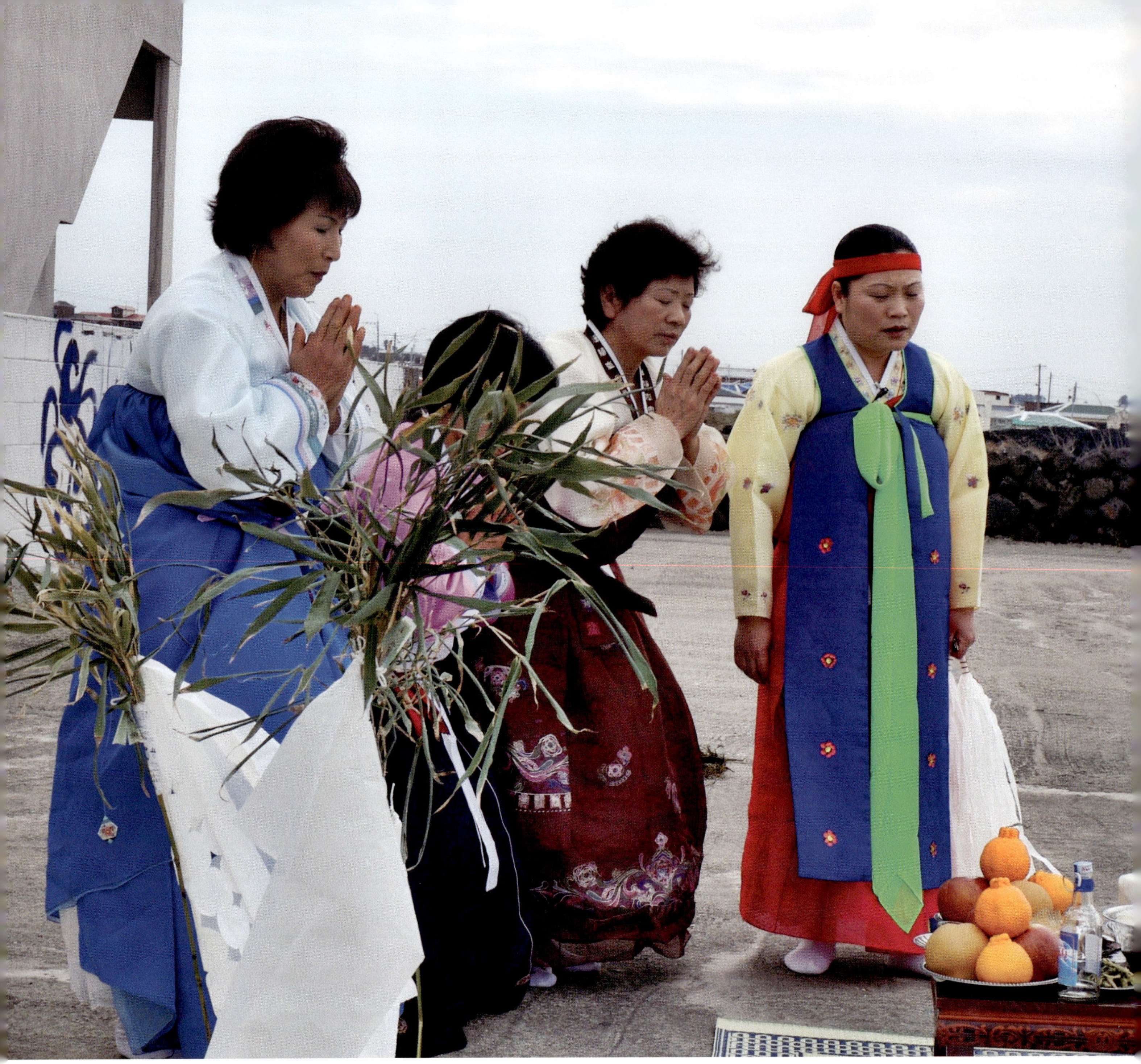

" I understand a *haenyeo*'s life is difficult and dangerous because they talk a lot about it. But I think there are healing and comforting elements in the water. I think water has a positive influence on *haenyeo*'s bodies and minds. Even old *haenyeo* like to go to the sea even though they can't catch very much. I think they enjoy being in the water."

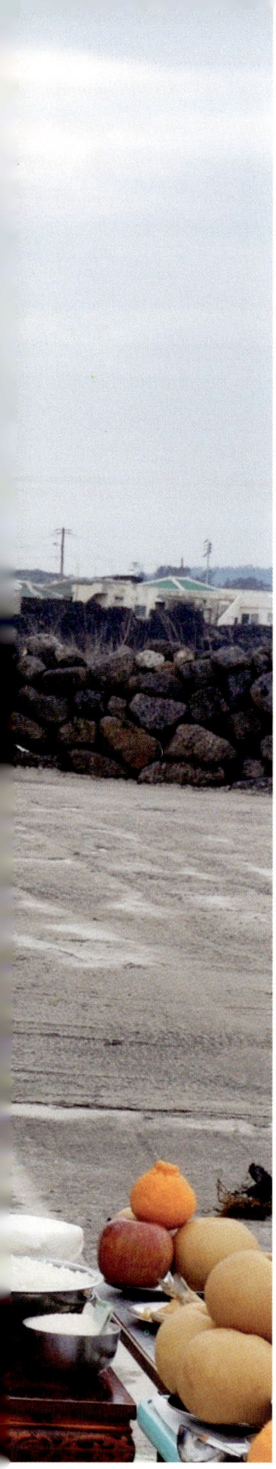

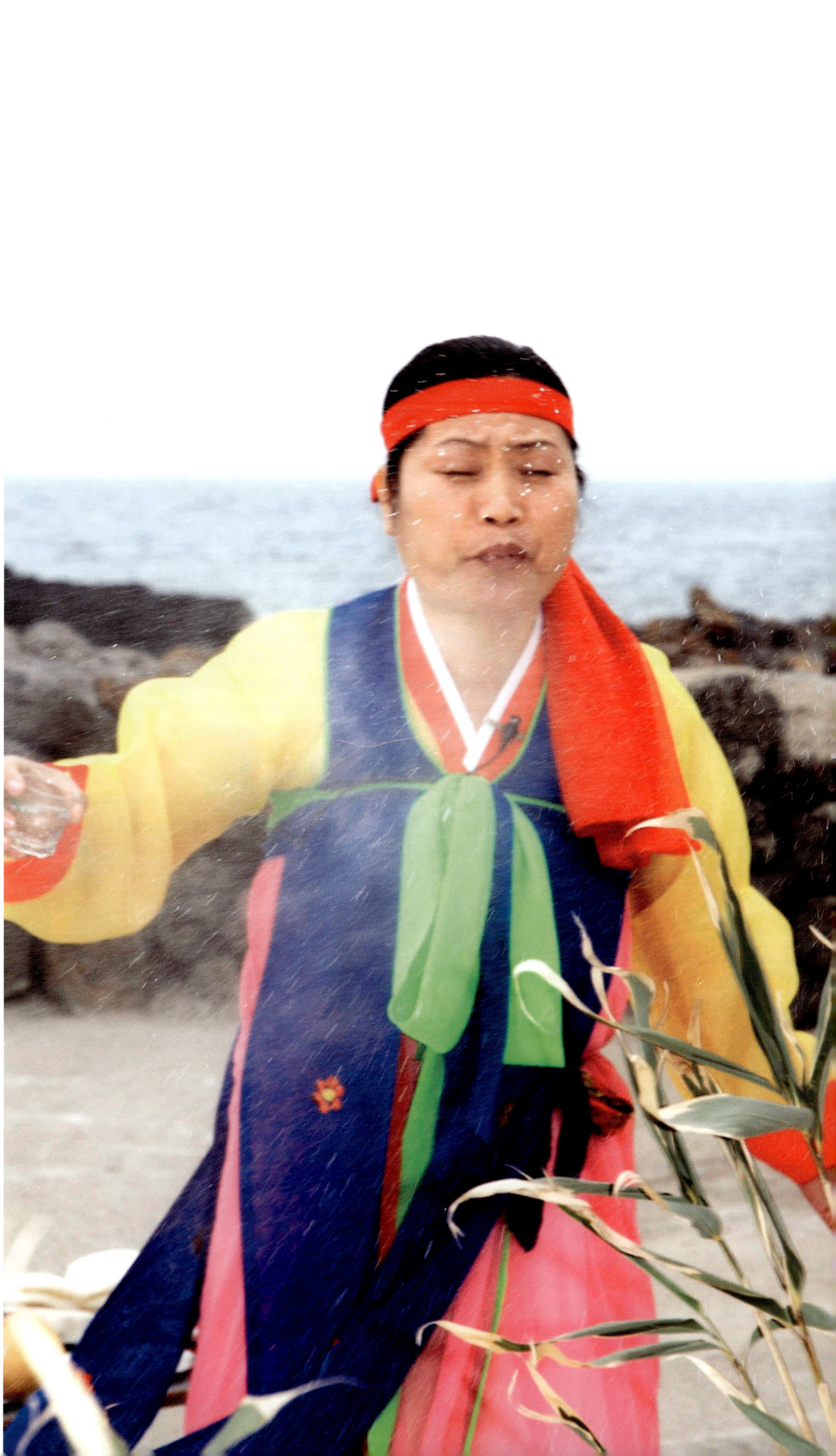

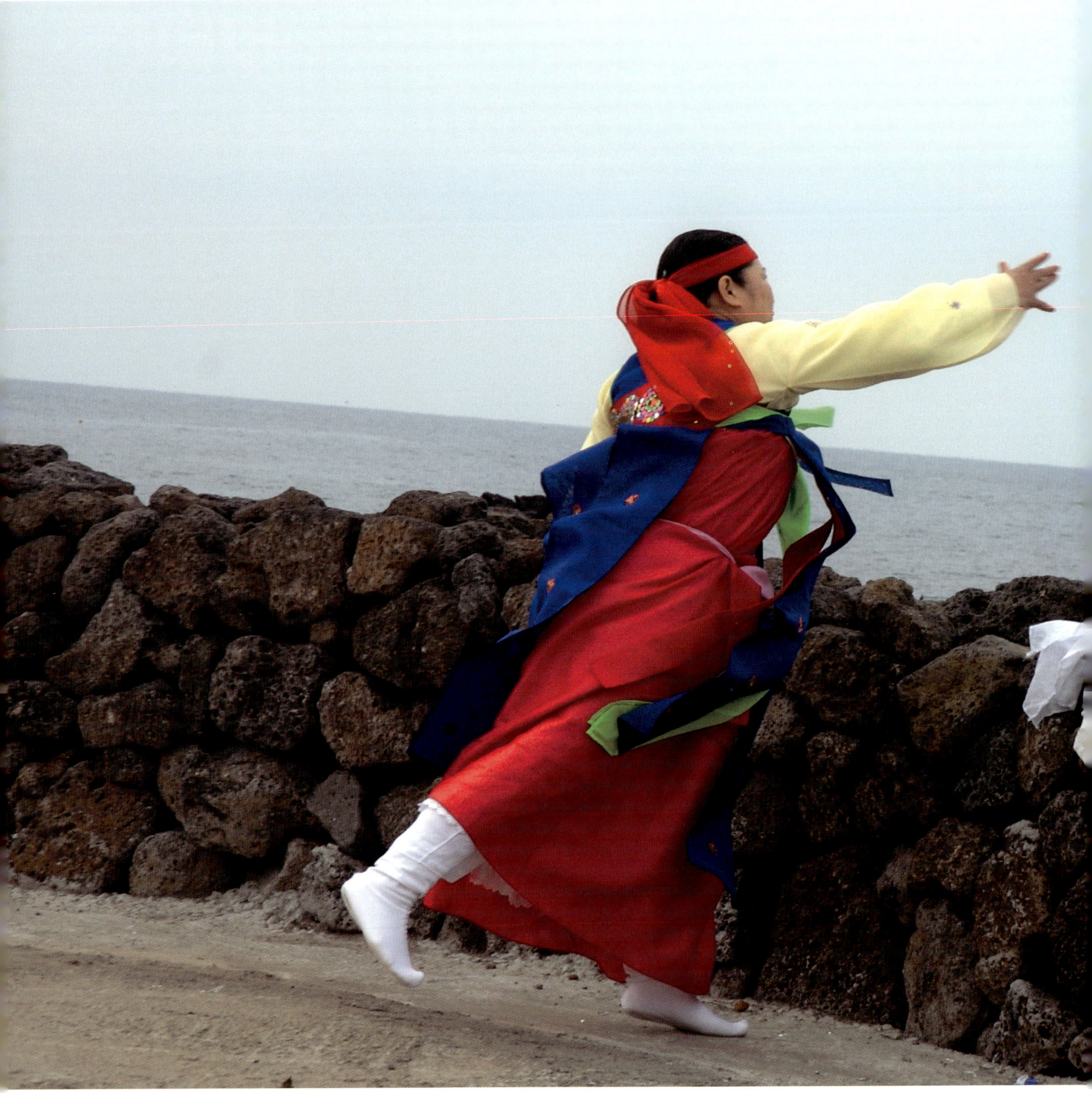

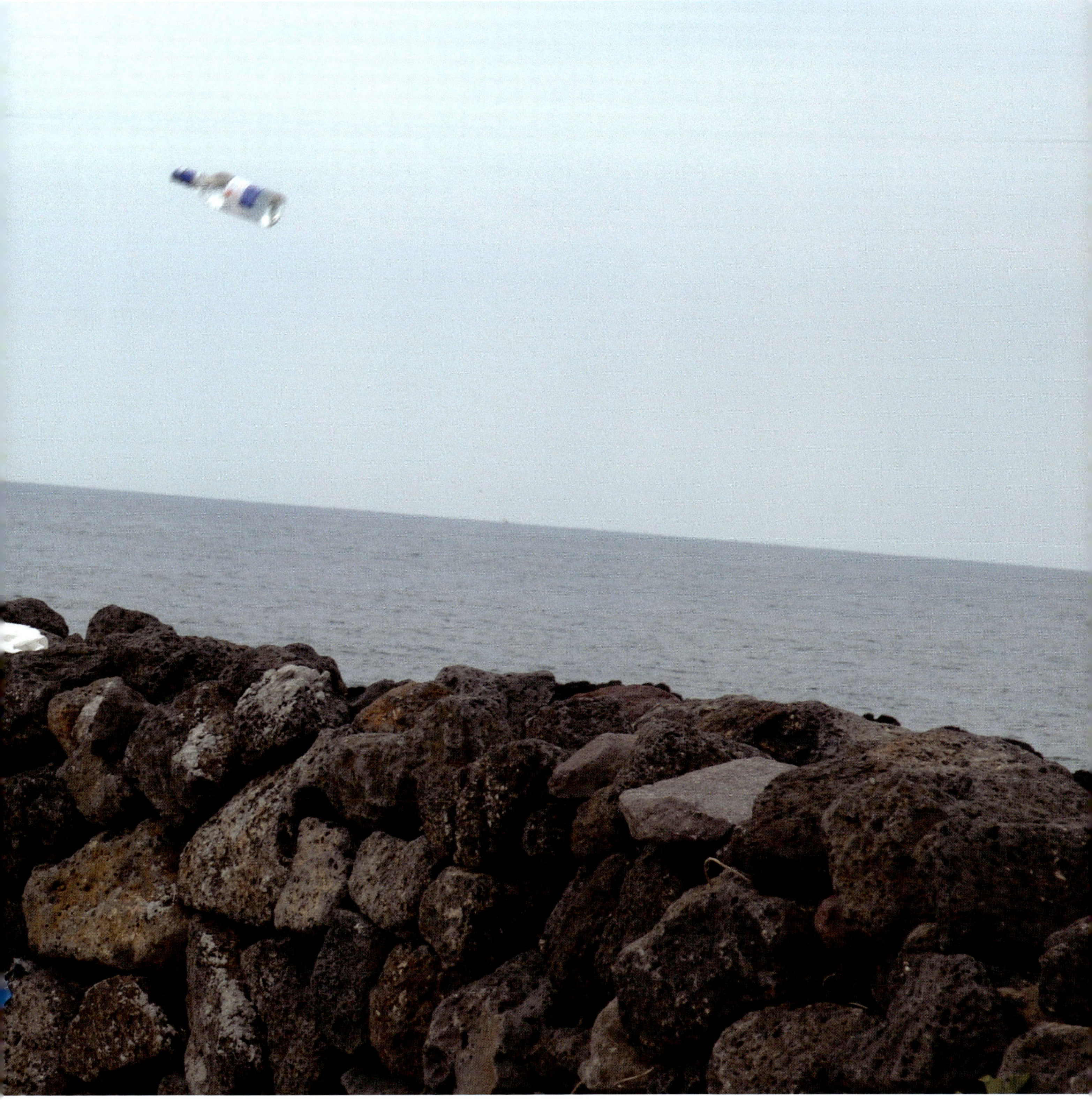

HADO-RI JAMSU GUT

Woman divers (also known as "*jamsu*") gather each year in Hado-ri for a shamanist ceremony to ask for protection and an abundant harvest. They set up an altar by the sea with offerings of rice cakes, fruit, and sea products for the gods of the sea and wind. Later, a local shaman reads the women's fortunes with coins and fresh rice grains brought from their homes.

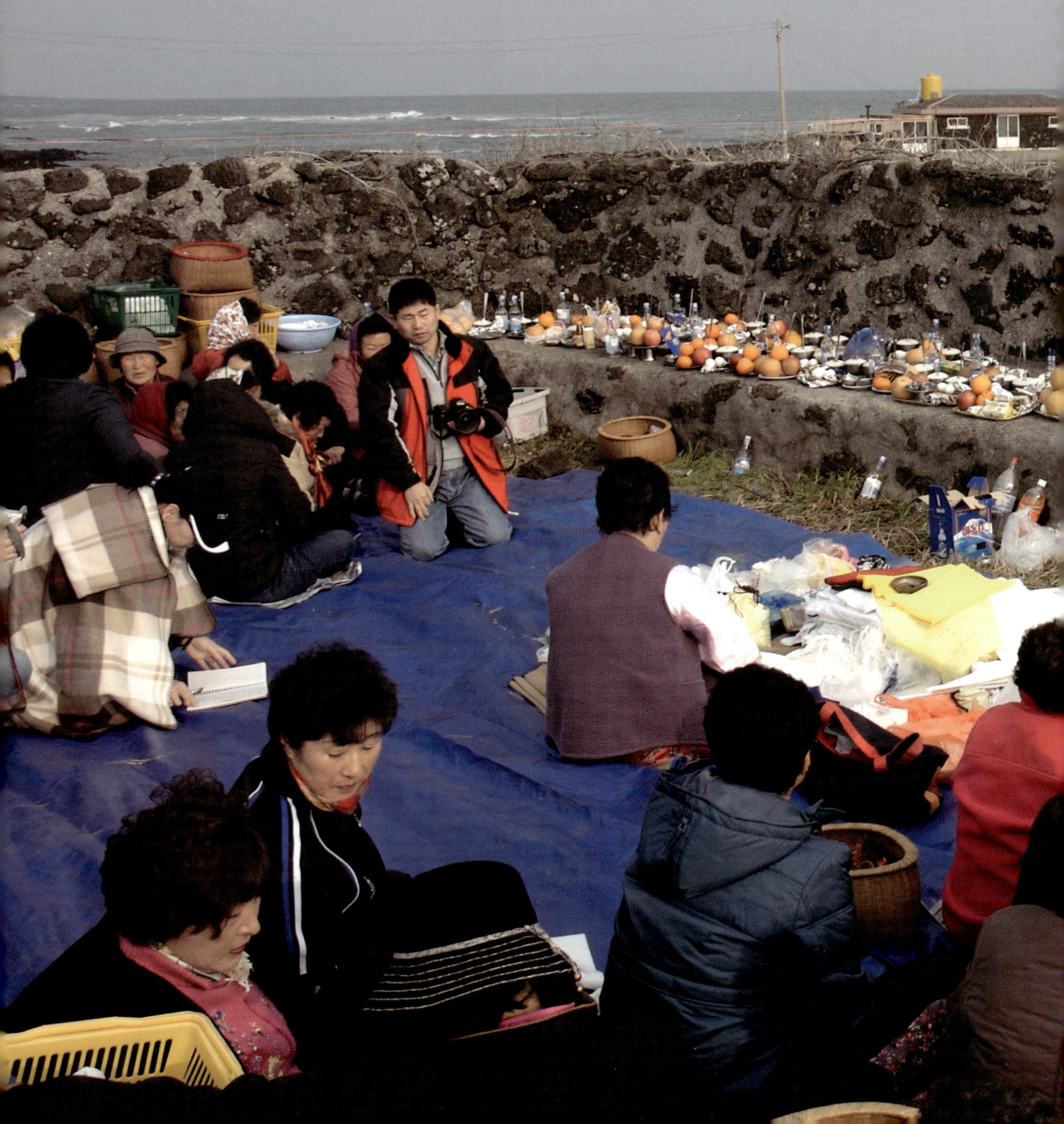

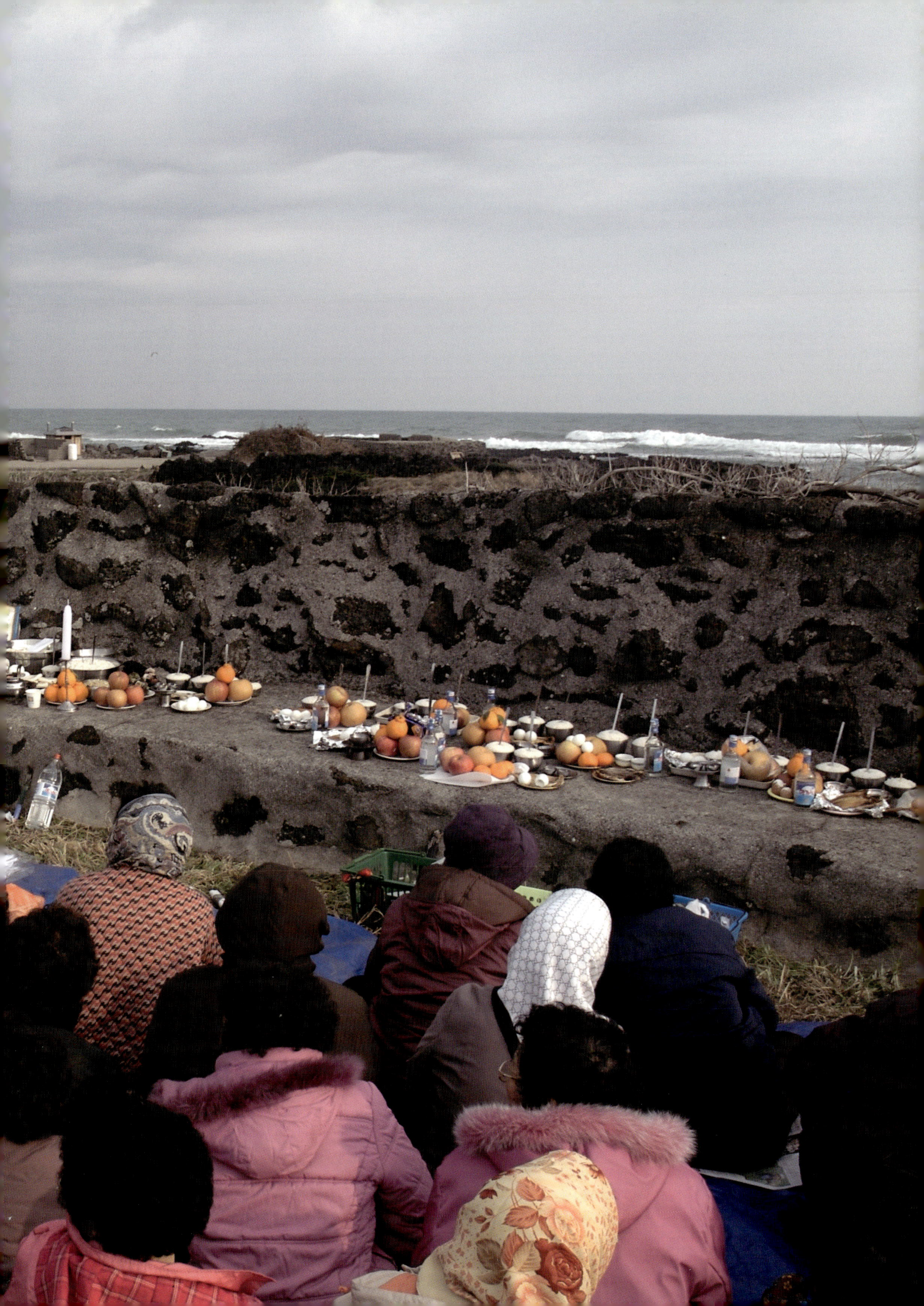

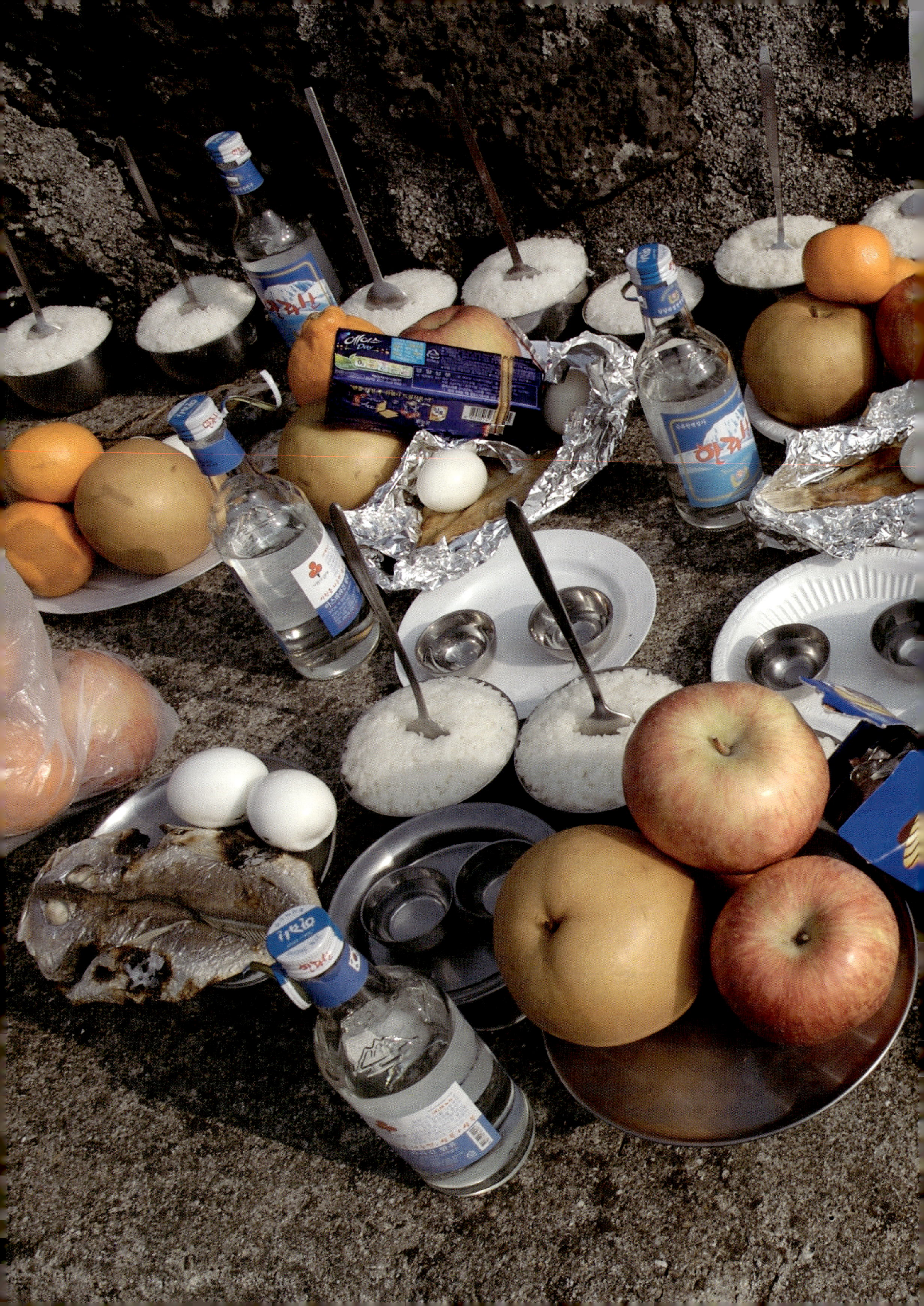
한라산

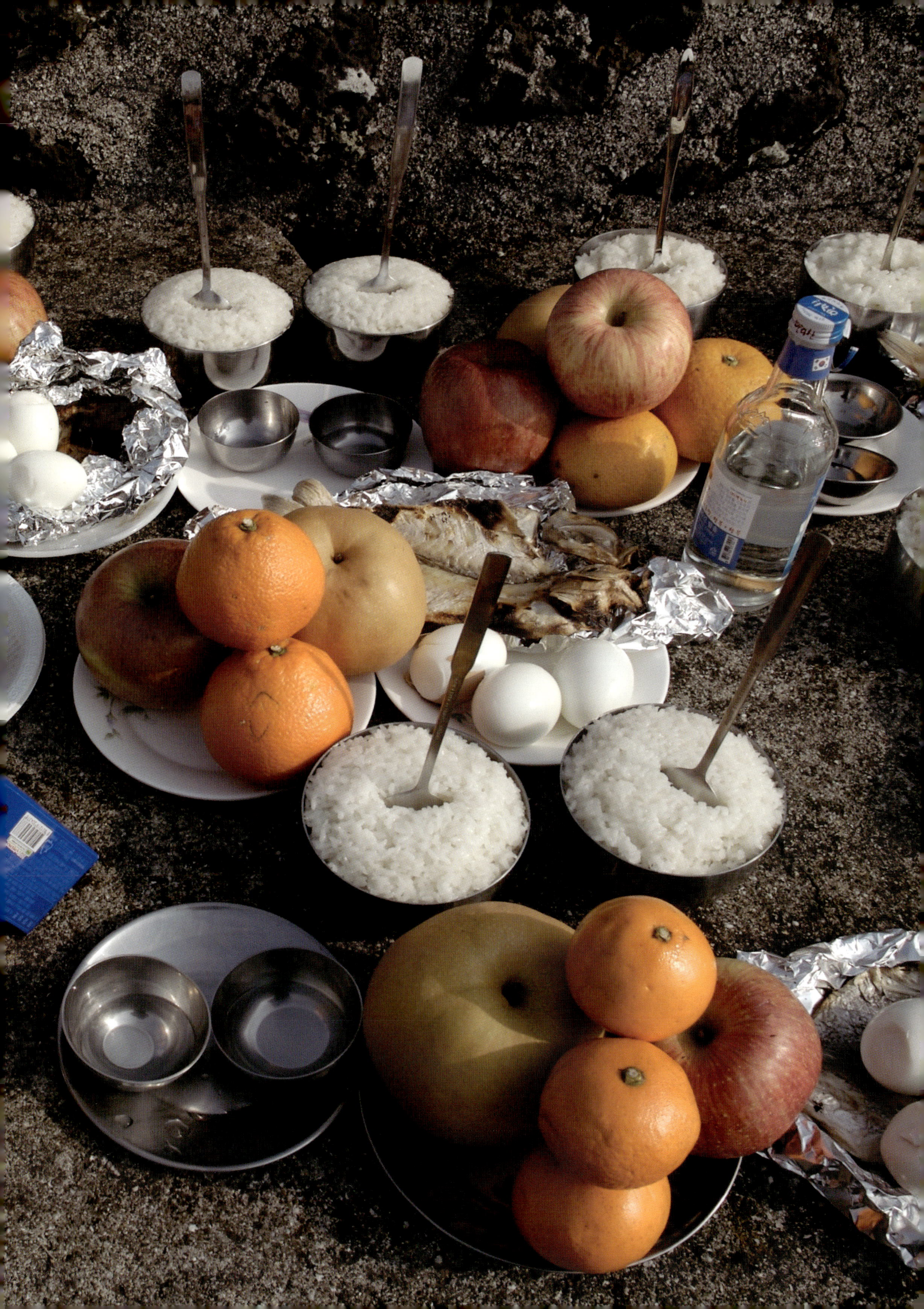

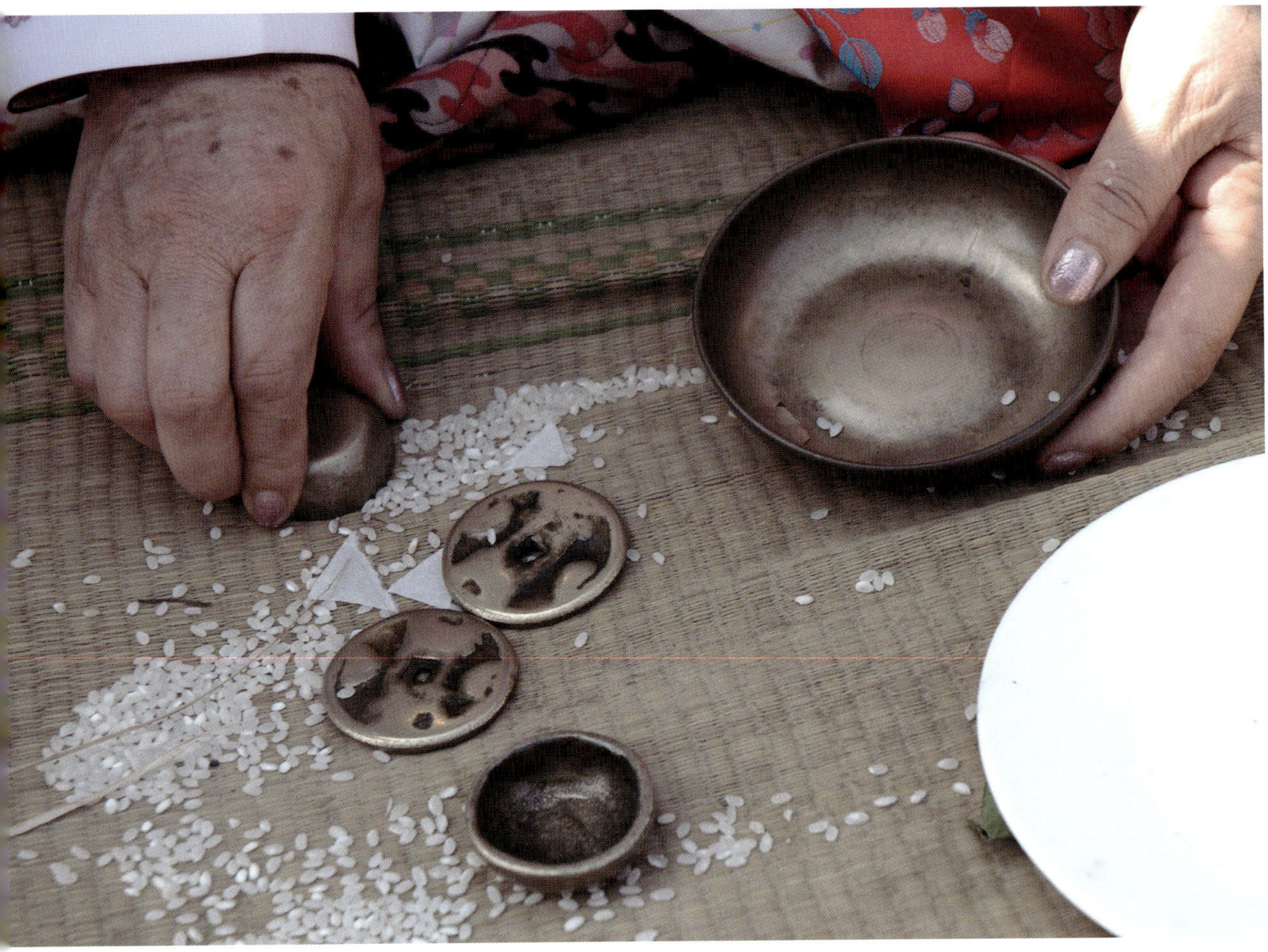

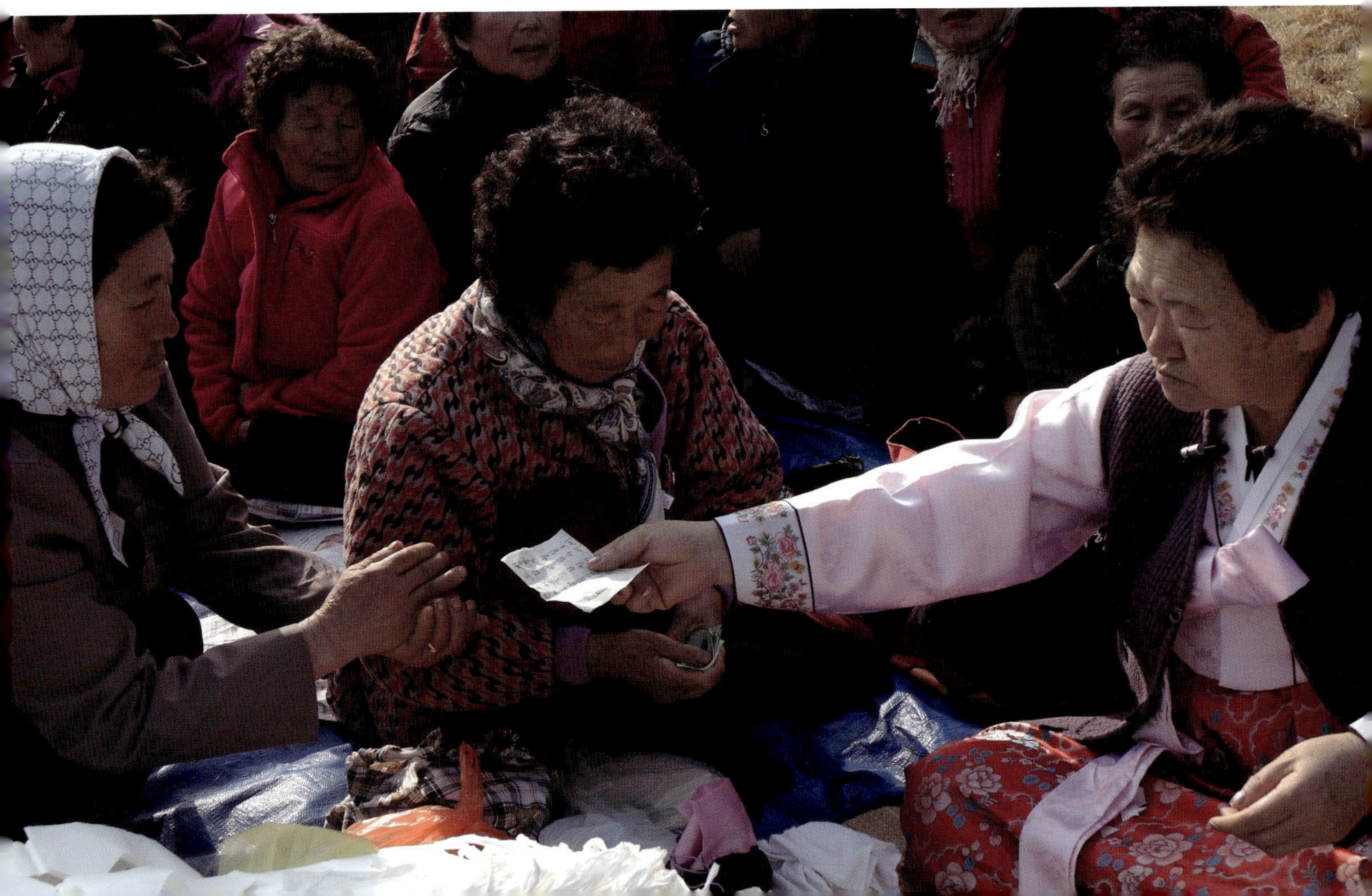

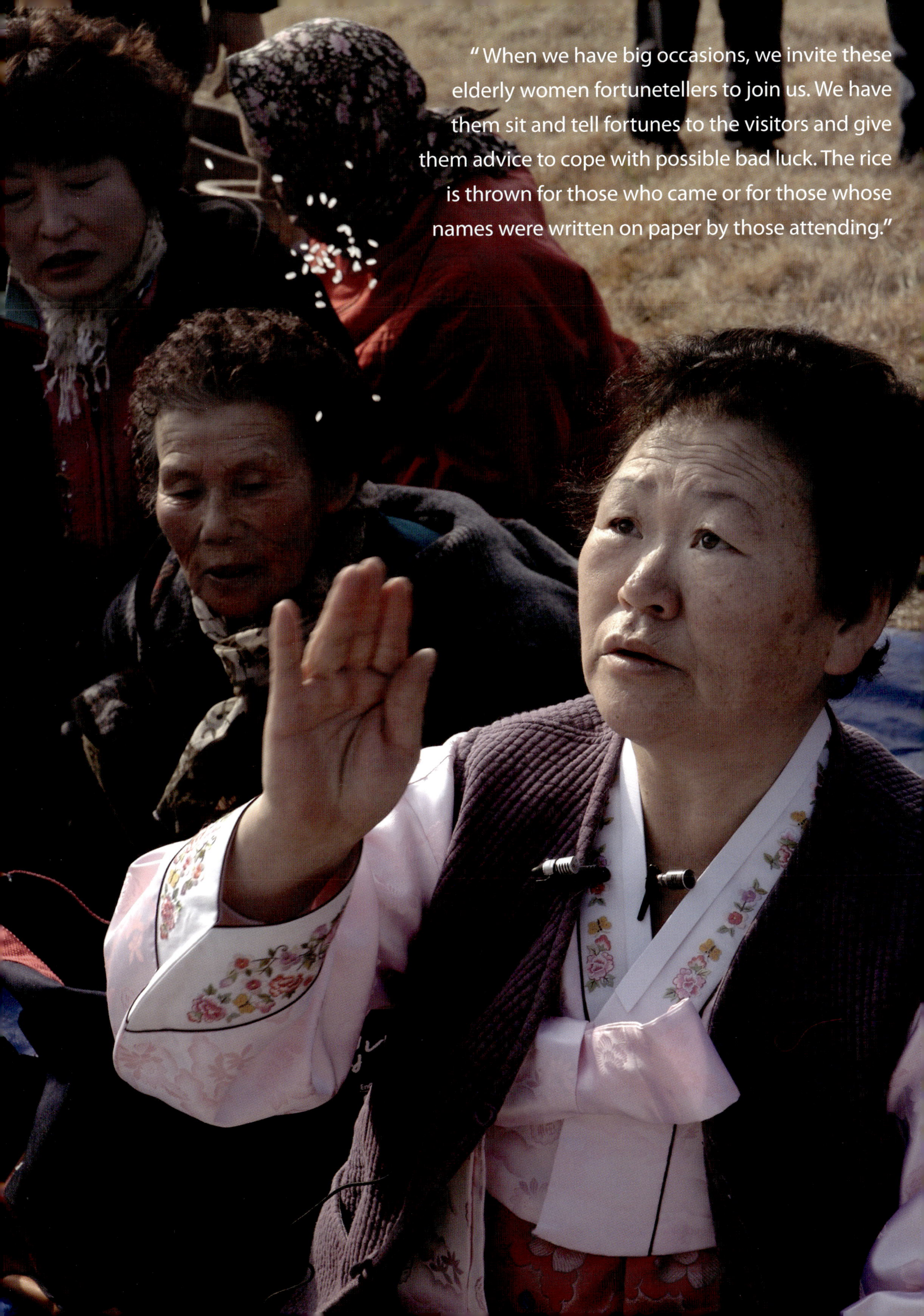

" When we have big occasions, we invite these elderly women fortunetellers to join us. We have them sit and tell fortunes to the visitors and give them advice to cope with possible bad luck. The rice is thrown for those who came or for those whose names were written on paper by those attending."

"The *gut* are important to the *haenyeo* because they feel they are risking their lives all the time while in the water. They need something to rely on and make them have a sense of security. *Gut* can give them that sense of security and make them feel at ease."

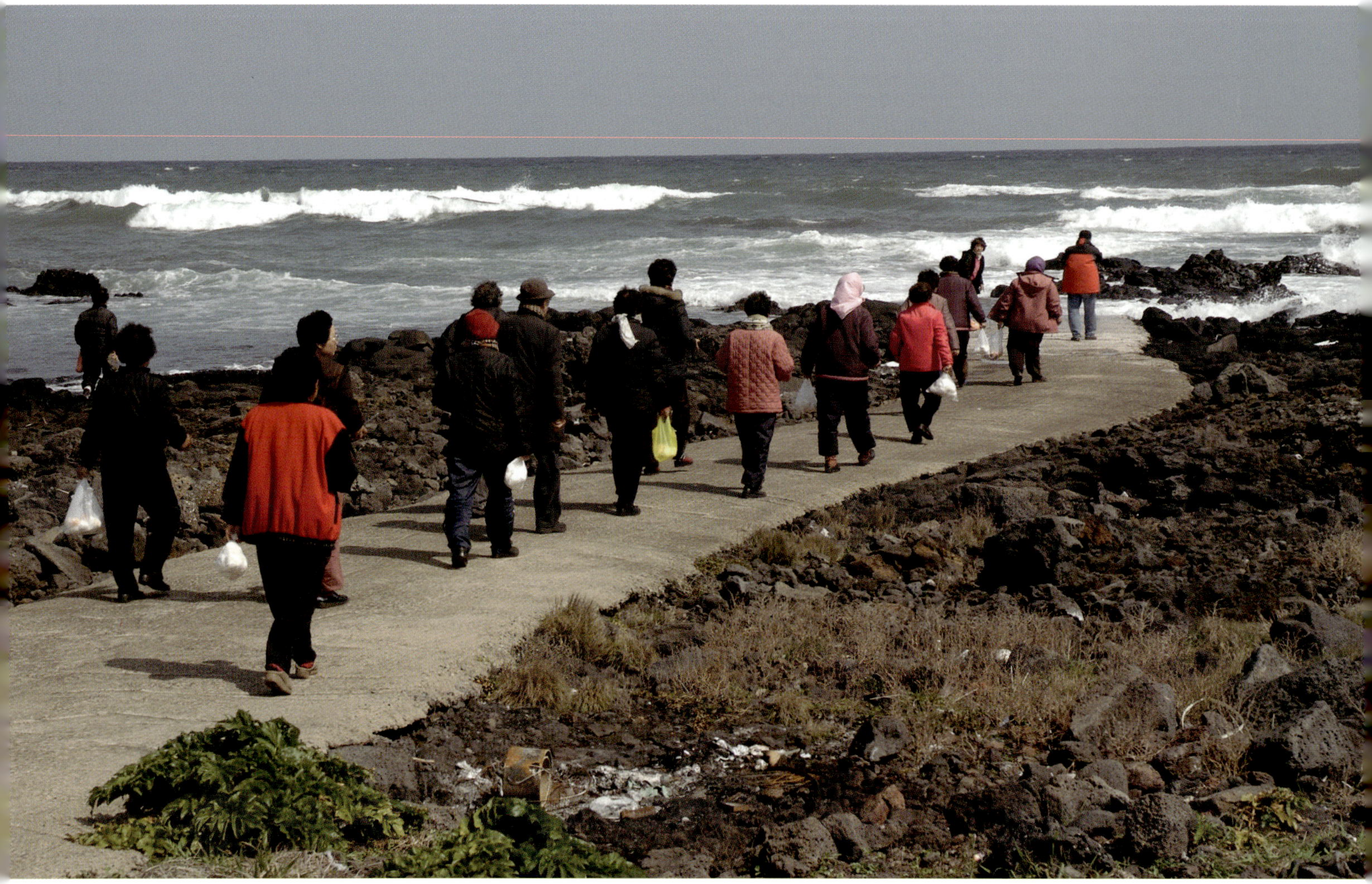

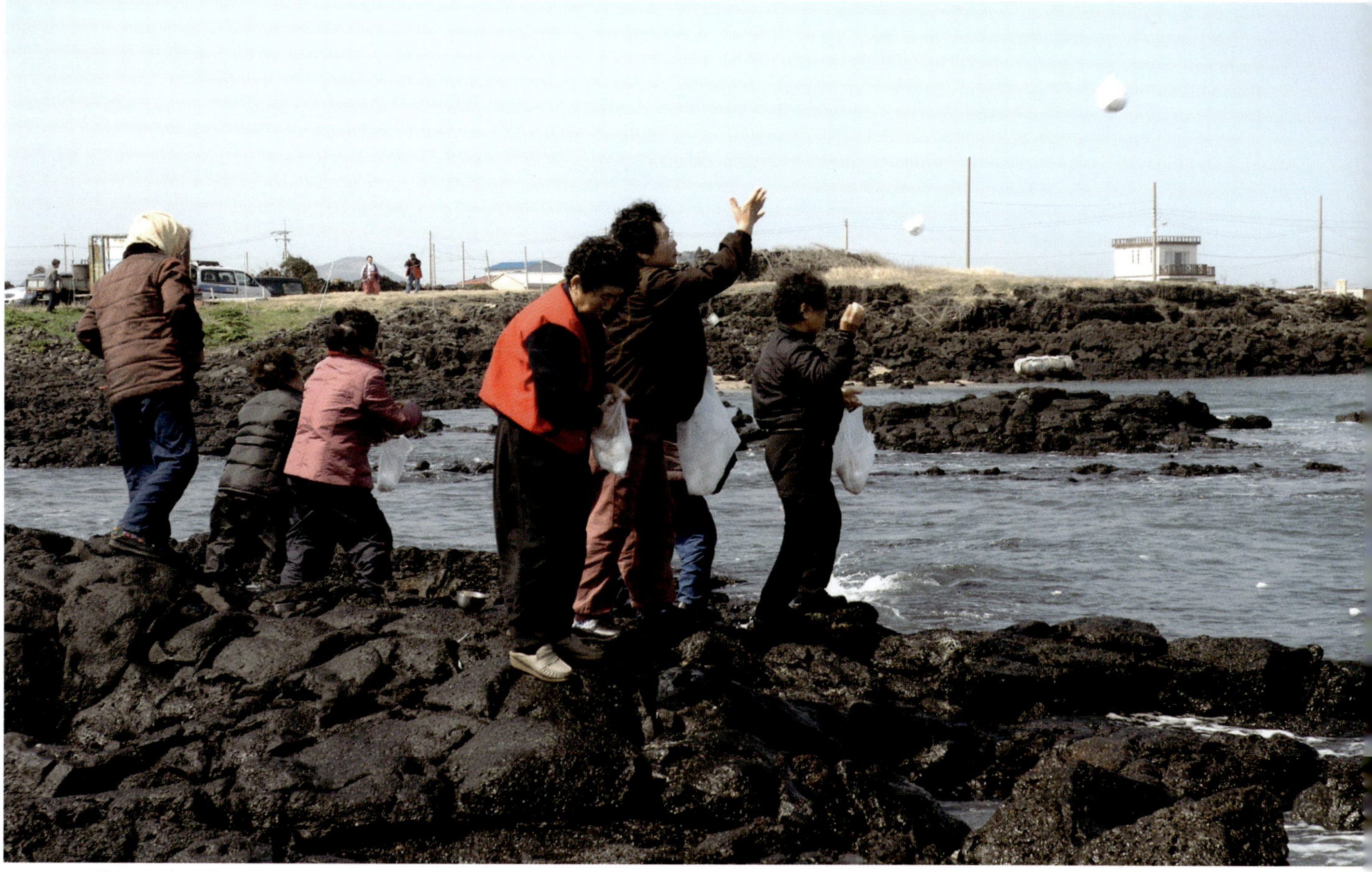

After eating the altar foods together, the *haenyeo* throw the rice cakes out as nourishment for the gods and goddesses.

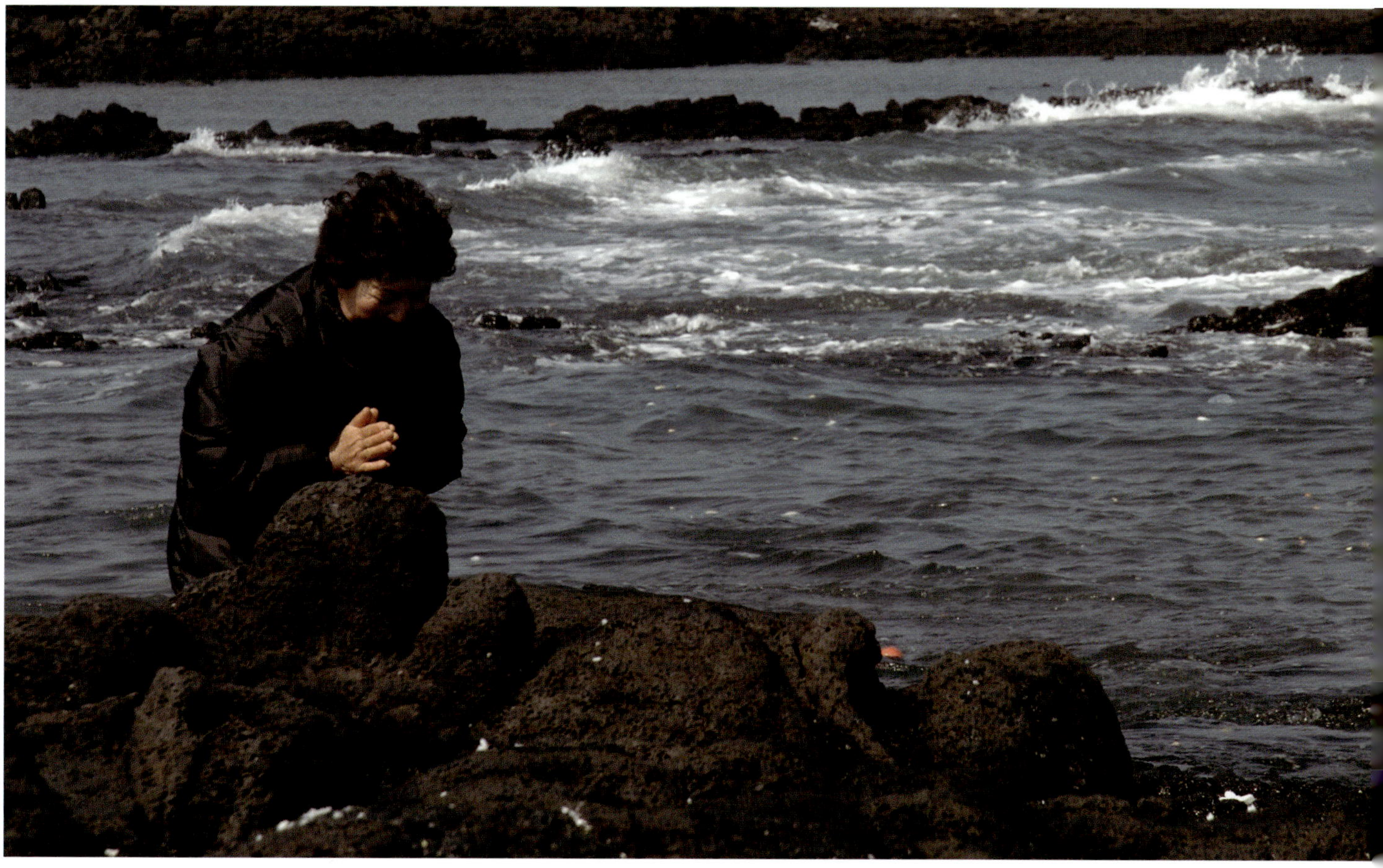

One *haenyeo* kneels in prayer for the divers' safety at sea.

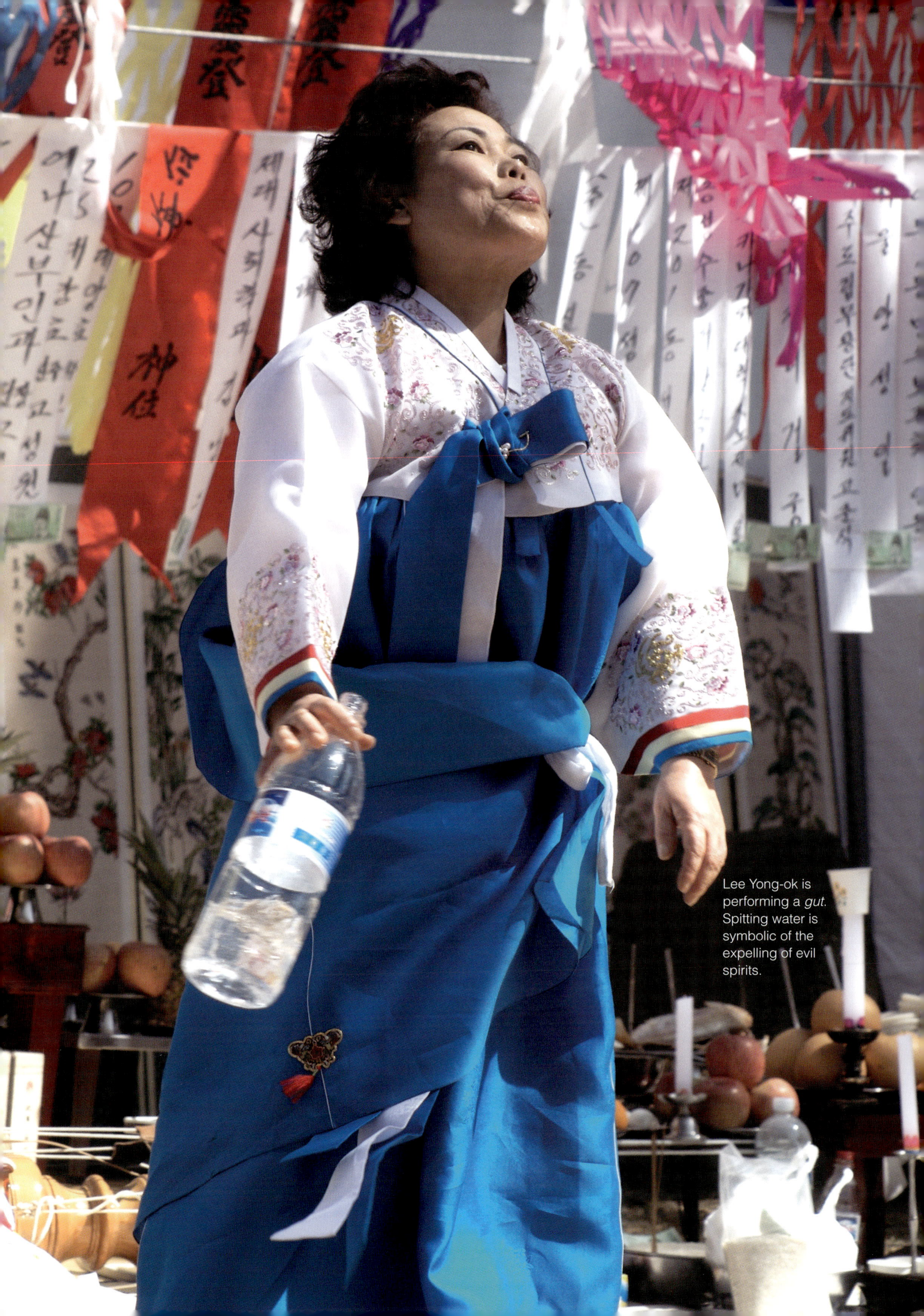

Lee Yong-ok is performing a *gut*. Spitting water is symbolic of the expelling of evil spirits.

" It's an honor for me to be a shaman.
I can comfort people and make them feel
better by performing *gut*."

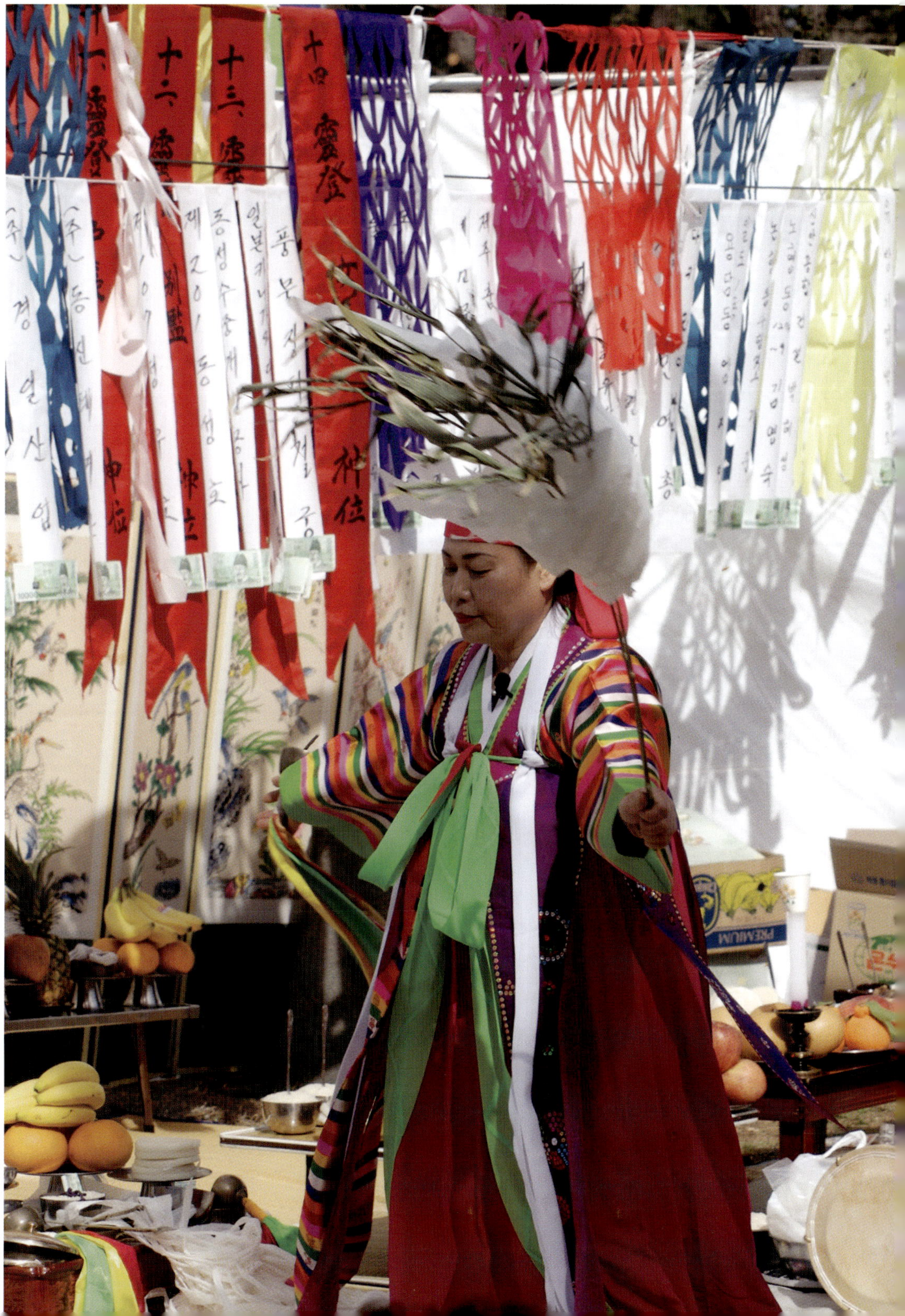

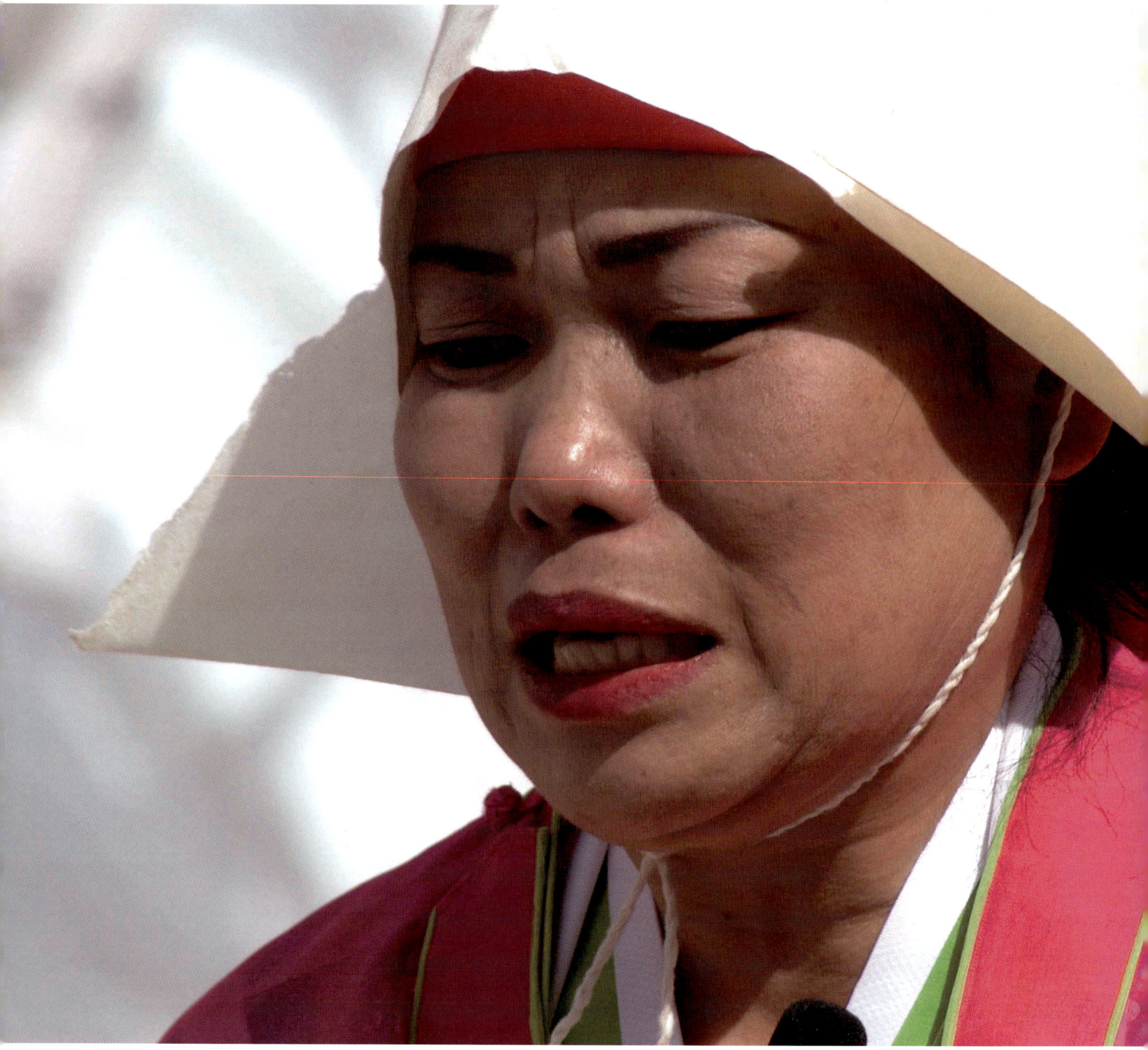

" When I am performing a *gut*, I try to relieve the host's sorrow and bad feelings inside their hearts. While relieving their suffering, I feel that I, too, become purified. I sometimes shed tears while conducting these private *gut*. It's a catharsis for me. I feel so happy and comfortable when I can forget my own pain and sorrows. "

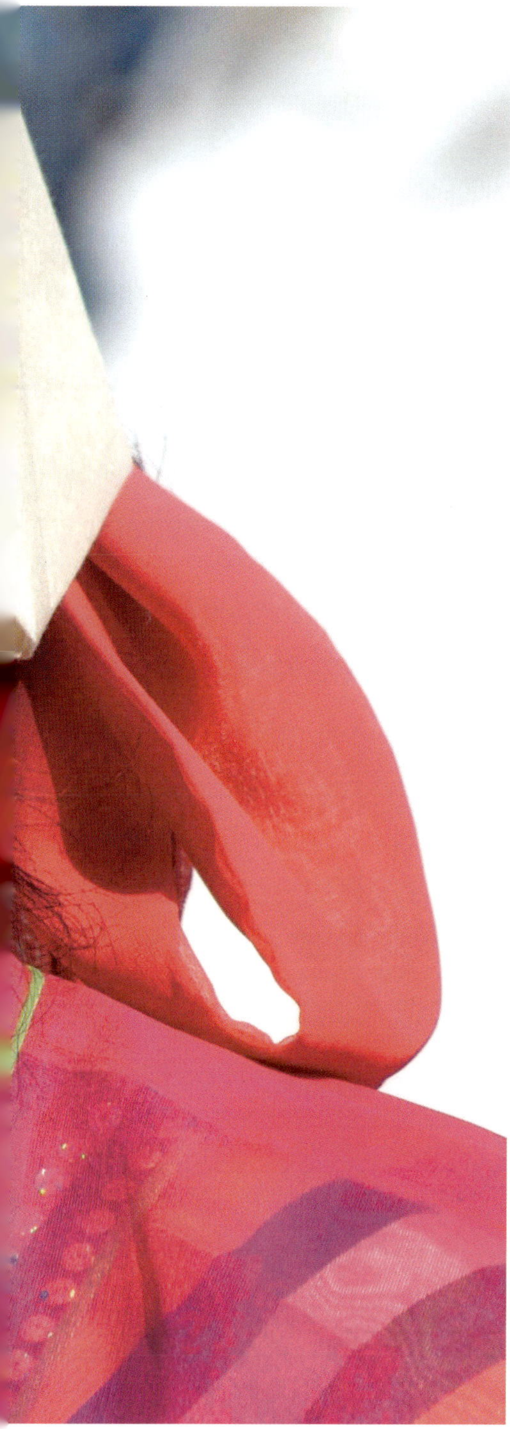

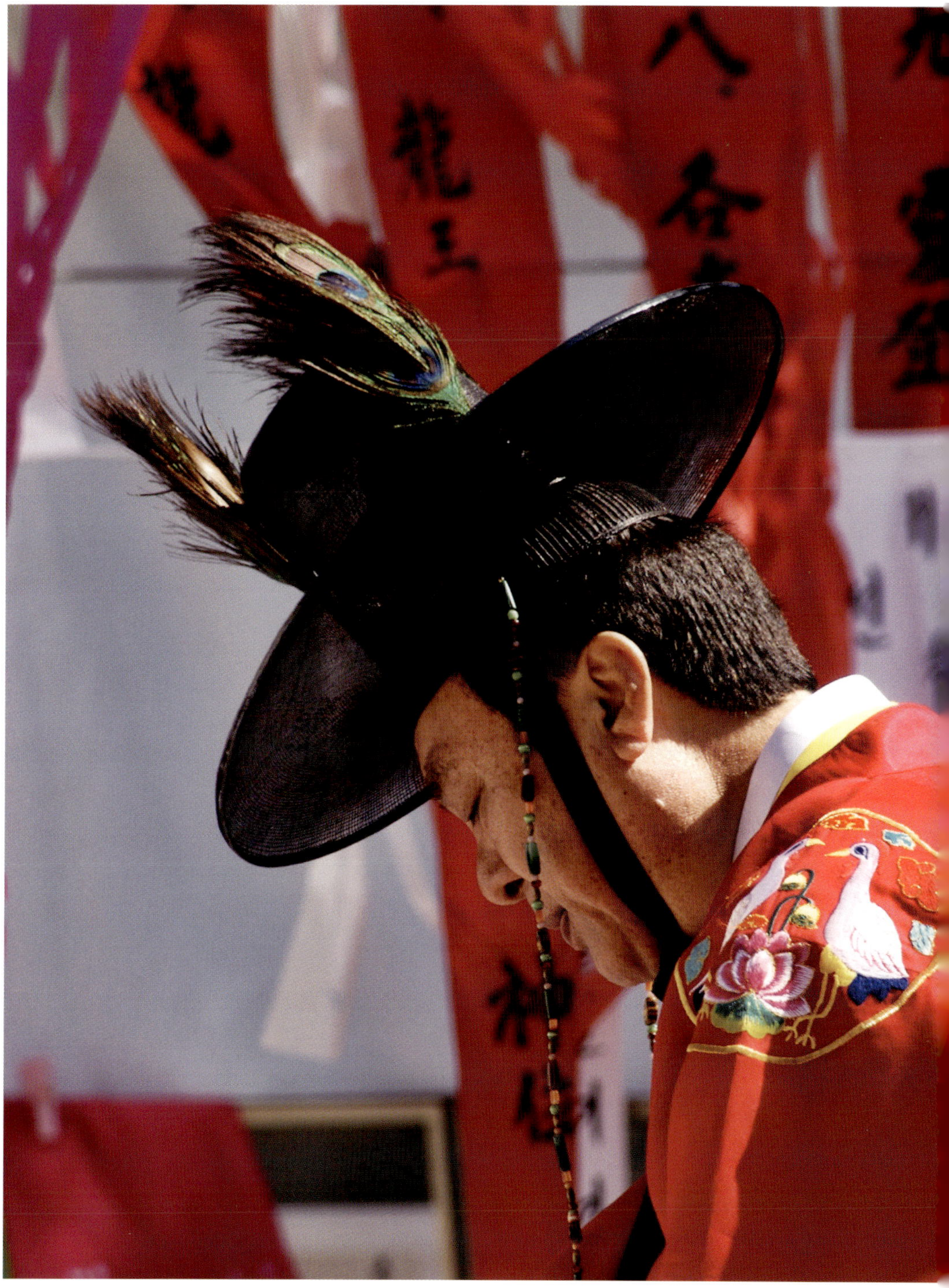

Lee Yong-ok's husband, Kim Yun-su, encouraged and trained her to become a shaman.

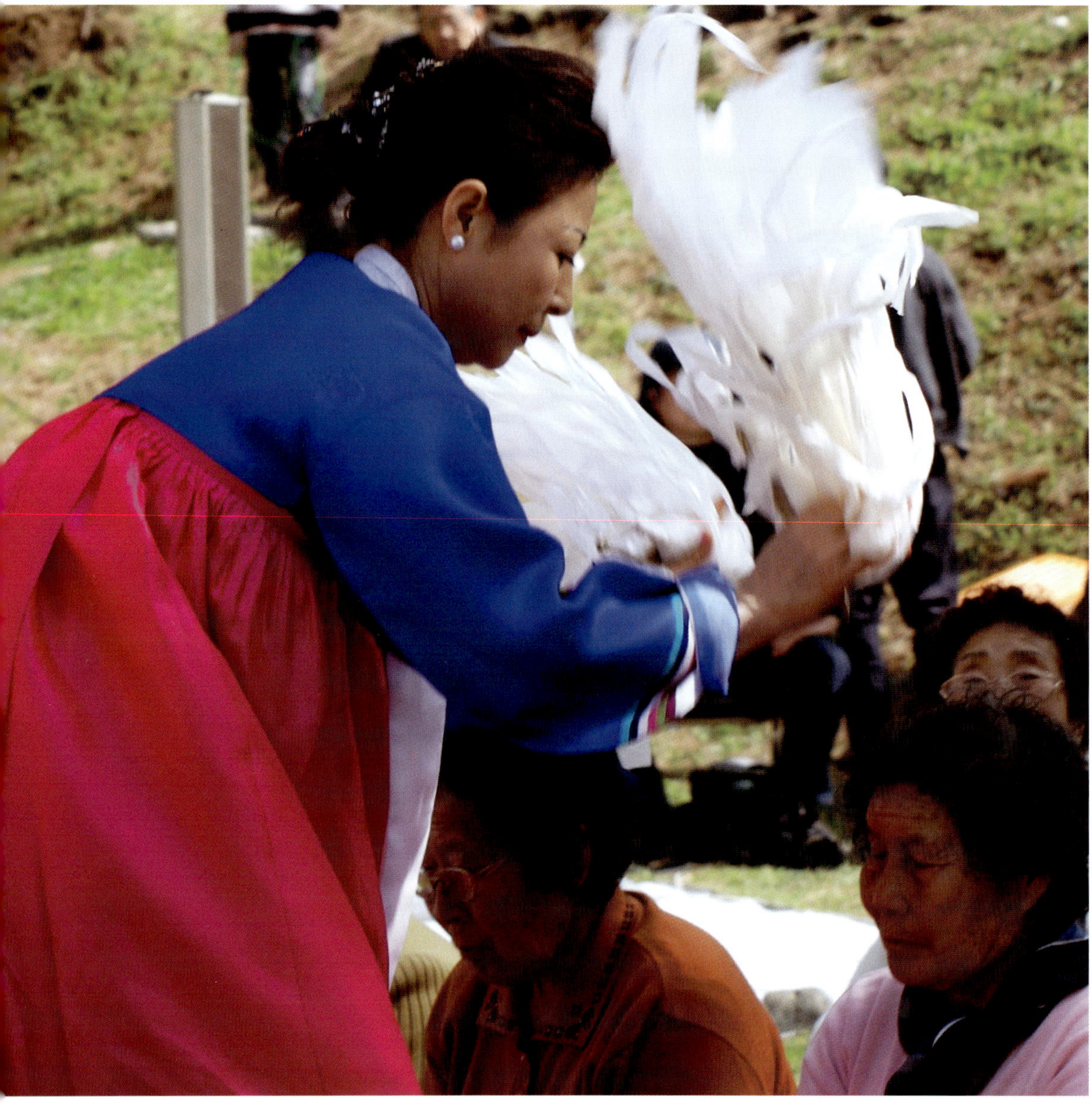

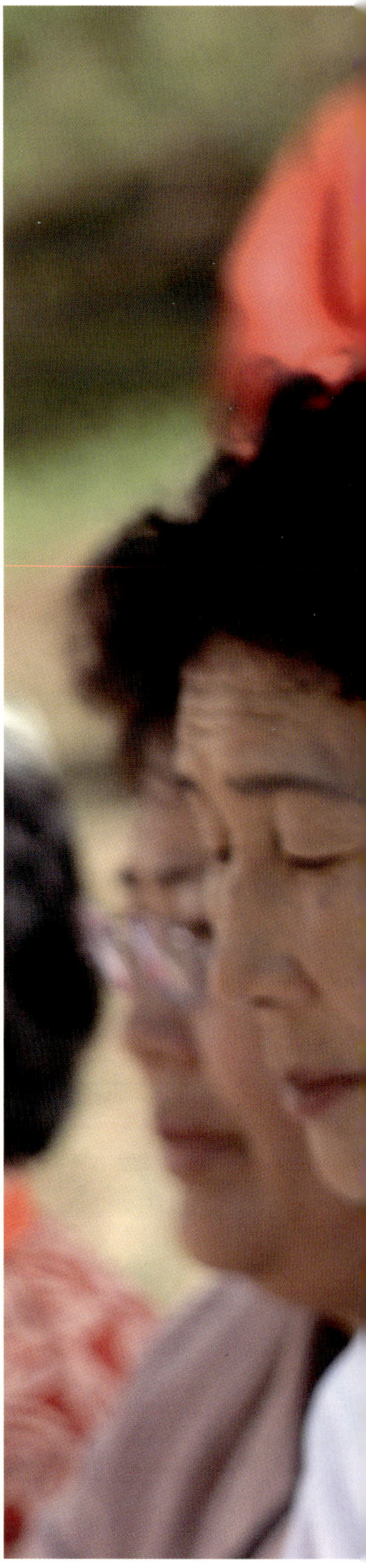

" Shamanism has been a practice in Jeju for centuries. Most of the people have always relied on the sea for a living. They needed to have some belief they could rely upon and pray to for their safety and abundant catch. "

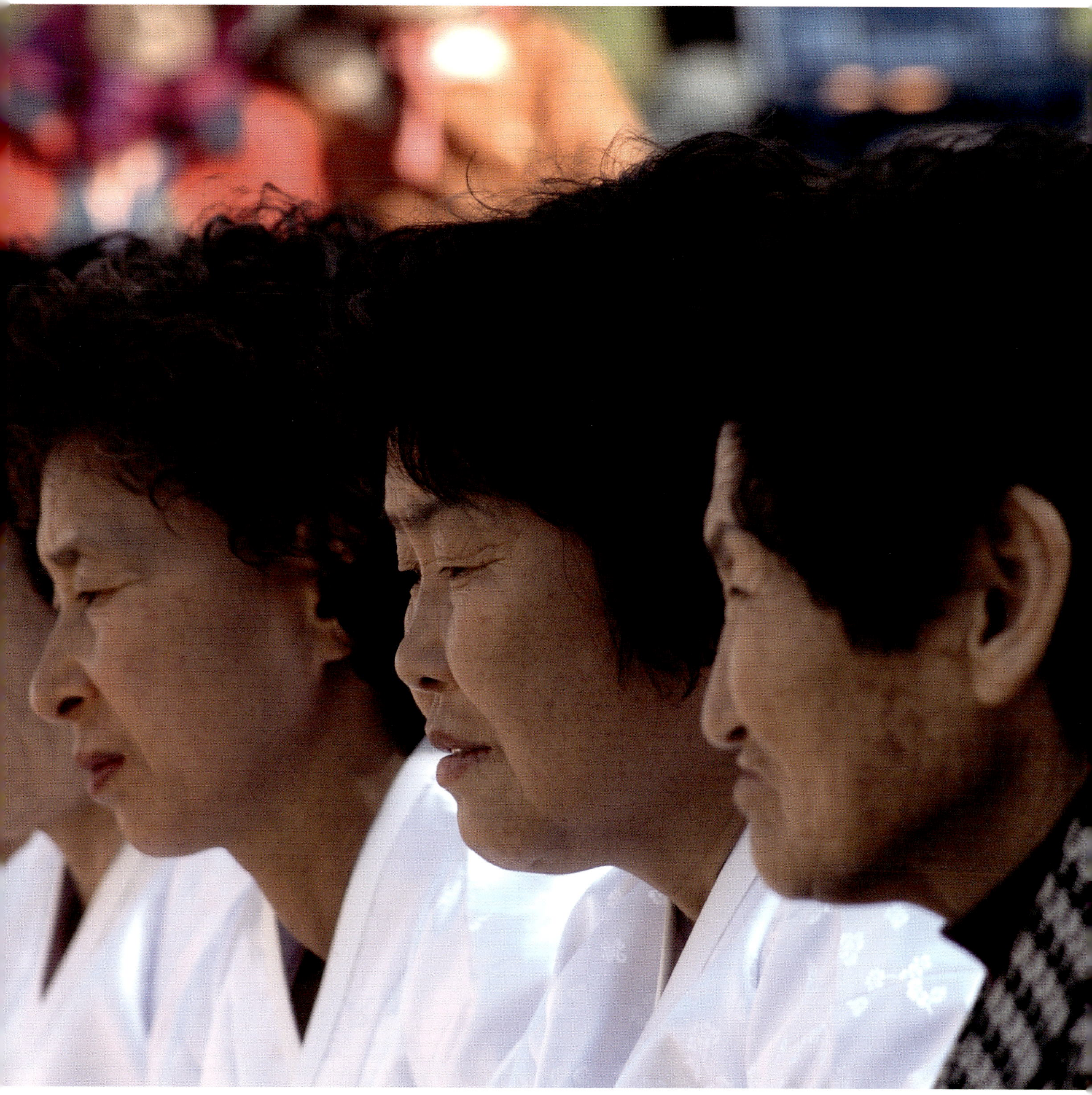

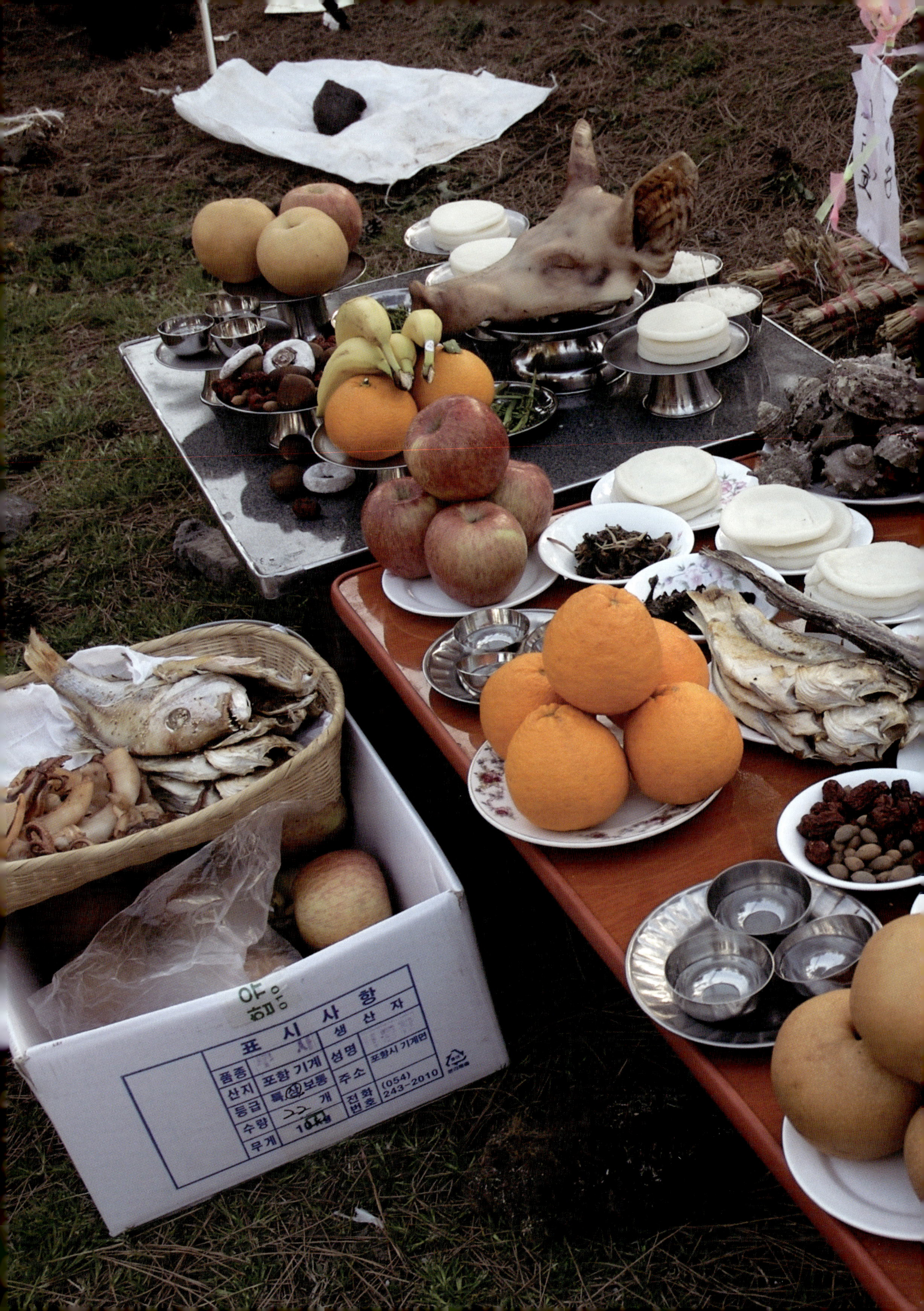

표시사항
생산자
품종
산지
포항 기계
등급
특 상 보통
수량
개
무게
성명
주소
포항시 기계면
전화번호
(054) 243-2010

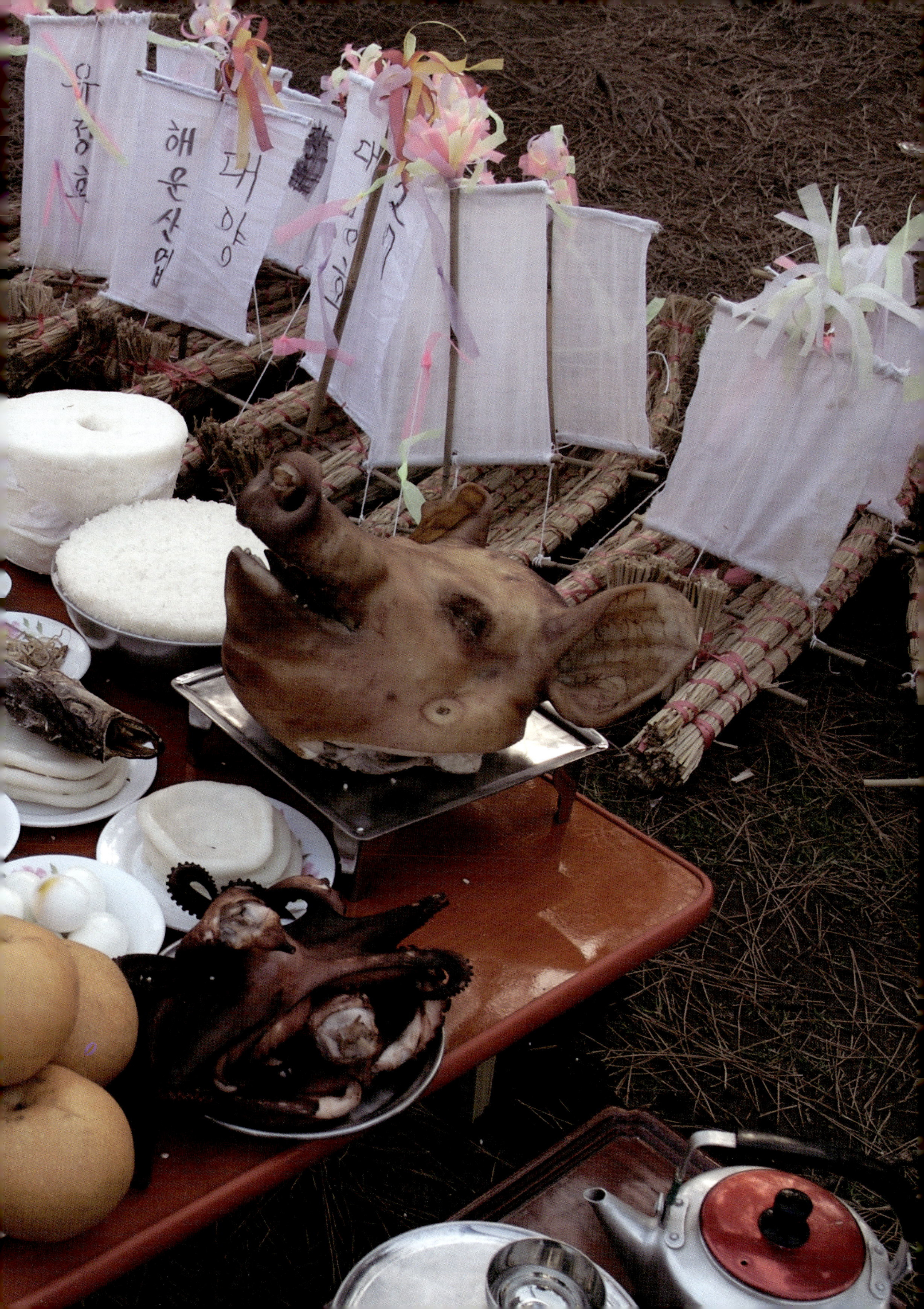
유정호
해운산업
대양

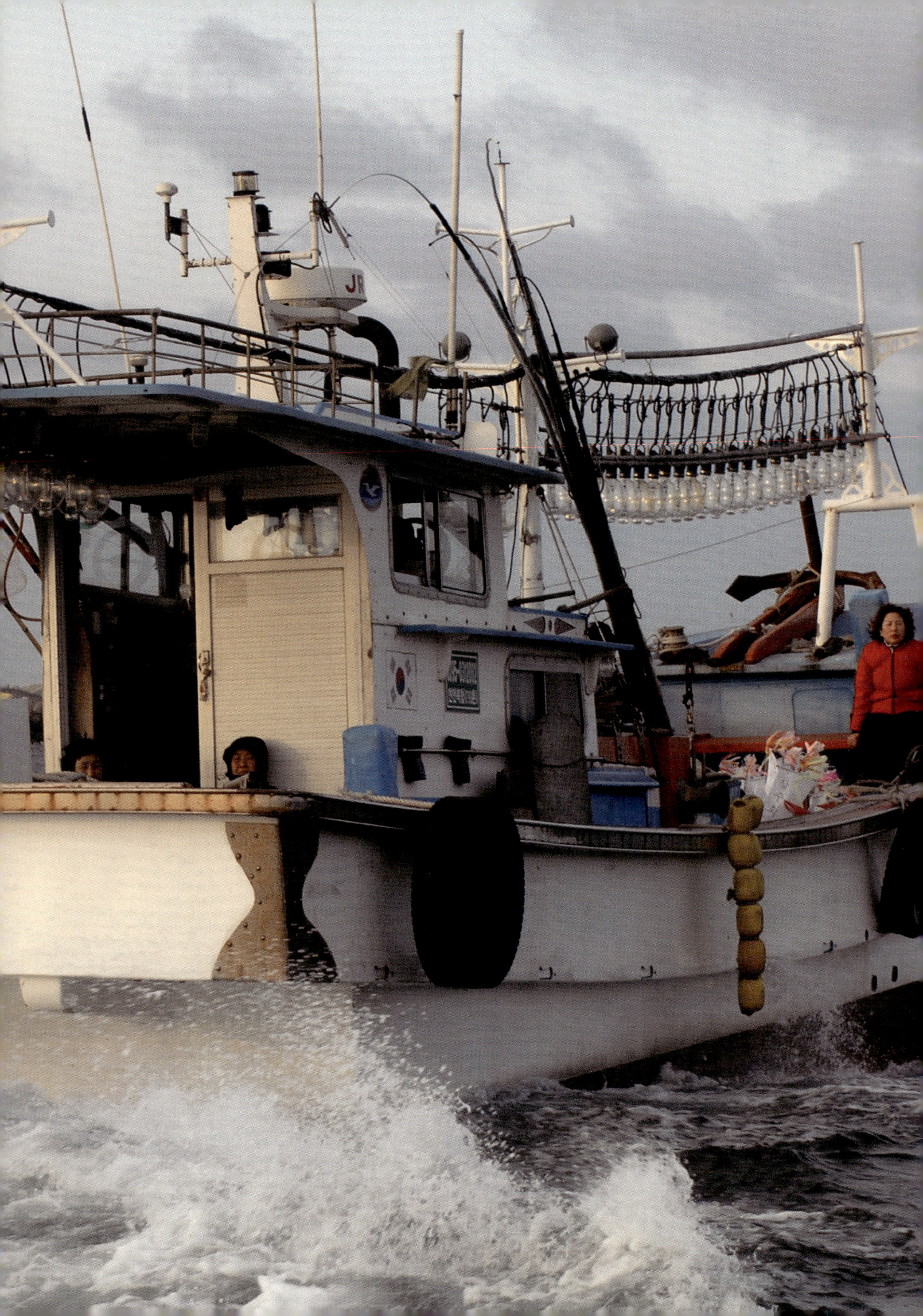

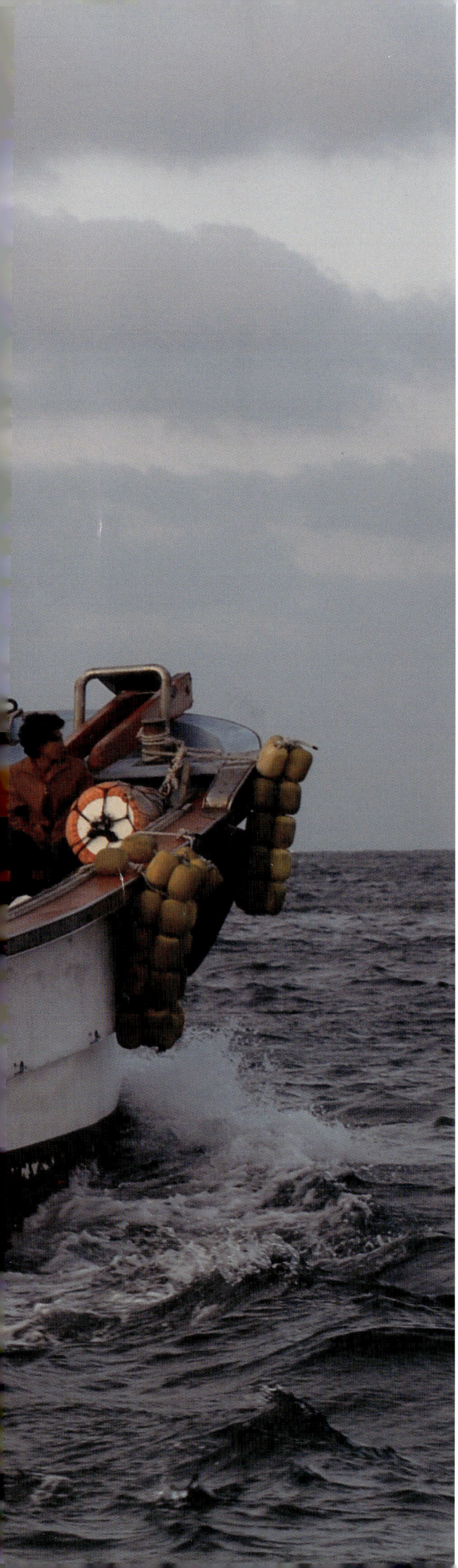

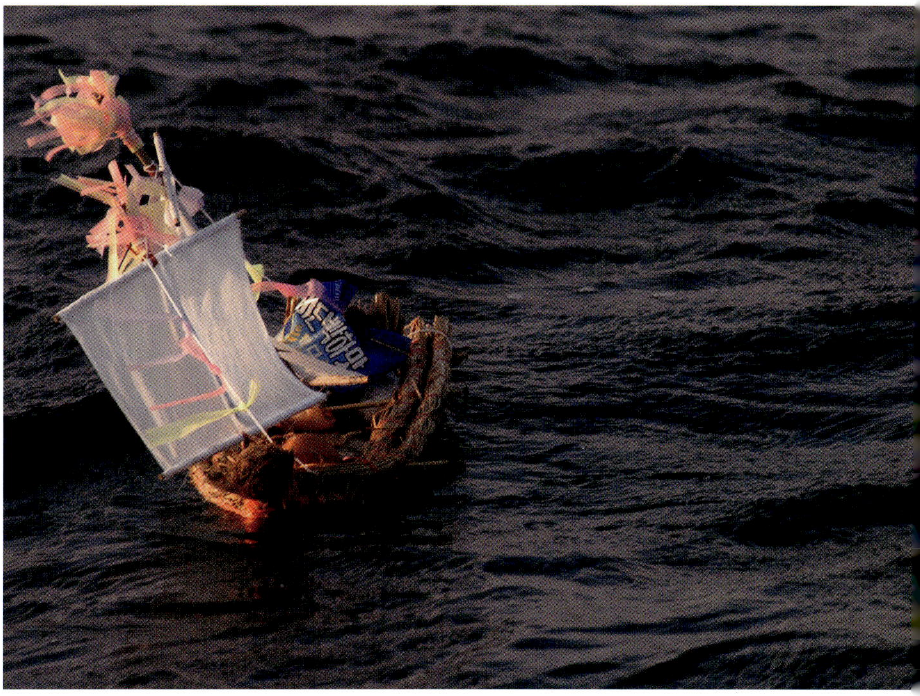

" We bid farewell to all of them by sending them off on little straw boats. The Yeongdeung goddess rides on the first boat. The rest follow on the other boats."

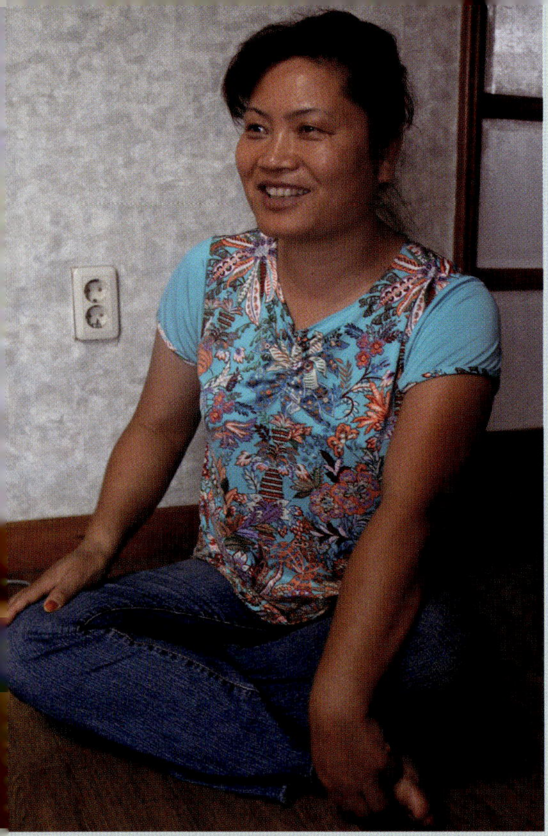

SUH SUN-SIL

b.1961 | Gimnyeong-ri

I was born in Goheung County, Jeolla Province, Korea, where my father was born. My mother was born in Gimnyeong, in Jeju. We came to Gimnyeong when I was four years old. I went to school here. I have three brothers and three sisters. I'm the second youngest and was born in 1961.

My path as a shaman began as a child. I was sick very often when I was young. At that time, a fortuneteller came to visit our village once a year from the mainland. My mother visited the fortuneteller and had my fortune told. The fortuneteller said that I would be dead at the age of 17 if I didn't become a shaman like my mother. So my mother persuaded me to learn the *gut* (shaman ceremony). She thought I had better be a shaman than die at an early age. After I graduated from elementary school, my mother took me whenever she was performing a *gut*.

Shamans have been despised and considered ignoble for a long time. But I am lucky because our work has been better recognized since the 1980s.

At the Haenyeo Museum Anniversary in 2007, I was asked to perform a *gut* related to *haenyeo* life. Usually, it takes an entire day to perform all parts of the *gut*. But I only performed two parts because it was part of a larger ceremony. The two parts included the "Welcome of the Dragon King, God of the Ocean" and "Paving the Road for the Dragon King." This *gut* is excerpted from the *gut* usually held on the eighth day of the third lunar month.

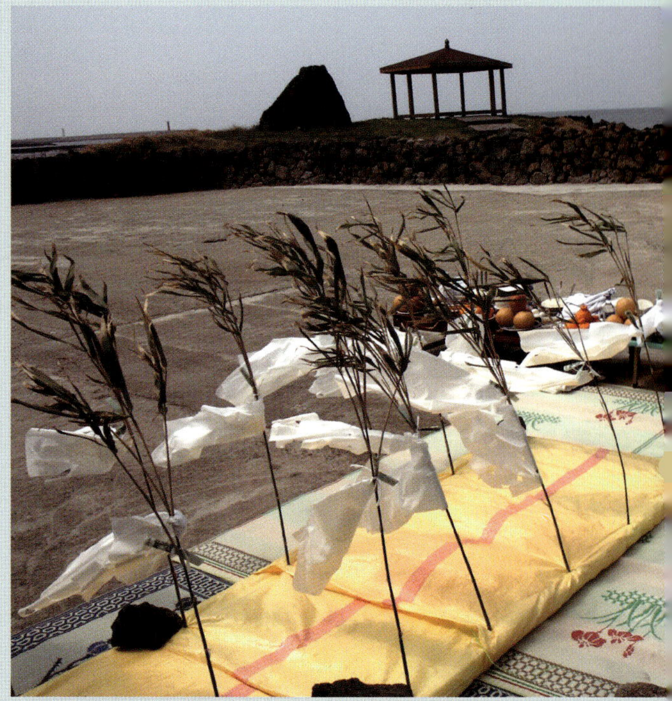

There are 18,000 gods and goddesses in Jeju. During the welcome of the Dragon King, we invite all of the gods, our ancestors, and spirits to join us. And then we pray for the *haenyeo*'s safety while they dive for their products. The bamboo symbolizes the gates through which the Dragon King and the gods enter. There are 16 short bamboo poles, two of which represent one gate. So there are eight gates.

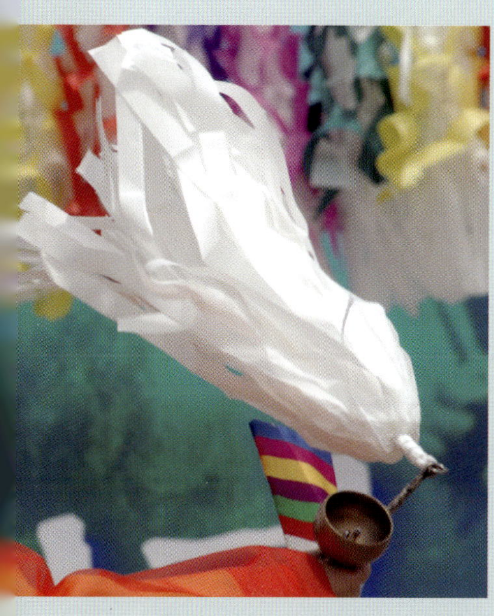

When we perform a *gut*, there are five of us. The main shaman performs the *gut*. Three others play three instruments to open the doors of the heaven and the Dragon King's palace. Another person assists the main shaman. When I perform the first part, I wear the red *hanbok* (Korean dress) over another one. Then I take it off when I finish the first part. It is a ceremonial robe.

We invoke the spirits of our ancestors and the gods/goddesses from four directions in the course of the *gut*. We also place various foods on the altar for the spirits. In addition to the foods, we place cloth and paper money on the altar to represent items we're offering for their use in the other world. I also shake a bell and make a sound to open heaven's door. It takes about 20 minutes to invoke them. I can then judge what is good and what is bad with the sacred sword, and use it to expel the evil spirits. The sword is called *sinkal*, and the bell is called *yoryeong*.

Three days before the *gut*, the *haenyeo* voluntarily come and make rice cakes and foods for themselves because this *gut* is for their safety and abundant harvest. No outsiders can intrude on this process. They cleanse their bodies and minds a few days before the *gut*.

When I am communicating with the spirits—through dancing and talking—I tell the gods what the *haenyeo* want and need for the year, and get the messages from the gods. Then I can tell them when is a good time or bad time for diving and when to be careful. When we *simbang* perform a *gut*, we concentrate with our whole heart, mind, and body. Sometimes, we cry.

I recall that during the Haenyeo Museum *gut*, when I was opening the doors to the heaven, they weren't opening widely. That's why I called the director of the museum and other men to come bow in front of the altar. It was not right that no one was greeting the gods and ancestors at the point of

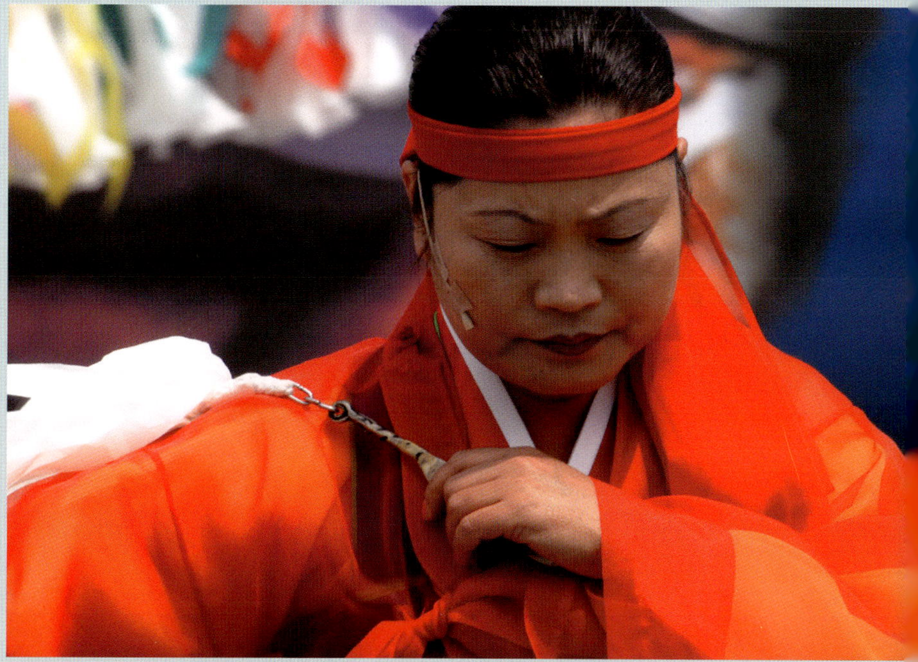

their invitation. Upon their bowing, the doors of the heavens opened widely.

After the doors opened, I paved the way for the Dragon King to come. When he arrives, a lot of other gods follow him, but not all of them come to the altar. During the *gut*, I prayed for the *haenyeo*'s safety and abundant harvest as well as their health and the prosperity of the museum.

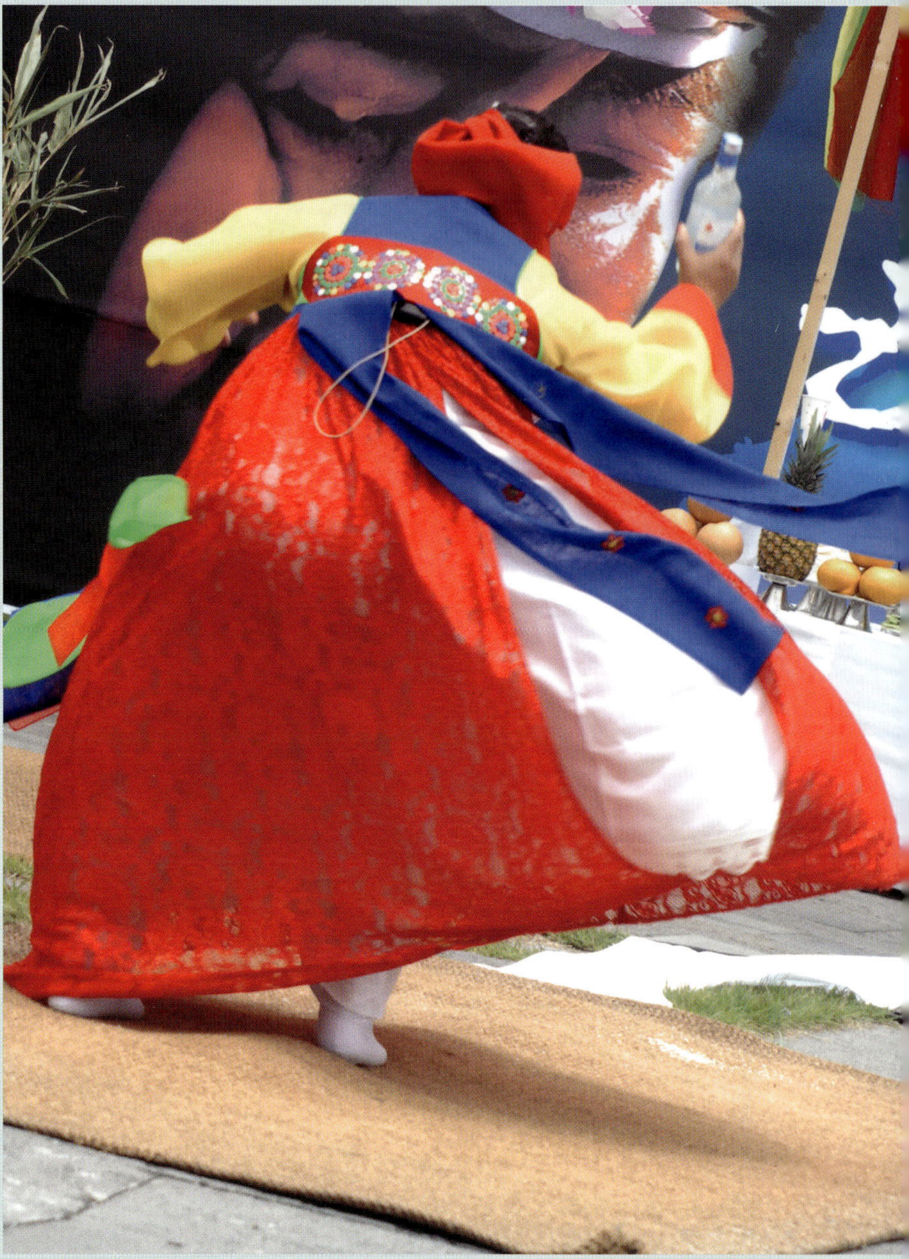

The *gut* are important to the *haenyeo* because they feel they are risking their lives all the time while in the water. They need something to rely on and make them have a sense of security. *Gut* can give them that sense of security and make them feel at ease. So many of them request cooperative *gut* as well as private ones in their respective homes.

Since long ago, the *haenyeo* have believed that the *gut* are the only thing they can rely upon. They are afraid that if they don't arrange for *gut*, they will have accidents. For example, during the Japanese occupation the Japanese authorities forbade all shamanist activities. At that time, many *haenyeo* reportedly died while working in the sea. During the Park Chung-hee government (1961–1979), shamanist rites were also forbidden. But *gut* were still secretly conducted. Since the 1980s, however, traditional cultural practices have been encouraged. The *haenyeo* rely on *gut* because they are always conscious of death.

(August, 2007)

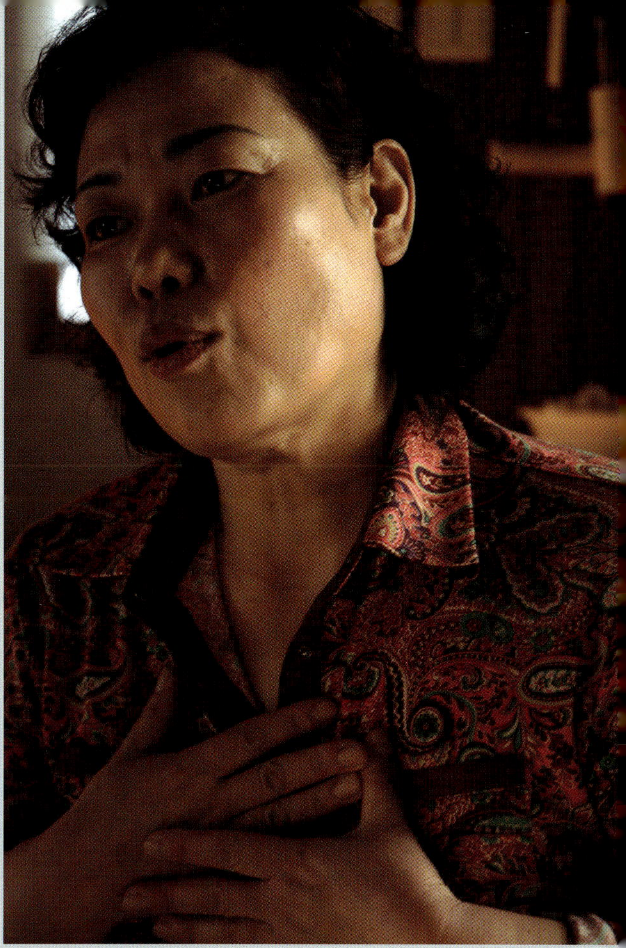

LEE YONG-OK

b.1955 | Jocheon-ri

My father was a shaman, as were my mother and her relatives. When I was nine years old, my mother's sister asked me to do some errands. It was the third day of the first lunar month. Along the way, I found a red pouch lying on the snow in an alley. I picked it up and shook it. It sounded like some coins were inside. I hid it without telling my mother. But one day my aunt took off my clothes to remove some lice and found out that I had kept this pouch. Upon opening it, we realized they were old coins that shamans used in their *gut*. They are called *ganjaebi*. When my aunt told my family what I had found in the streets, my grandfather predicted that I would become a shaman.

I attended Hamdeok Elementary School here in Jeju. But for some reason I wasn't interested in studying. Very often, I wandered on the beach without going to school. At that time, my father was a well-known shaman who performed *haenyeo gut* and Yeongdeunggut in Hamdeok, so everyone knew me as well. Whenever *haenyeo* saw me, they'd ask me to predict if their sea products would bring them good money at the markets. And I was only nine years old. Later, they would tell me that my predictions were accurate and had resulted in good or bad luck, as I had told them.

One day, my mother asked me why I didn't study. Suddenly, I asked her to get the *ganjaebi* that my aunt had taken from me. After I obtained them again, I began my fortunetelling. Then my family moved to Jeju-si, and I attended another elementary school. But I still got sick and couldn't attend school regularly. I stopped when I was a third-grader. At age 13, my uncle—also a shaman in Hamdeok—offered to teach me how to perform *gut*. So I learned quickly by following him and assisting him during the ceremonies. He was very strict, and I began doing some parts of the *gut* by myself at age 17. The more I did, the more I preferred to become a spiritual mediator instead of a fortuneteller. However, I thought that if I married a common person I would quit the role of a shaman. But at age 20, I married a shaman, and so my decision to remain a shaman became even firmer.

I am in my 50s. When I stay at home, I feel pain all over my body due to backaches and arthritis. It's because I have done too much work, not only the *gut* but also family-related tasks and raising children without much help at home. But when I perform *gut*, all of my pain disappears.

Shamanism in Jeju

There was a time when the practice of shamanism was suppressed by the government. But even during those periods of time, people still asked us to perform *gut*. So we performed these *gut* secretly in the mountains and caves so we wouldn't be discovered.

Shamanism has been a practice in Jeju for centuries. Most of the people have always relied on the sea for a living. They needed to have some belief they could rely upon and pray to for their safety and abundant catch.

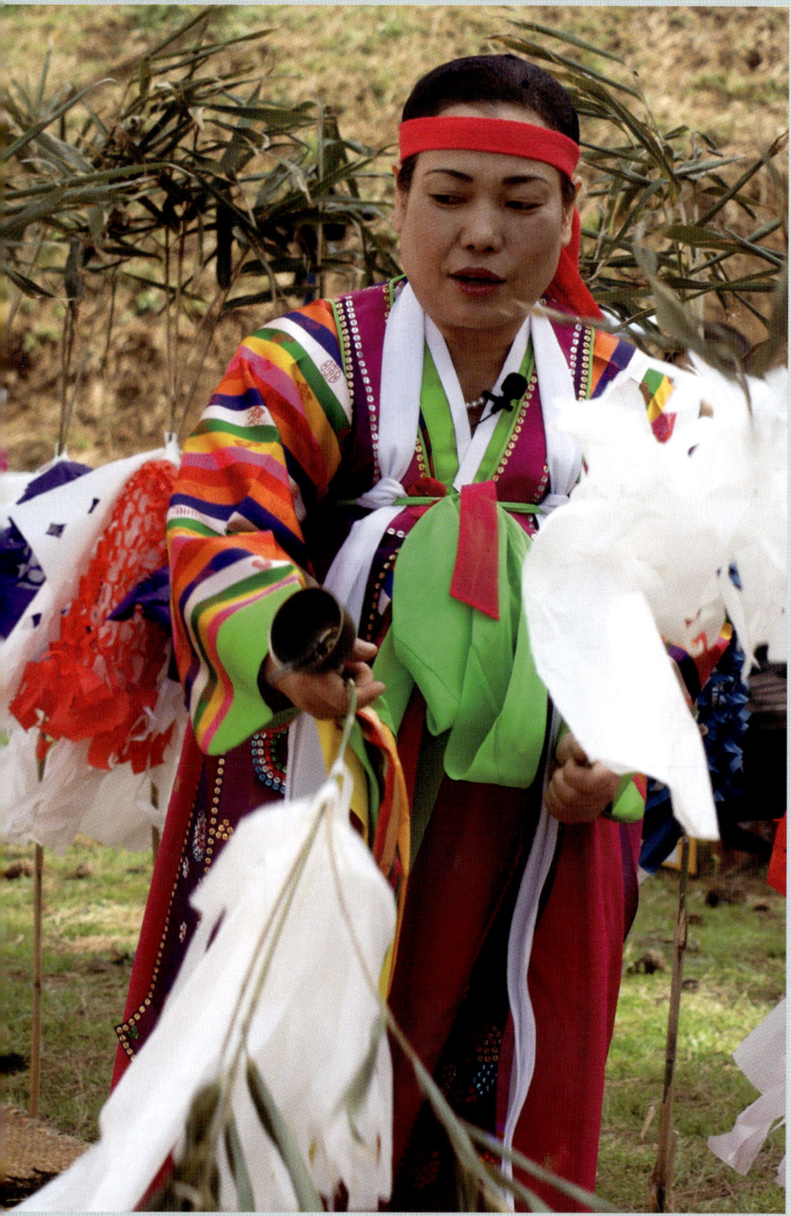

Shamanism and Buddhism have some attributes in common. Even I sometimes visit a Buddhist temple to pray for our family. I know *haenyeo* still believe in shamanism very strongly. They especially believe in the praying to the Dragon King because they rely on the ocean for their living, and the Dragon King is considered god of the ocean. They pray for their safety while diving in the sea, and for an abundant harvest. Sometimes, they even pray to have sons. Many of them not only participate in the public *gut*, but ask me to conduct private *gut* in their own homes.

I think Suh Sun-sil in Gimnyeong and I are among the last generation of Jeju shamans. In my opinion, young people

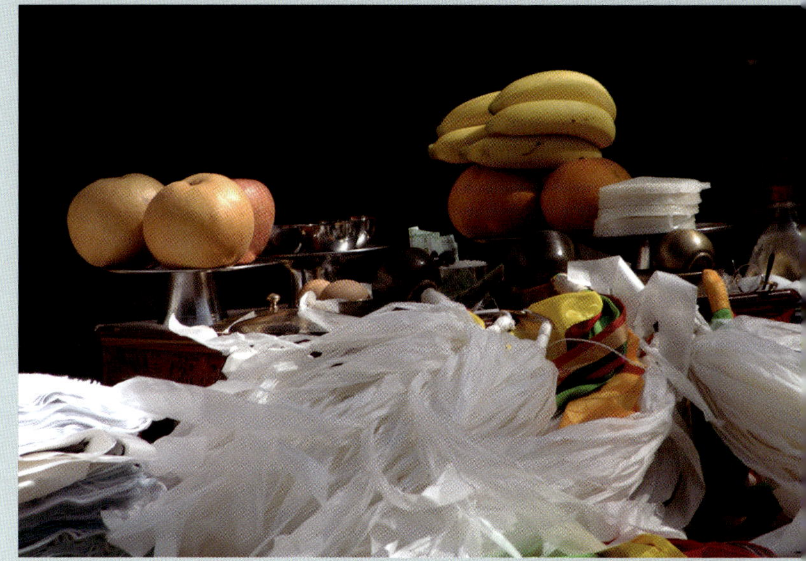

who want to learn to become shamans are only interested in doing so in order to make money. I can tell they don't feel the gods and goddesses in front of them when I watch them perform the *gut*. I feel we have a responsibility to train successors to properly continue our shamanist tradition.

As science and medicine have developed, people have tried to solve many of their problems according to these disciplines. But I am sure there are things that can't be helped simply through science and medicine. We've seen many people recover from their illnesses after performing a *gut* for them, even though their condition was considered incurable.

It's an honor for me to be a shaman. I can comfort people and make them feel better by performing *gut*. Every year, for example, around the time of the April 3 Massacre, we conduct special *gut* for the dead spirits at the various massacre sites. I have also been asked to perform *gut* and pray for babies and ancestors. I want to be remembered as a good person even after I die. If I am just a housewife, who will remember me or consider me as a person who has helped troubled people?

However, when I am performing a *gut*, I try to relieve the host's sorrow and bad feelings inside their hearts. While relieving their suffering, I feel that I, too, become purified. I sometimes shed tears while conducting these private *gut*. It's a catharsis for me. I feel so happy and comfortable when I can forget my own pain and sorrows. Sometimes I joke with my husband that we should divorce so I can conduct *gut* full-time.

Yeongdeunggut

There are two Yeongdeunggut. The first one is to greet and welcome the Goddess of the Wind (Yeongdeung). It used to be simple and short, but these days we've been asked by the Jeju Fishermen's Cooperative Association to conduct a rite for a good harvest along with the welcoming ceremony. So it is a bigger event these days. The last one is to bid farewell to the gods. That's the larger event.

Yeongdeung is a visiting goddess who comes annually to Jeju on the first day of the second lunar month. When arriving, Yeongdeung comes with a lot of attending spirits. They remain for two weeks, going around the island scattering seeds in the sea and in the fields to ensure an abundant harvest.

During the welcoming *gut*, the gods are not allowed to enter the actual site of the *gut*; they stay outside. That's why we treat them afterwards by throwing food into the sea. While talking to the gods, we use a scripture. But in the middle of a *gut*, we receive messages from the gods and ancestors which we share with the people. So our words can be different depending upon what we hear. We always try to make the people happy and joyful. So when I narrate between the gods and the people, I ask for good luck; I utter words humorously. We are like mediators. Then the *halmeoni* (grannies) give friendly responses to my words...often yelling and laughing. After a *gut*, we shamans are exhausted. But I teach my shaman novices that whenever we perform these ceremonies, we should do it sincerely and comfortably.

We use several tools, such as a drum, a gong, and a bell. The drum is used to notify the gods that we are initiating the *gut* because the sound resonates far away. The gong is used to awaken the gods in heaven. And the bell is used to awaken the gods of the earth. Bamboo is also used because it is evergreen and abundant in Jeju. But the significance of it is more complicated to explain.

We wear different clothes depending on the gods whom we greet and to whom we

In 2009, UNESCO inscribed Jeju Chilmeoridang Yeongdeunggut on the Representative List of the Intangible Cultural Heritage of Humanity. The ritual is held in the second lunar month to pray for calm seas, an abundant harvest, and a plentiful sea catch, at Chilmeoridang in the village of Geonip.

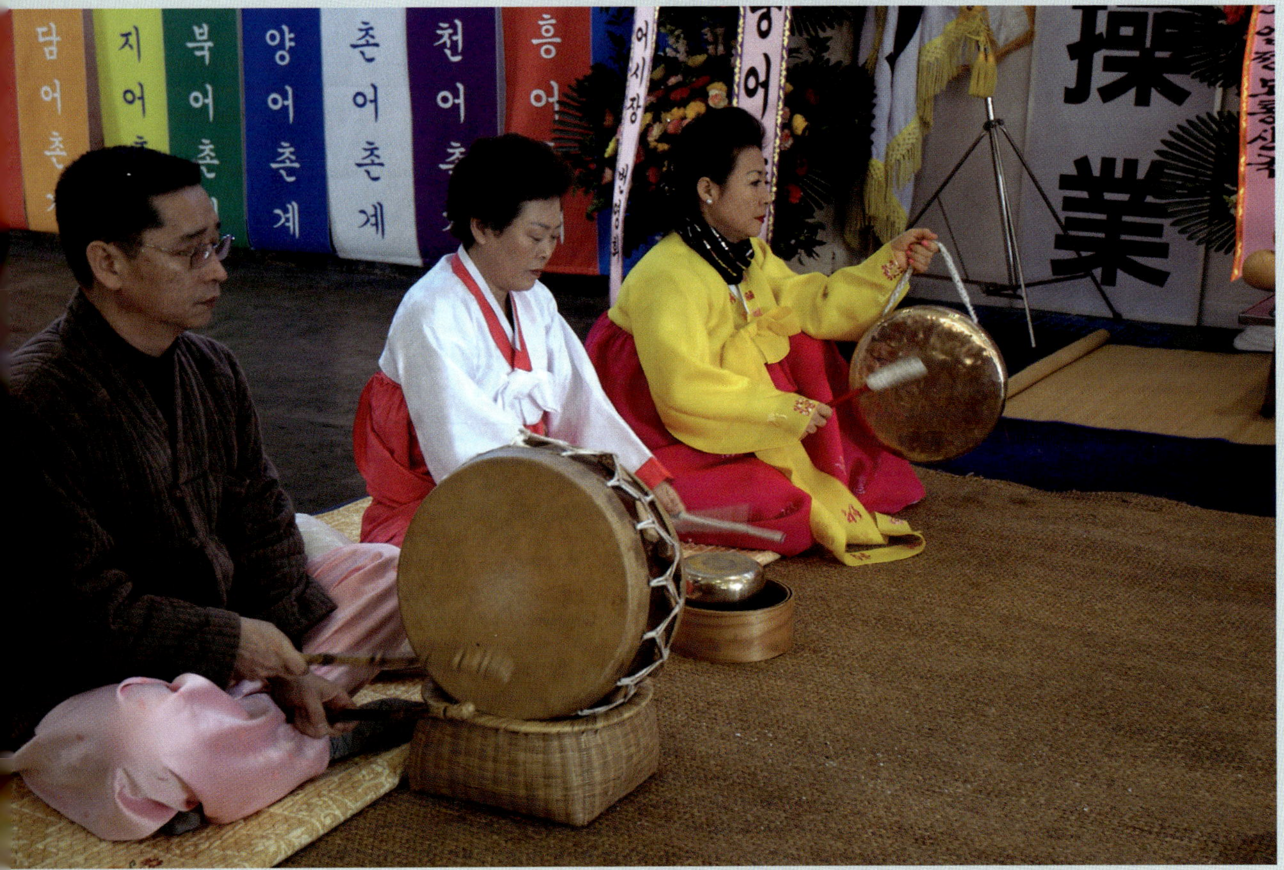

pray. We wear more ornate clothes when we pray to the gods representing a higher level, such as the Dragon King of the Ocean. As for our swords, they have such a long, detailed history. Basically, every part of the knives has a specific meaning. But if you don't understand the history of shamanism, it's hard to grasp the significance. It is a tool that we use when we inform the gods about what the people want.

On the altar we place food such as *tteok* (white rice cake), cooked rice, eggs, liquor, and vegetables. But on the day of the *gut*, we place fresh sea products there as well because the Goddess of the Wind is also involved in the work of the sea. We always share the food that was received by the gods after the *gut*. It is like the food we share after *jesa* ceremonies (Korean annual memorial rites honoring one's dead ancestors).

Many *gut* are happy festivals. Each year, we speak different narratives depending on the weather. If the weather is good, people think the gods are happy. If the weather is terrible, it's not a good sign. Then the people are more serious.

The scale of Yeongdeunggut has been getting smaller and smaller. I think it's because the owners of boats and ships have been fewer. Also, these days young people don't realize the importance of *gut*. They simply ignore them. Nevertheless, we do everything properly during the ceremony. Even though our clients are decreasing, many representatives from the media, scholars, and students come to view the *gut*. I think we should keep this tradition and never abandon it.

Some people believe that the major parts of the *gut* should only be performed by males. I don't believe this. In fact, my husband, who is also a shaman and designated as one of Jeju's "Human Cultural Assets," lets me take part even though I have some trouble dancing due to arthritis. Because I have conducted many private *gut* for the spirits of the dead after funerals and other occasions, I was confident about doing the public Yeongdeunggut. The *haenyeo* who attended the Yeongdeunggut said they were very satisfied with my ceremonial work.

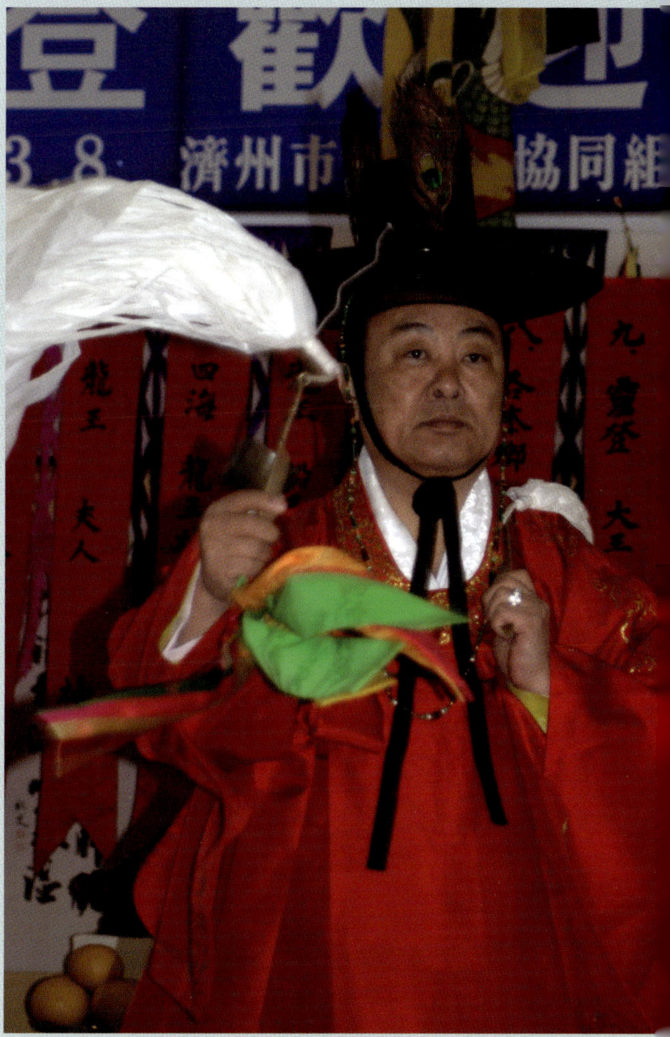

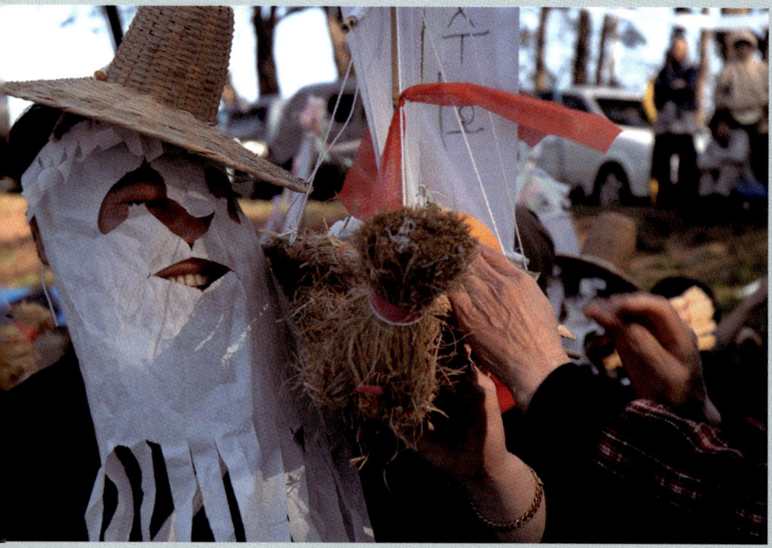
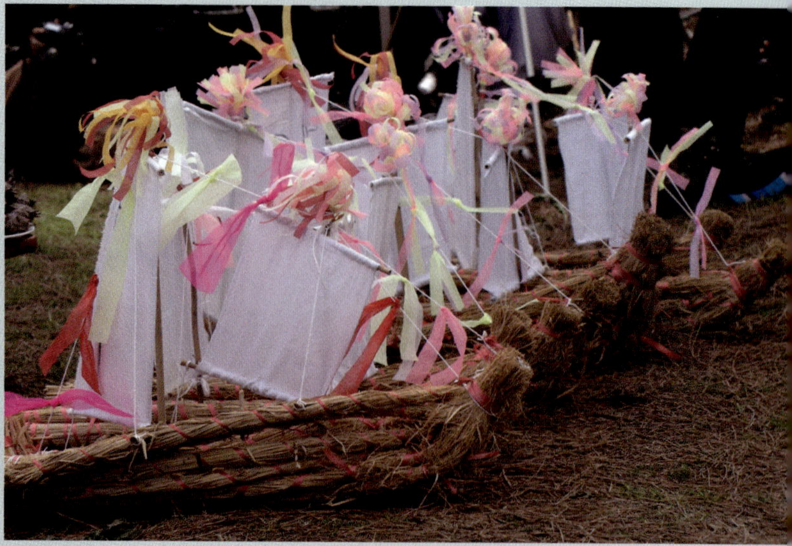

Locals avert harm by cajoling and entertaining "goblins" represented by men wearing paper masks at a Yeongdeunggut.

There's a point during the *gut* when people bring bags of rice to attending fortunetellers. When we have big occasions, we invite these elderly women fortunetellers to join us. Because they can't do very active things, we have them sit and tell fortunes to the visitors and give them advice to cope with possible bad luck.

The visitors usually bring rice to place on the altar. At the end of the *gut*, the fortunetellers fill their palms with rice from the sack and throw them up in the air and then catch them again in their hand and see how many they have left. Six grains of rice is the best luck. Eight also signifies a good harvest. Ten and twelve are good, too. But the numbers four, seven, and nine are considered bad luck. When they catch good numbers, the fortunetellers let the guest swallow the grains. When they catch bad numbers, they just throw them away.

After the two-week period in which the gods are present on the island, we prepare a farewell ceremony. As I mentioned before, there are many gods that come during this time. So at the end, there are dancers with masks that represent the goblins. We bid farewell to all of them by sending them off on little straw boats. The Yeongdeung goddess rides on the first boat. The rest follow on the other boats.

(March, 2008)

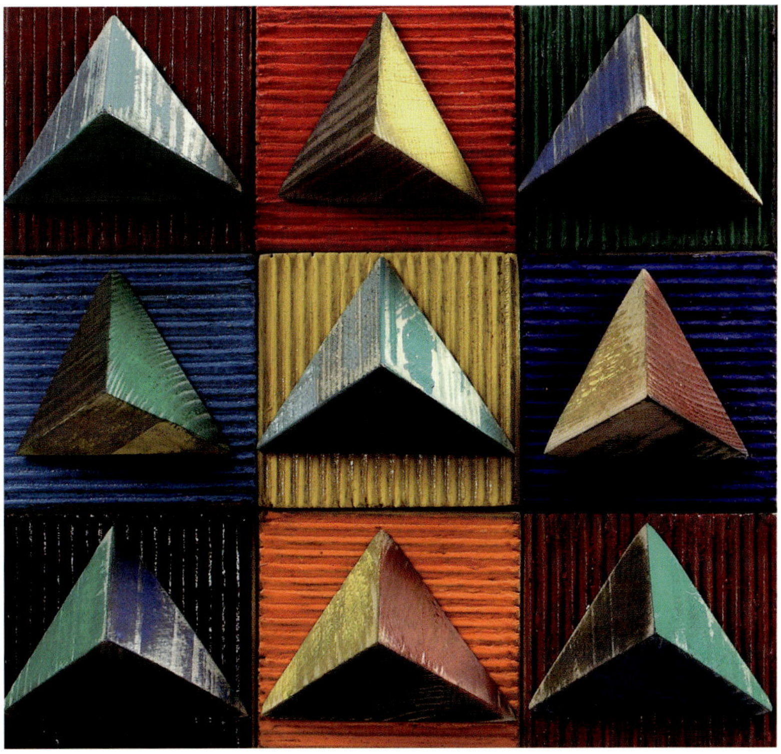

Abangguk & Eomeongguk 105x36 cm | paper, woods & pigment | 2008

Baekjo Halmang

(Grandmother of All Shamanist Gods)

Baekjo "Halmang" (grandmother) is the mother goddess of all shamanist gods in Jeju. Legend has it that she was blessed with 18 sons, 28 daughters, 78 grandchildren, and other relatives. Additionally, 368 gods are considered to be her nephews. Having been abandoned at childhood, and having survived many hardships throughout her life, she continues to be honored by many women for her courage, determination, and strong will.

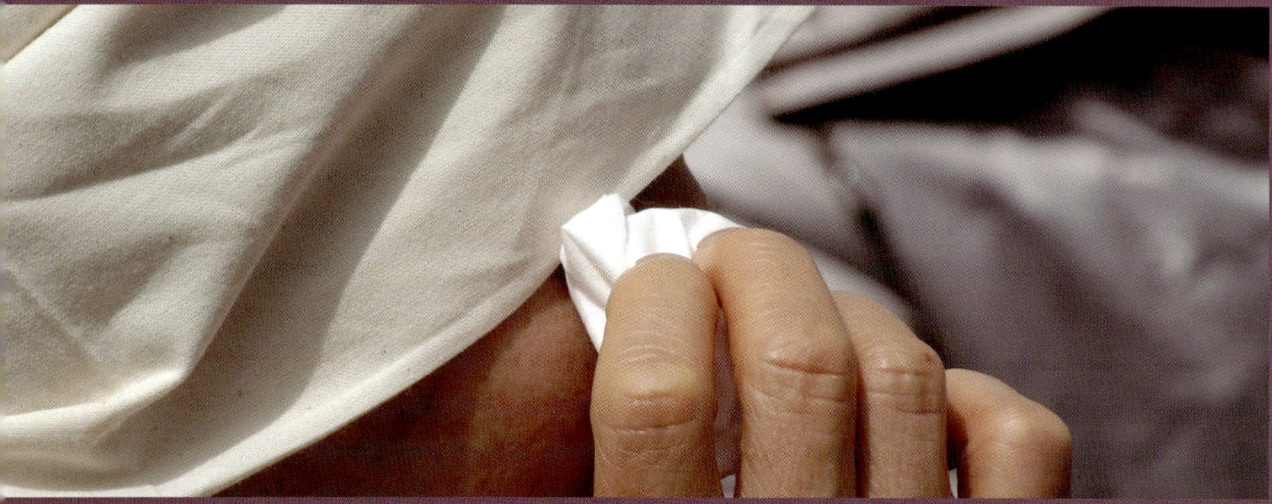

Chapter Three

LIFTED BY THE WIND & TIDES OF SUFFERING

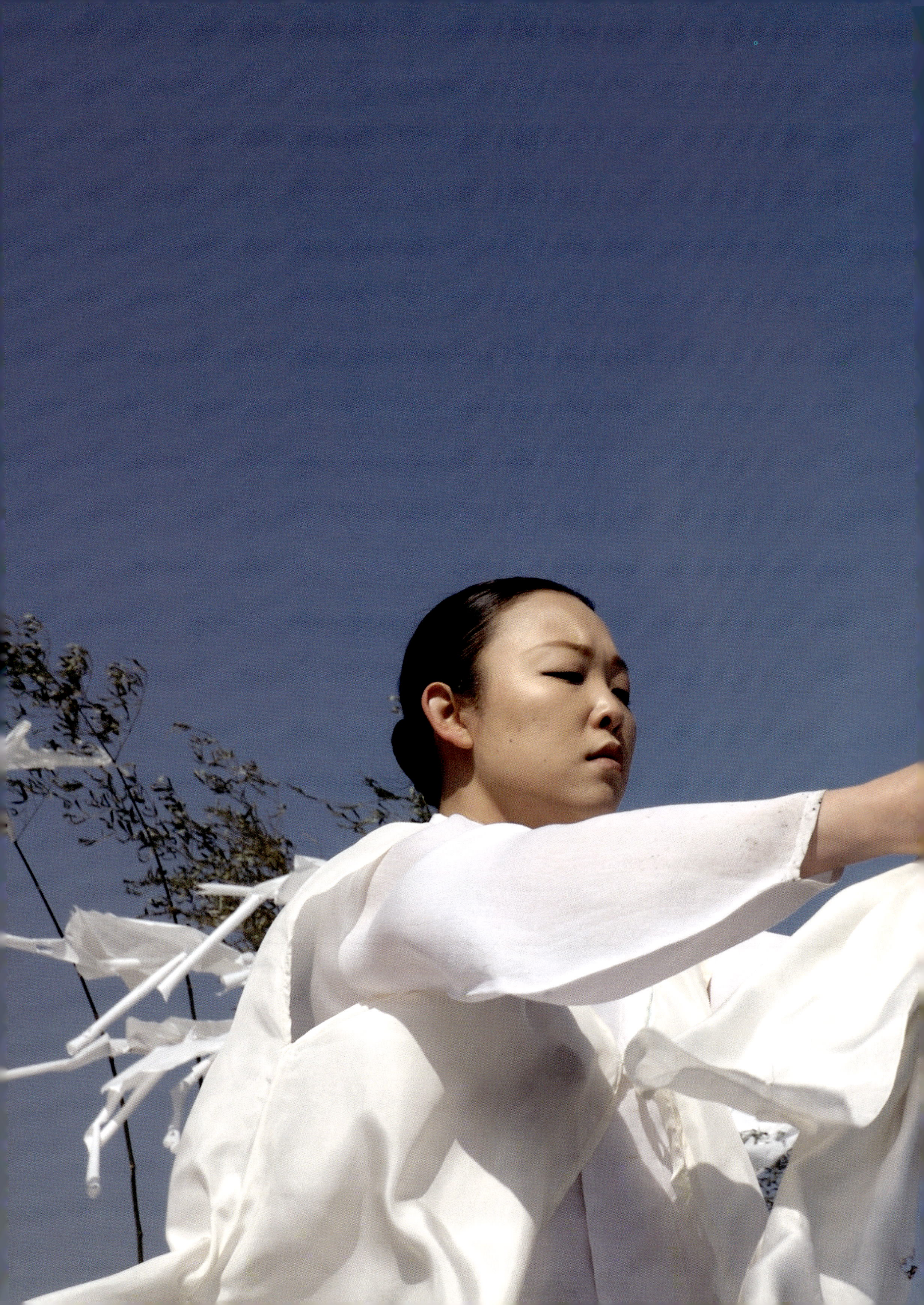

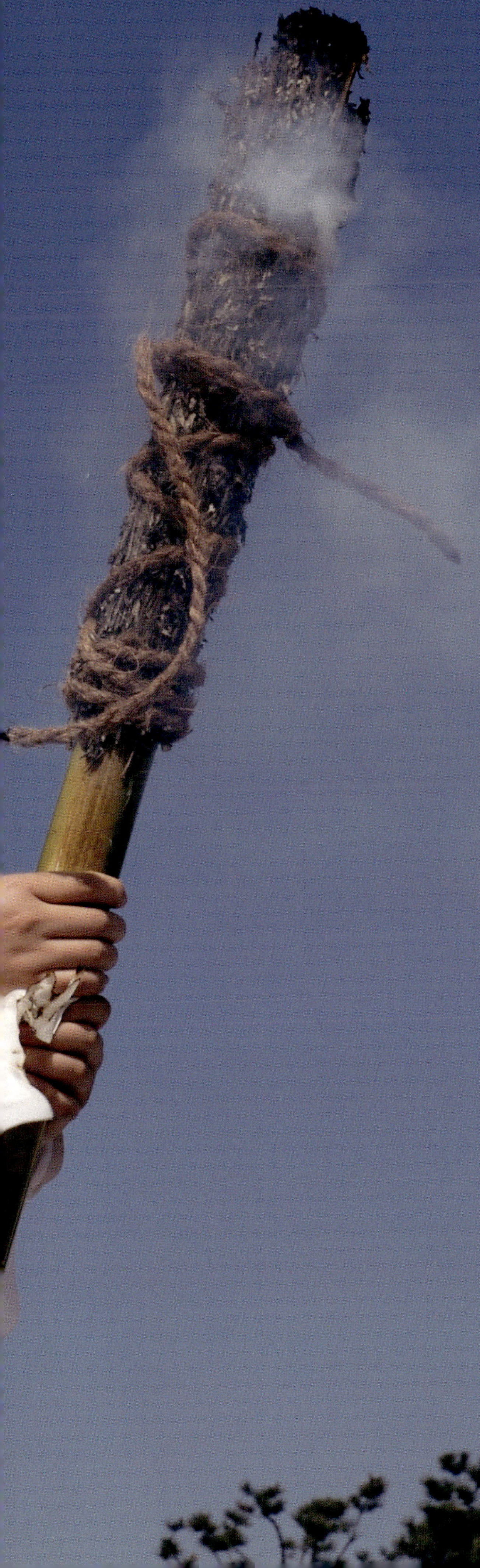

One by one, gray-haired seniors stepped forward with white chrysanthemums to remember the dead. Some of them wept—their kindred tears streaming down their cheeks as incense smoke vanished into the air. The altar was flooded that day by a fragrant sea of flowers.

Over 5,000 people had come by car and bus to Jeju April 3rd Peace Park to commemorate the 60th anniversary of Jeju's Massacre of April 3, 1948. After the Korean War (1950–1953), this was Korea's second largest massacre in its modern history. My cousin Peggy Choy—a dancer and social activist—first told me about this horrific chapter in Jeju history during my initial research. We both shared an interest in Korea's cultural and political history as third-generation descendants of patriotic grandparents.

The number of casualties has varied according to government and academic reports. Estimates, however, have ranged between 30,000 and 80,000—a significant loss in human lives given the total population of 300,000 at that time.

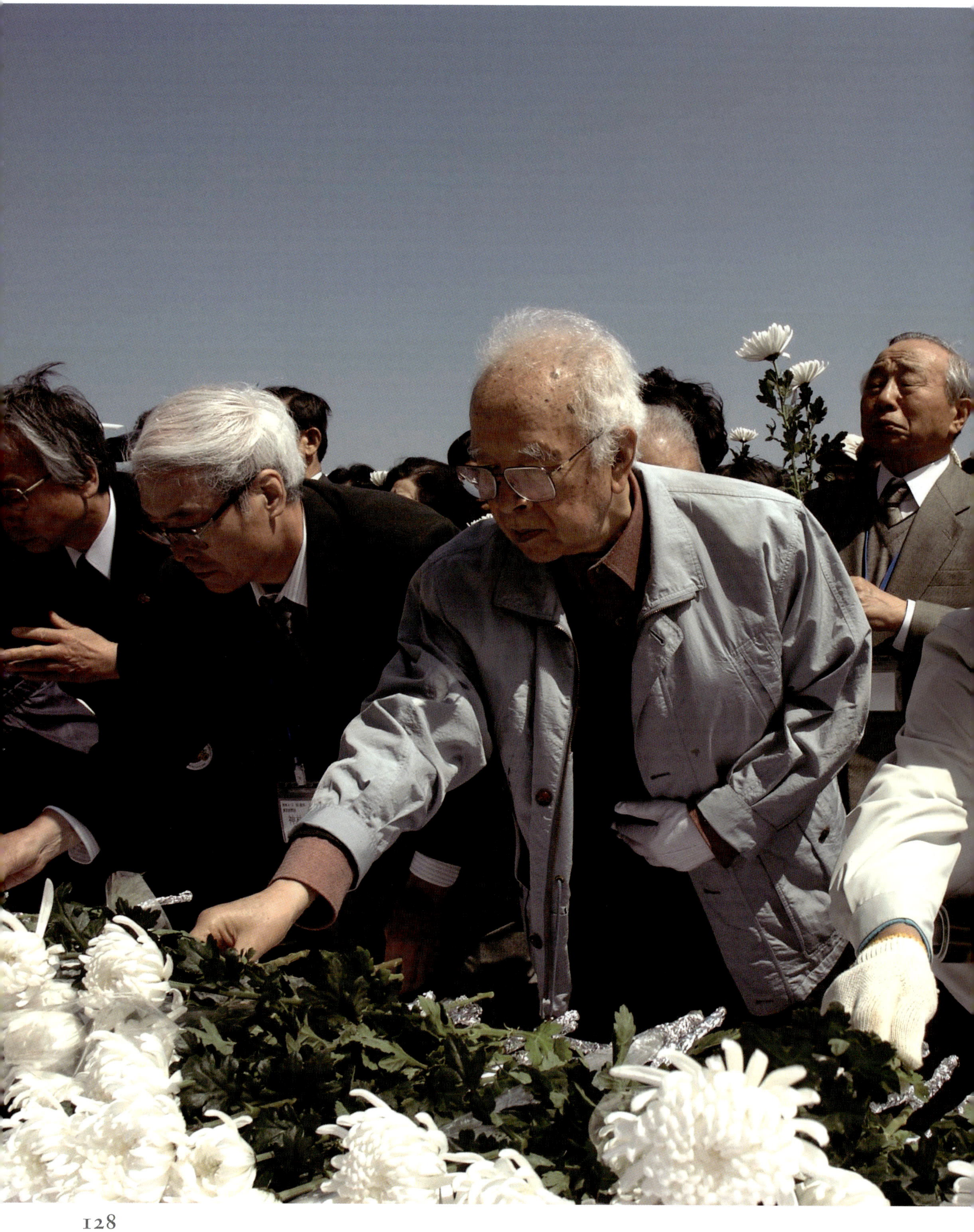

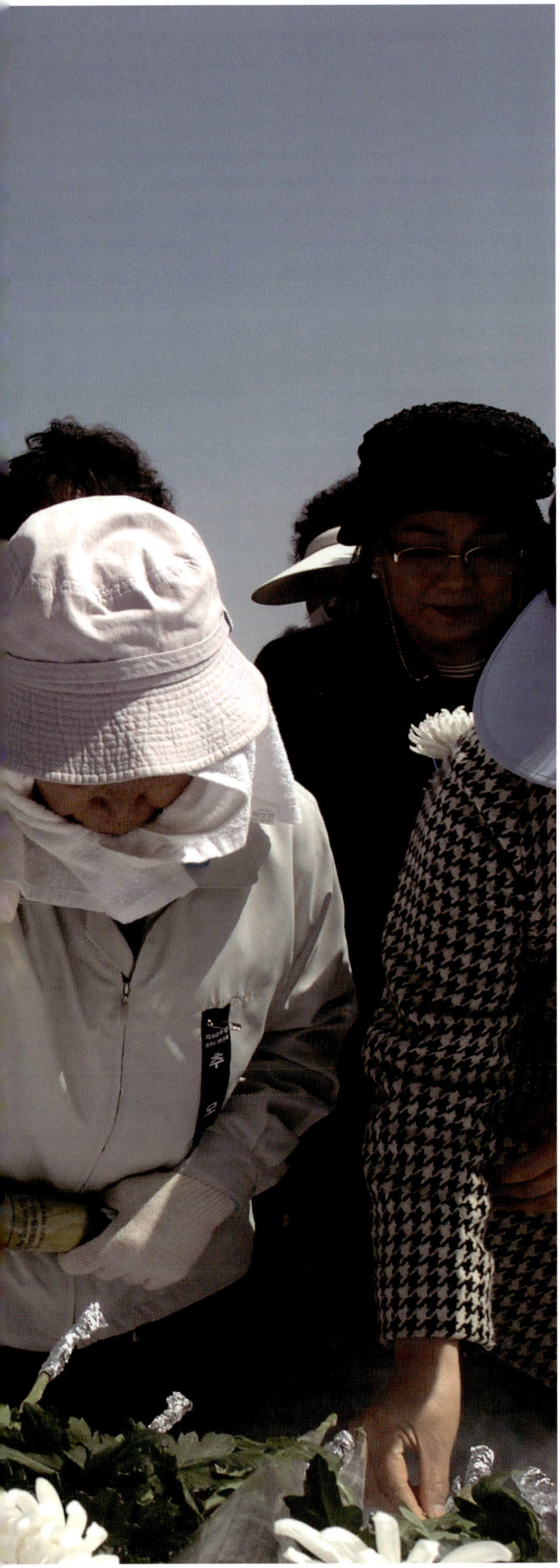

Although the massacre was perpetrated by the post-World War II Korean government and supported by a right-wing Jeju governor, and ultra-right mainland youth groups, the US Army Advisory Government was clearly the sole legal authority in Korea south of the 38th parallel at the time from August 1945 to August 1948. The tragedy was precipitated by protests against the holding of separate elections in South Korea planned for May 1948, and opposition to the installation of pro-American president, Syngman Rhee.

Researching the massacre has been illuminating, for sure. But bearing witness to the stories of its survivors is another matter. I first met Hong Suk-ja while living in Gimnyeong. She operated a minimart where I often bought my milk, noodles, and toiletries. Disabled for many years after a truck ran over her on the road, she also endured the mental anguish from others' memories of Sa Sam (which means "four" and "three," designating the third day of the fourth month). "Many of the grandmothers who survived the massacre used to visit me and recall their memories," she said. "One woman told me that four of her sons were killed, and she prepared no less than 17 *jesa* (ancestral rite) in one year because most of her family members were killed."

Such a degree of suffering is unimaginable to me. So when I hear that Jeju has been officially designated as an "Island of World Peace," a part of me responds hopefully. But another part of me can't help but wonder. Will the controversial presence of a proposed naval port benefit the island as some believe? How can Jeju Island remain harmonious and peaceful in the years to come?

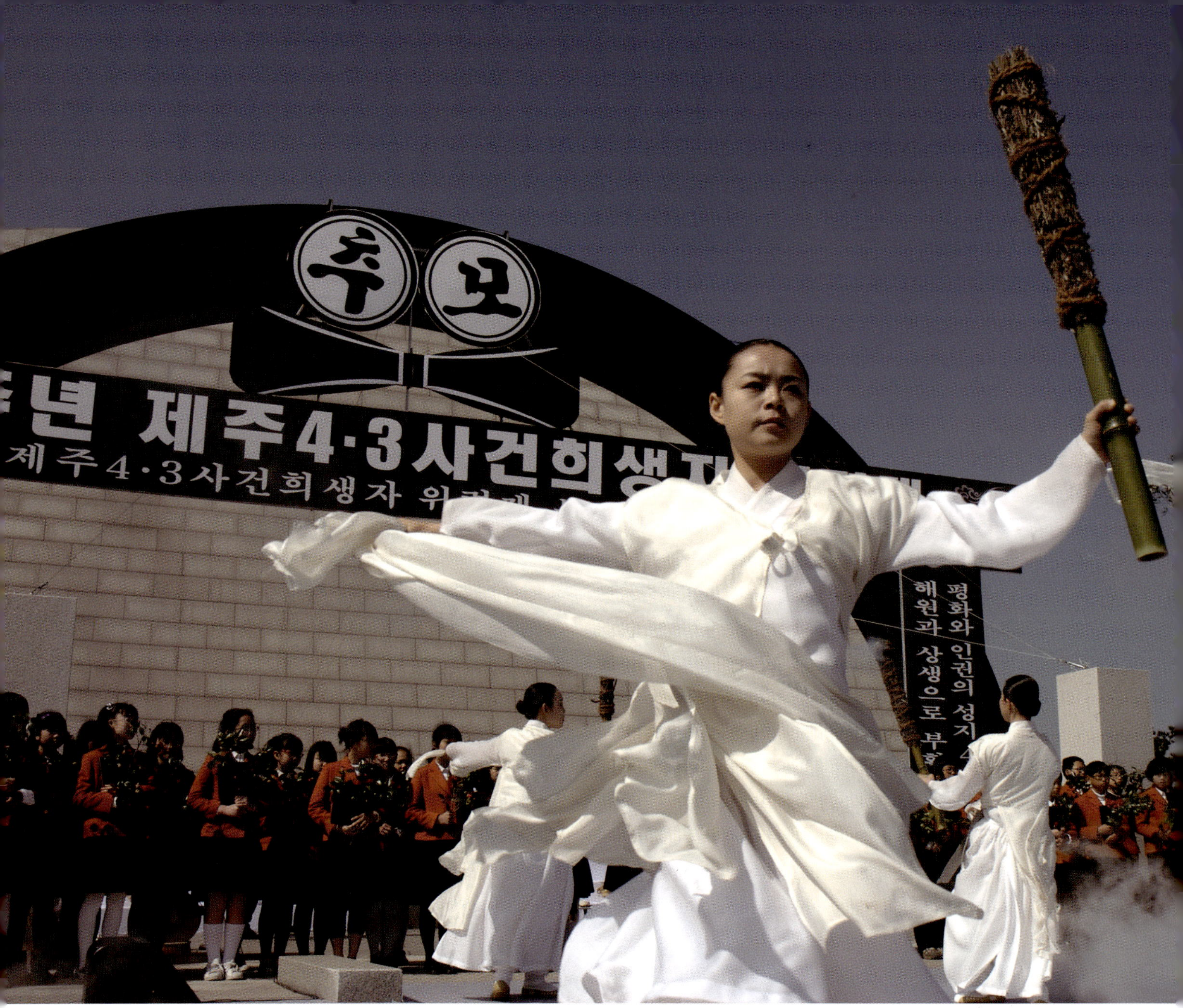

" During the massacre, a lot of young men were killed. Because of the Korean War, most young men had to join the army. So women had to do all the things that the men normally did. That's when they became involved in everything."

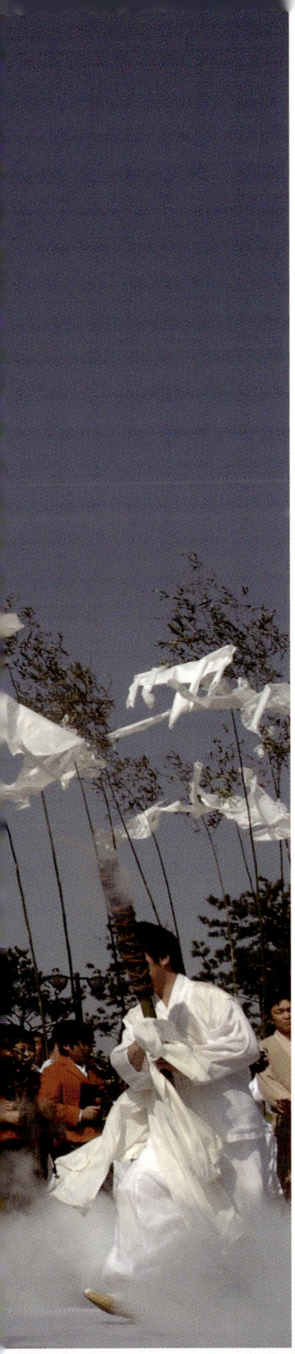

Sa Sam

The year 2008 marked the 60th anniversary of "Sa Sam" (April 3, 1948)—also known as the Jeju Massacre. At least 30,000 civilians were killed during a rebellion put down by the South Korean military. Every family in Jeju has been affected by a tragic loss that occurred during this time. Thousands attended a commemorative ceremony at Jeju April 3rd Peace Park on April 1, 2008, offering white chrysanthemums in honor of the dead.

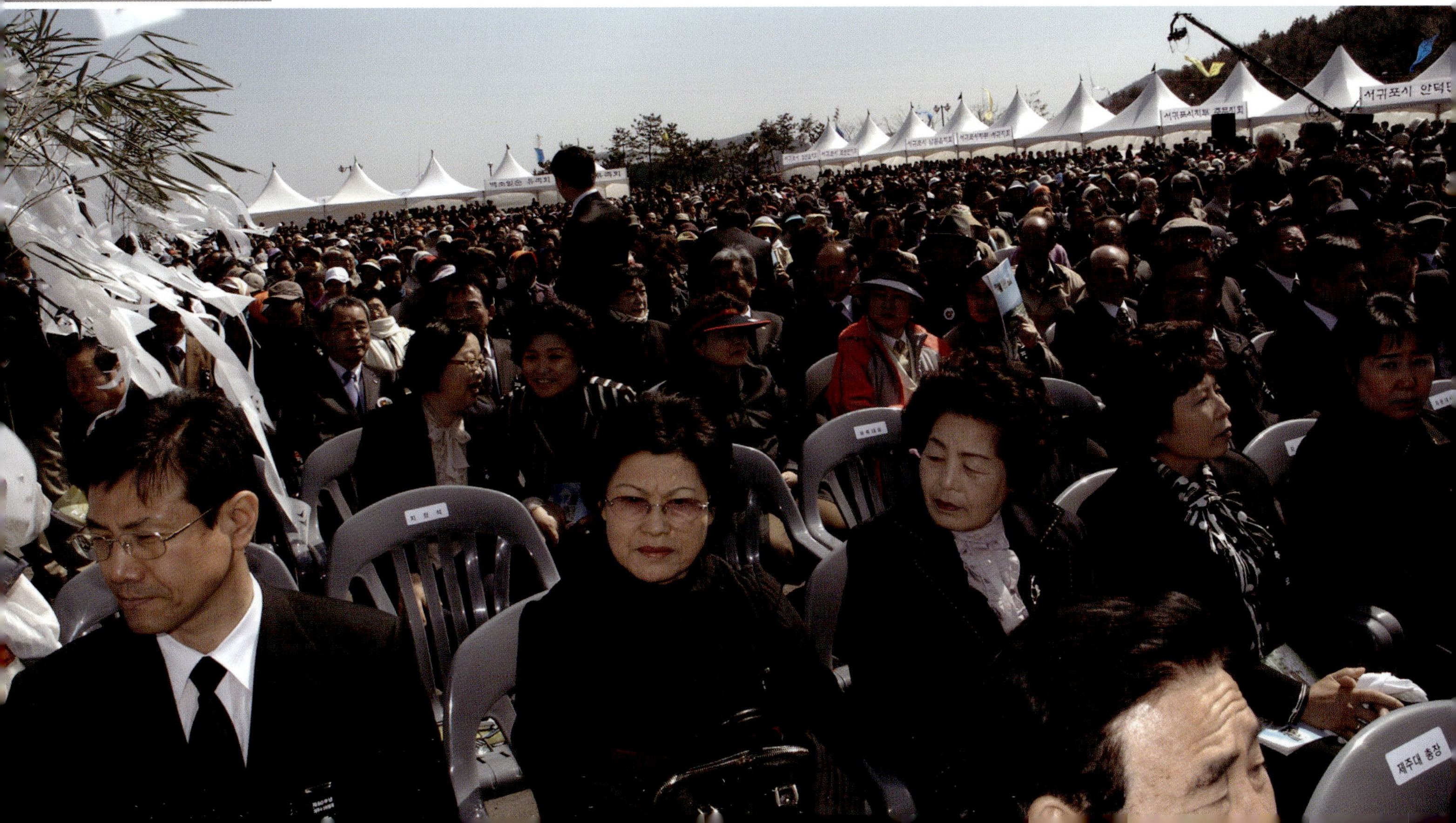

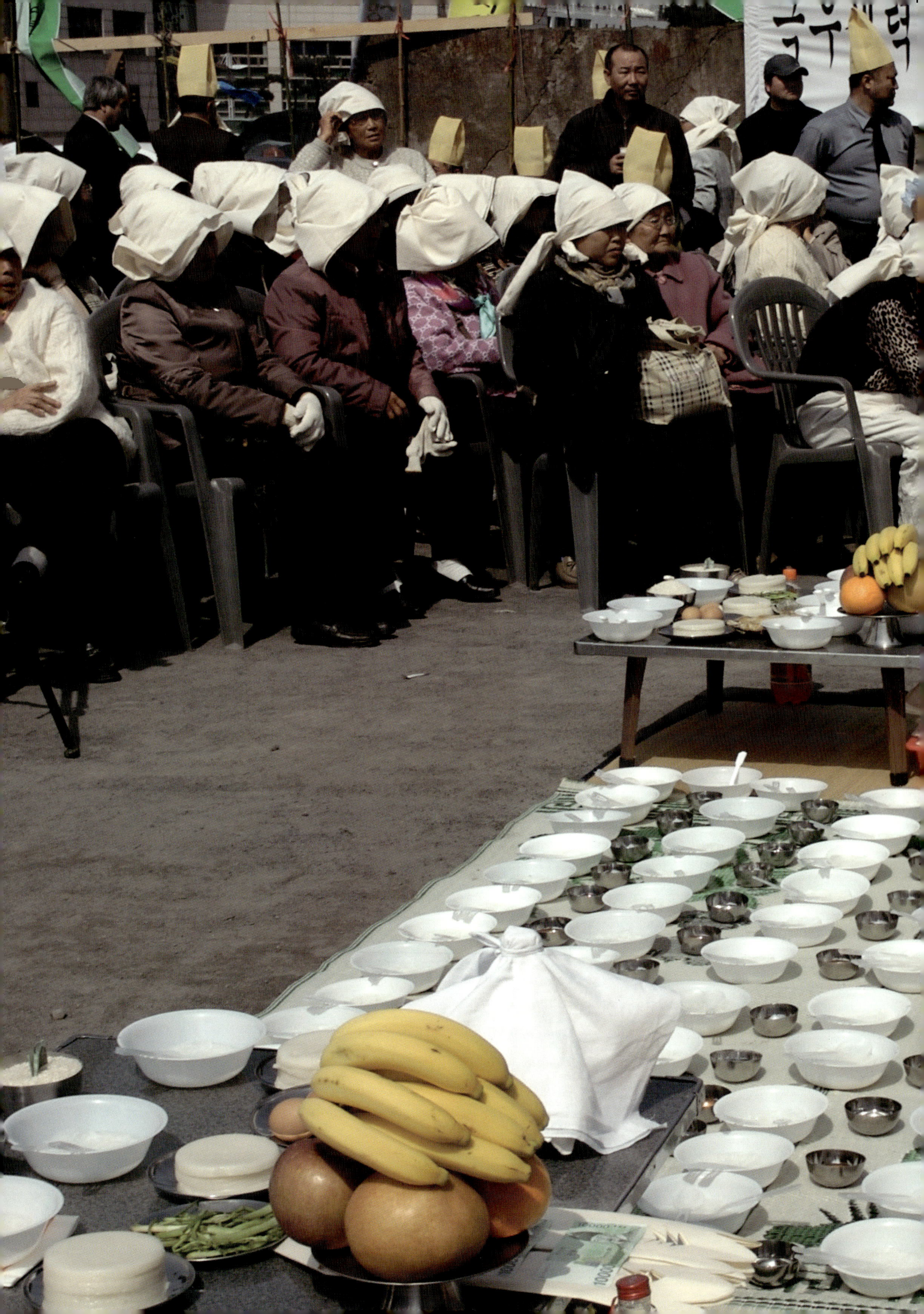

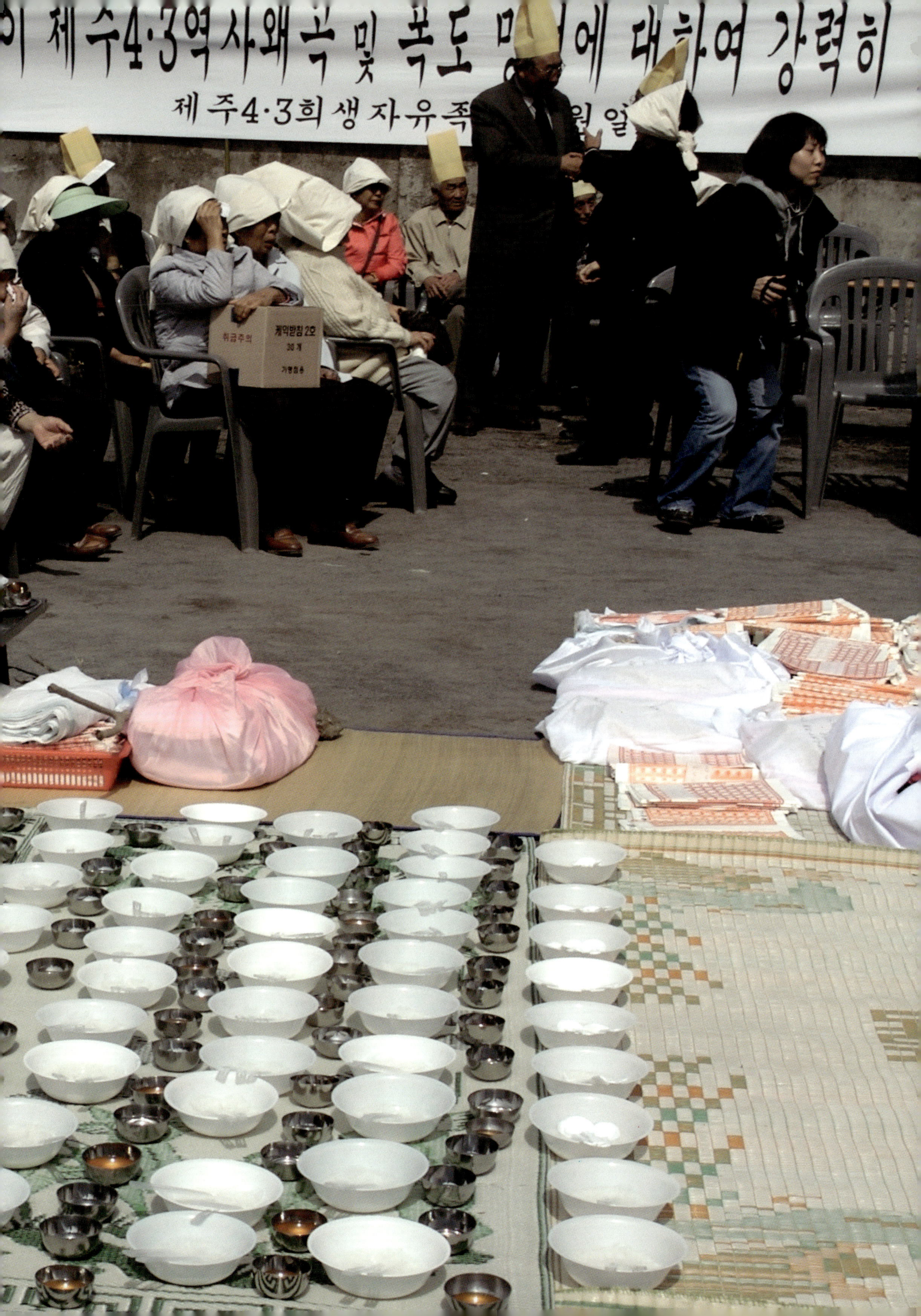
제주4·3역사왜곡 및 폭도
에 대하여 강력히
제주4·3희생자유족
취급주의

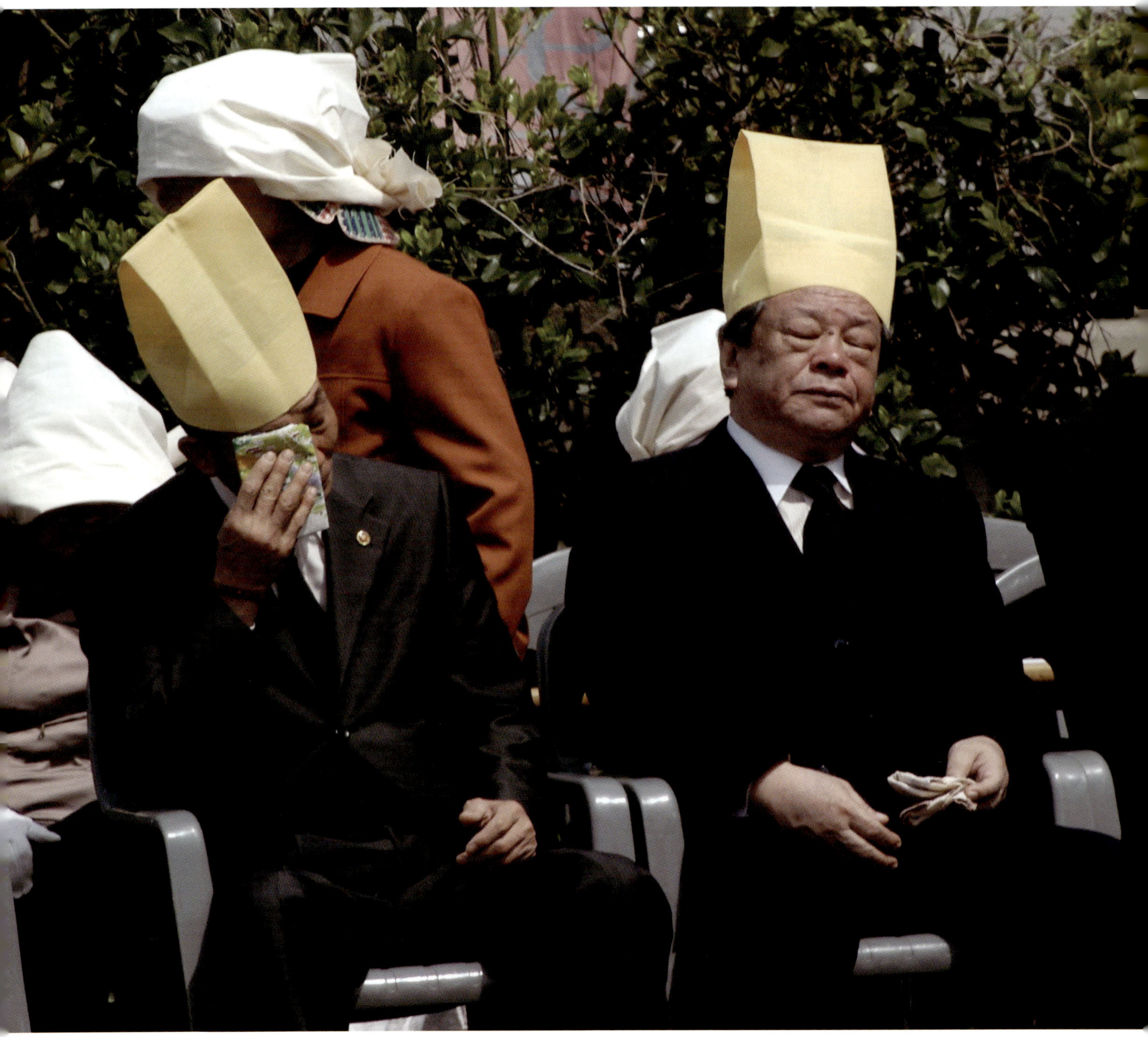

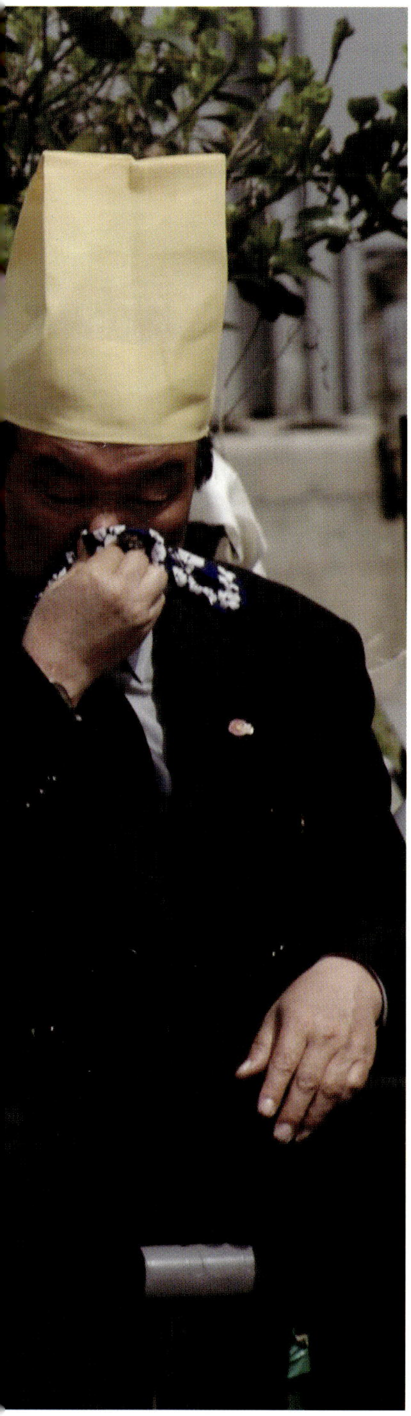

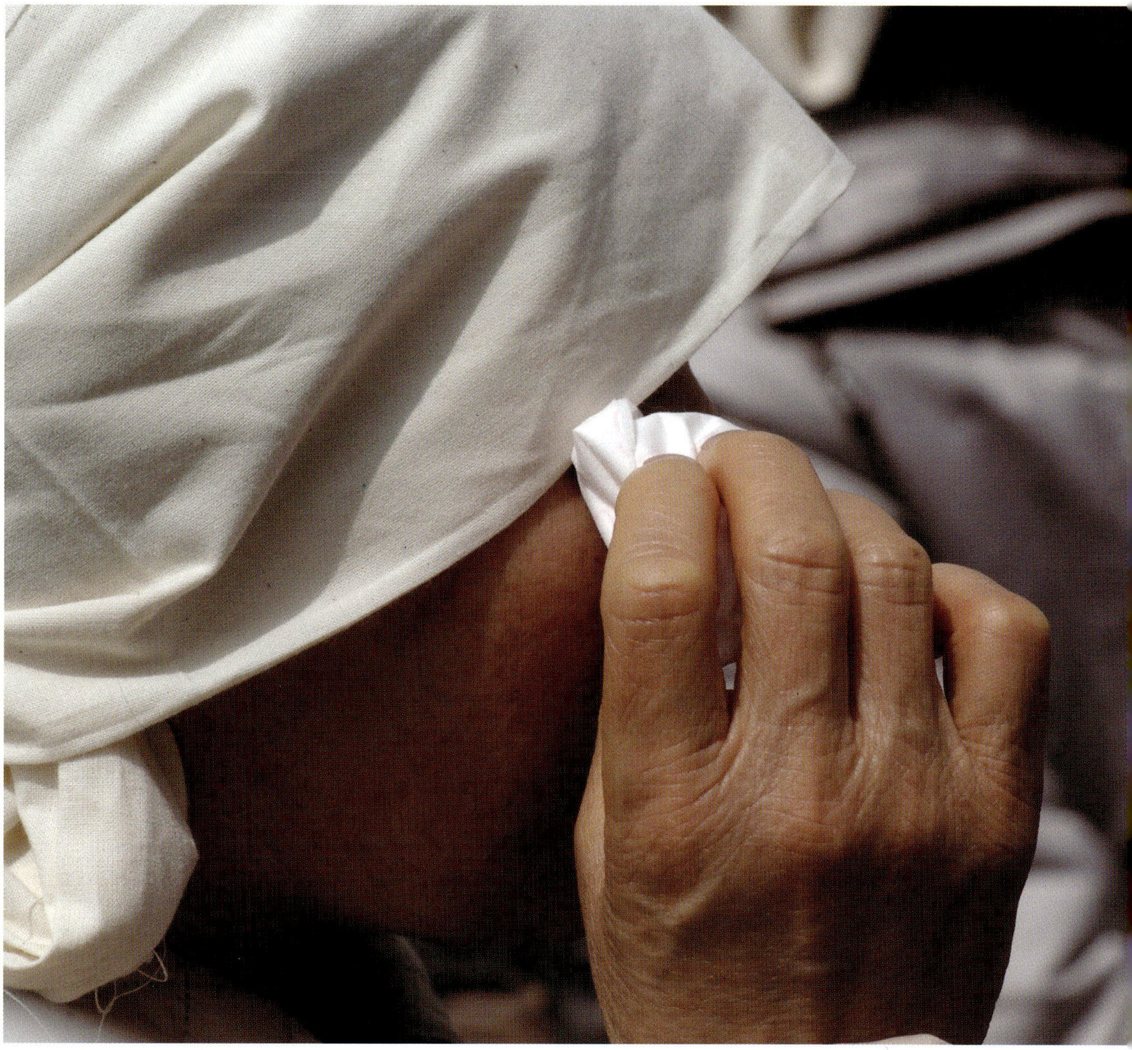

" Many innocent youths were killed by the police. The insurgents from the mountain tried to recruit smart youths and students. They were too young to have better judgment."

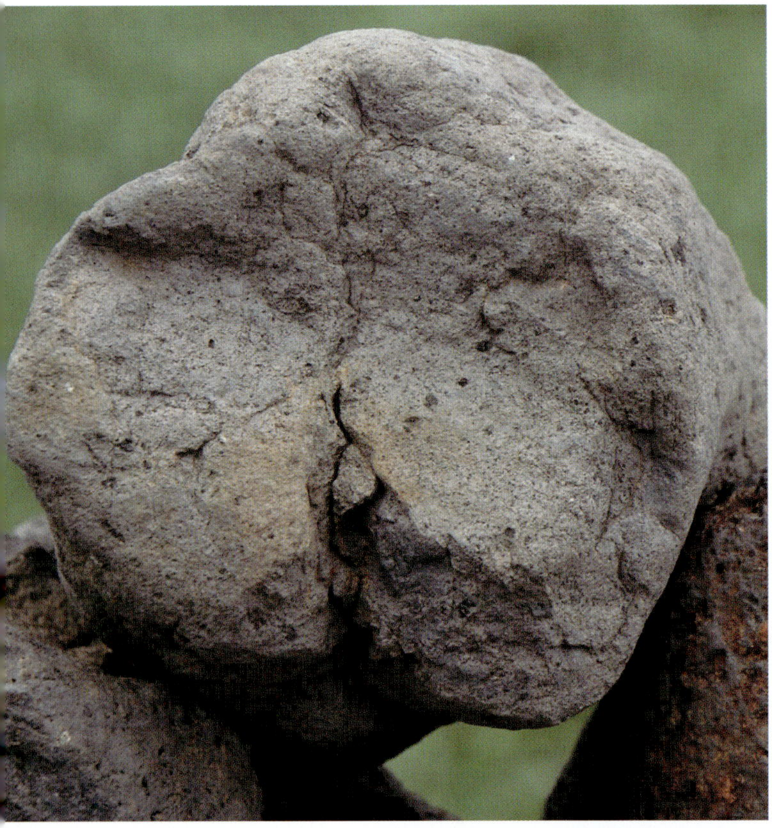
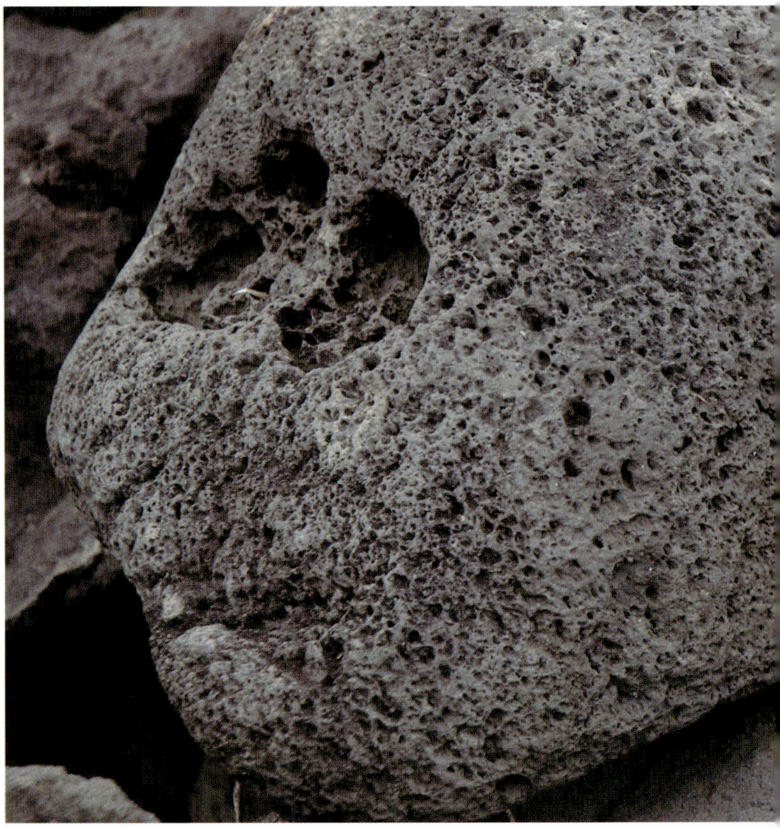
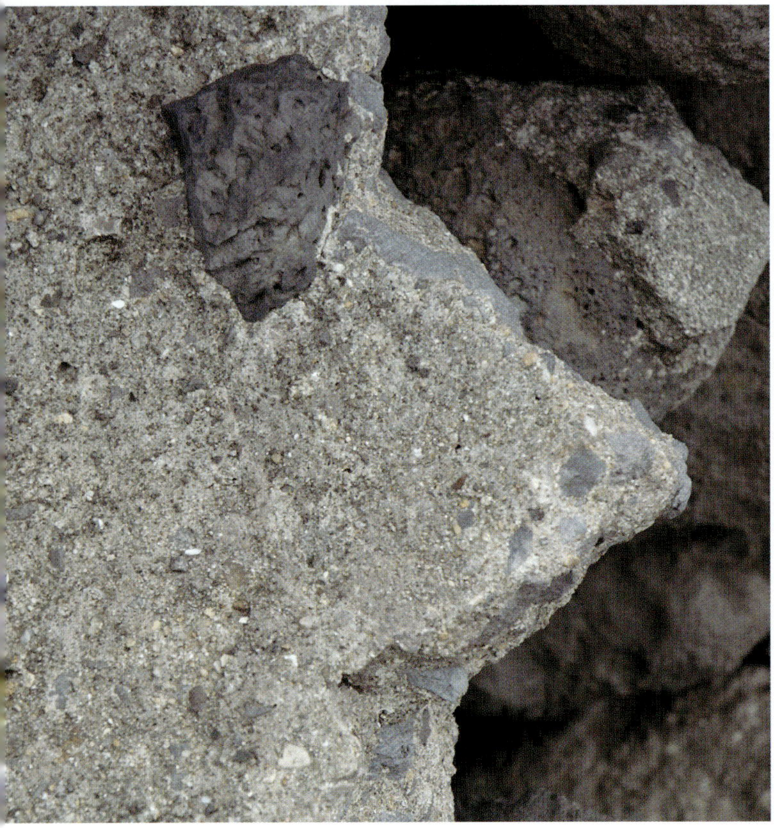
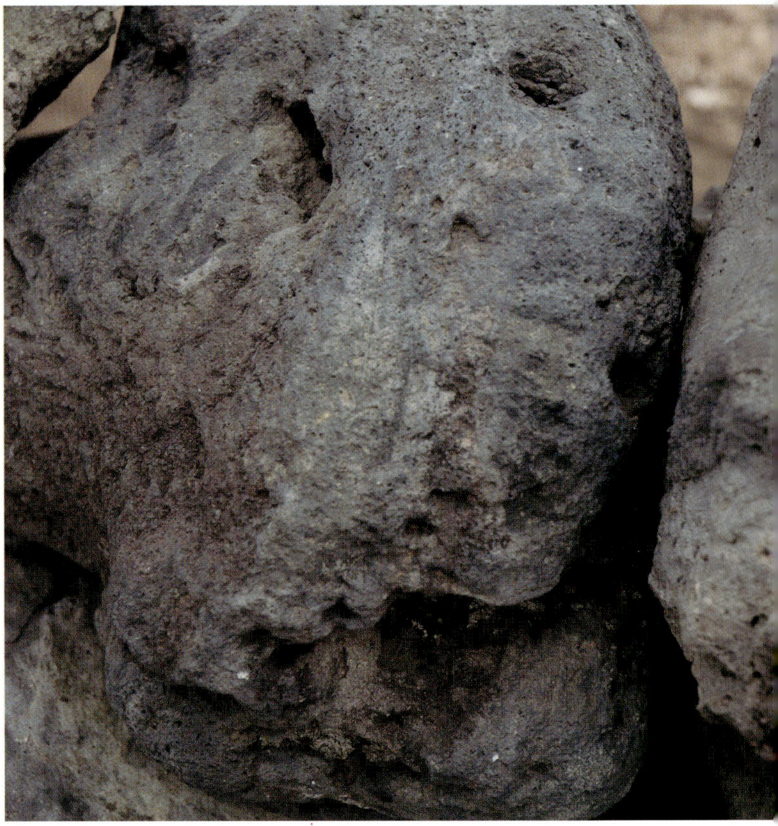

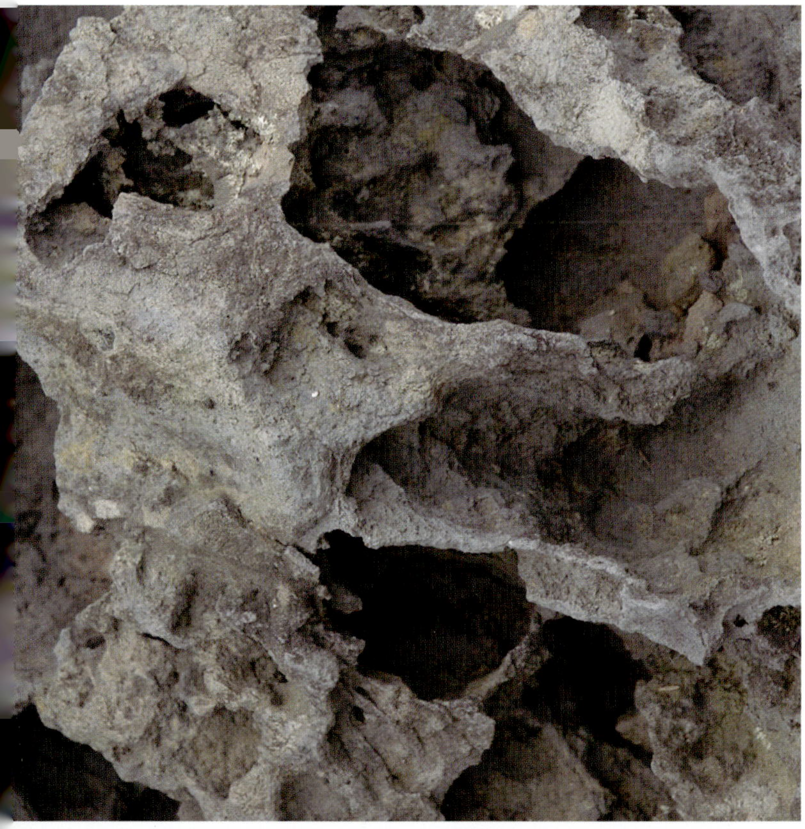

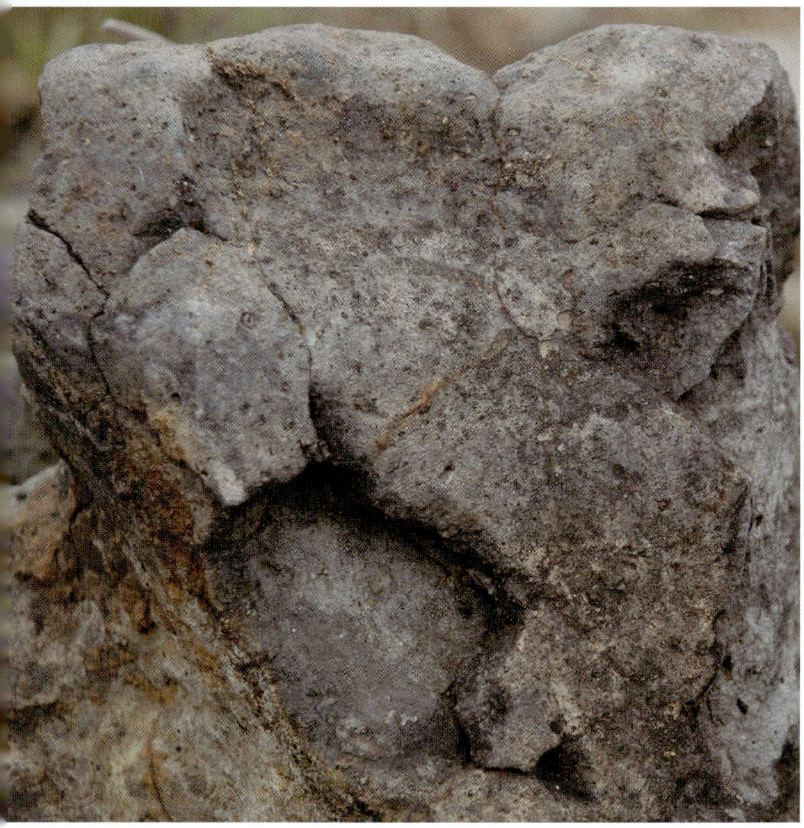

Faces of Suffering

One of Jeju's "three abundances" is its black volcanic rocks. The book *Young-shil, the Spirit Grove* by Paek Ja-won inspired further observations of spirits embedded within the basalt walls that distinctively mark the island.

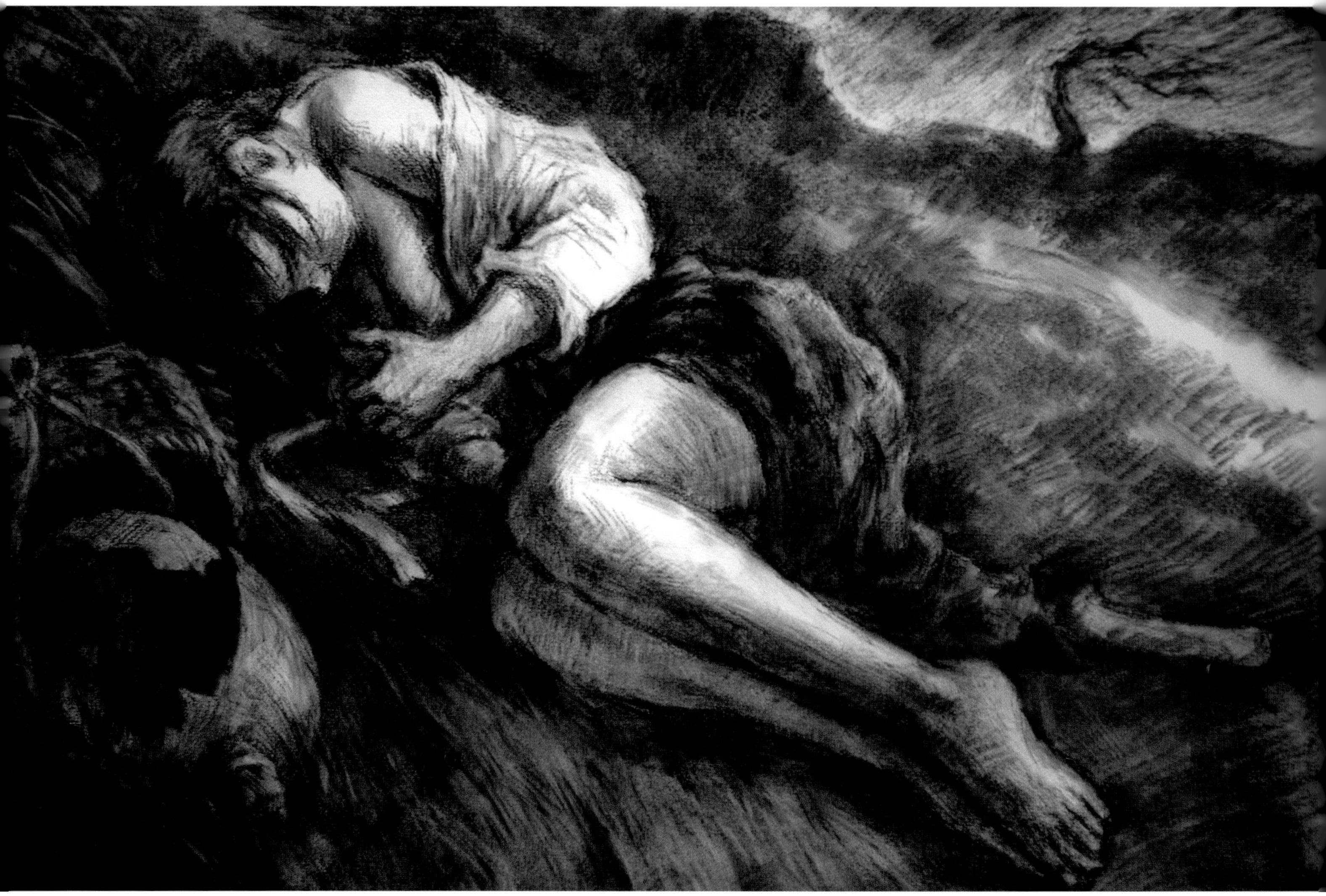

Rape by Kang Yo-bae
55.3x76.0 cm | paper, charcoal |1991

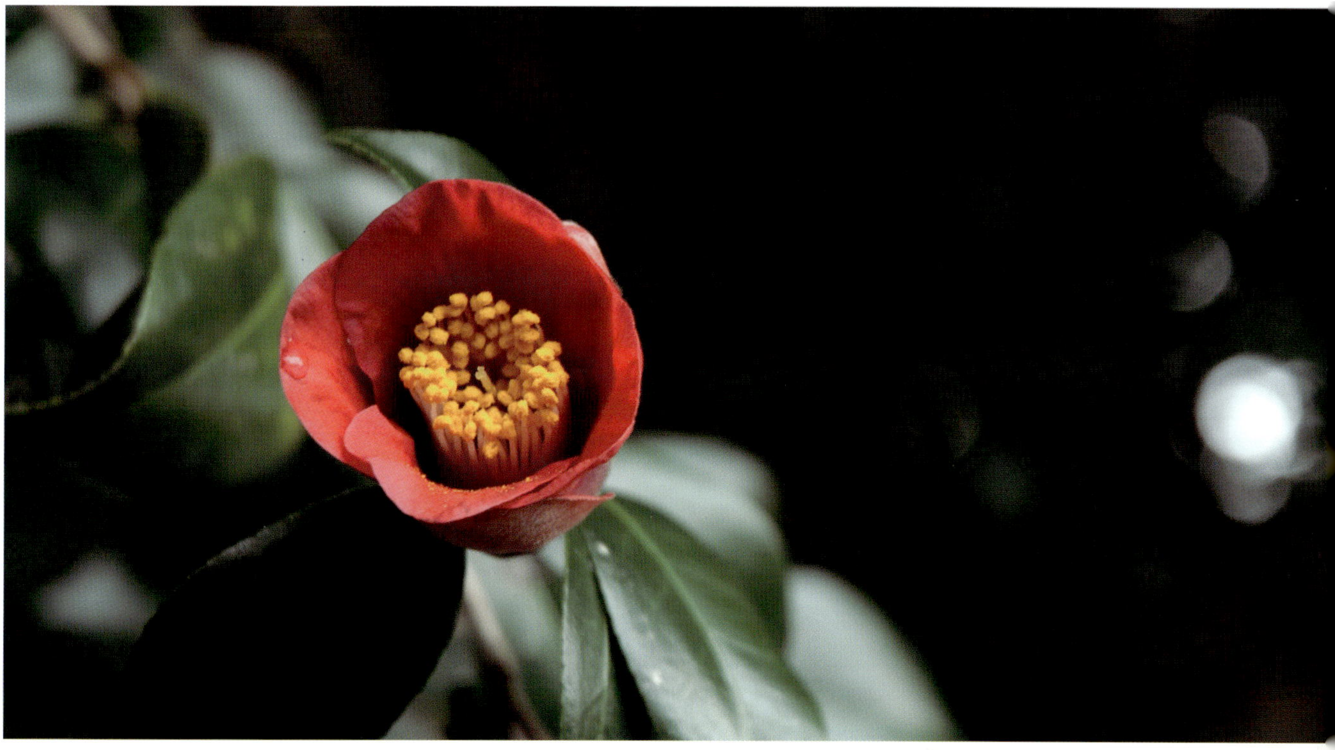

The camellia tree is characterized by flowers that fall suddenly and entirely off their stems, without shedding a single petal. Artist Gang Yo-bae has compared the cruel ordeals suffered by the people of Jeju to fall of a camellia flower.

" Jeju *haenyeo* were terribly exploited during Japanese occupation (1910–1945)...the Japanese authorities forced the *haenyeo* to work far away. But they never earned the money they deserved."

" The *haenyeo*'s anti-Japanese activities began in Seongsan-ri in 1931 when they realized they were being cheated out of their deserved earnings. The protest then spread to the Gujwa area. In 1932, the biggest protest was held in Hado-ri. About 17,000 people gathered together at that time. It was one of the biggest protests against Japan in the history of Korea's anti-Japanese movement. It was unique in that women spearheaded the protest."

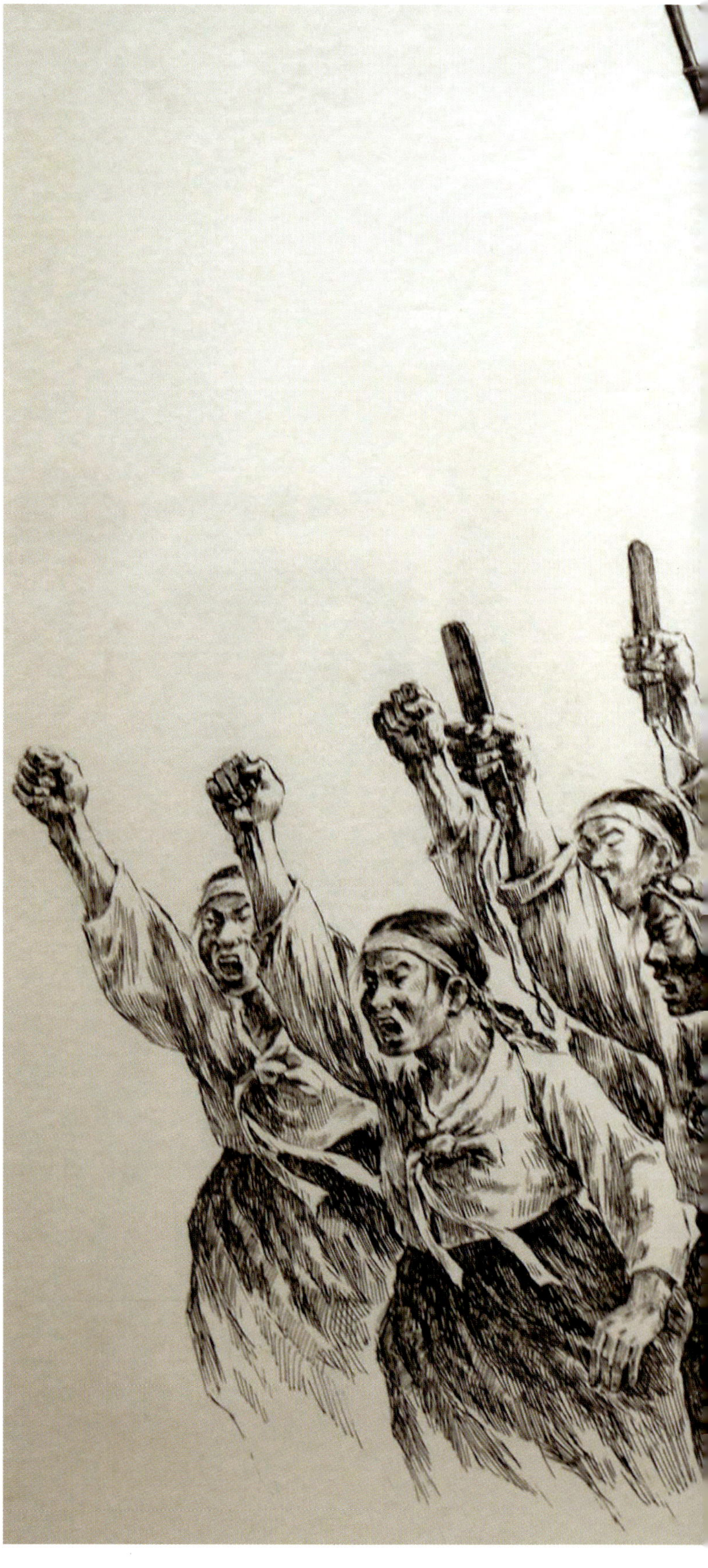

The Anti-Japanese Struggle of the Haenyeo by Kang Yo-bae
38.7x53.2 cm | paper, pen, brush, chinese ink |1989

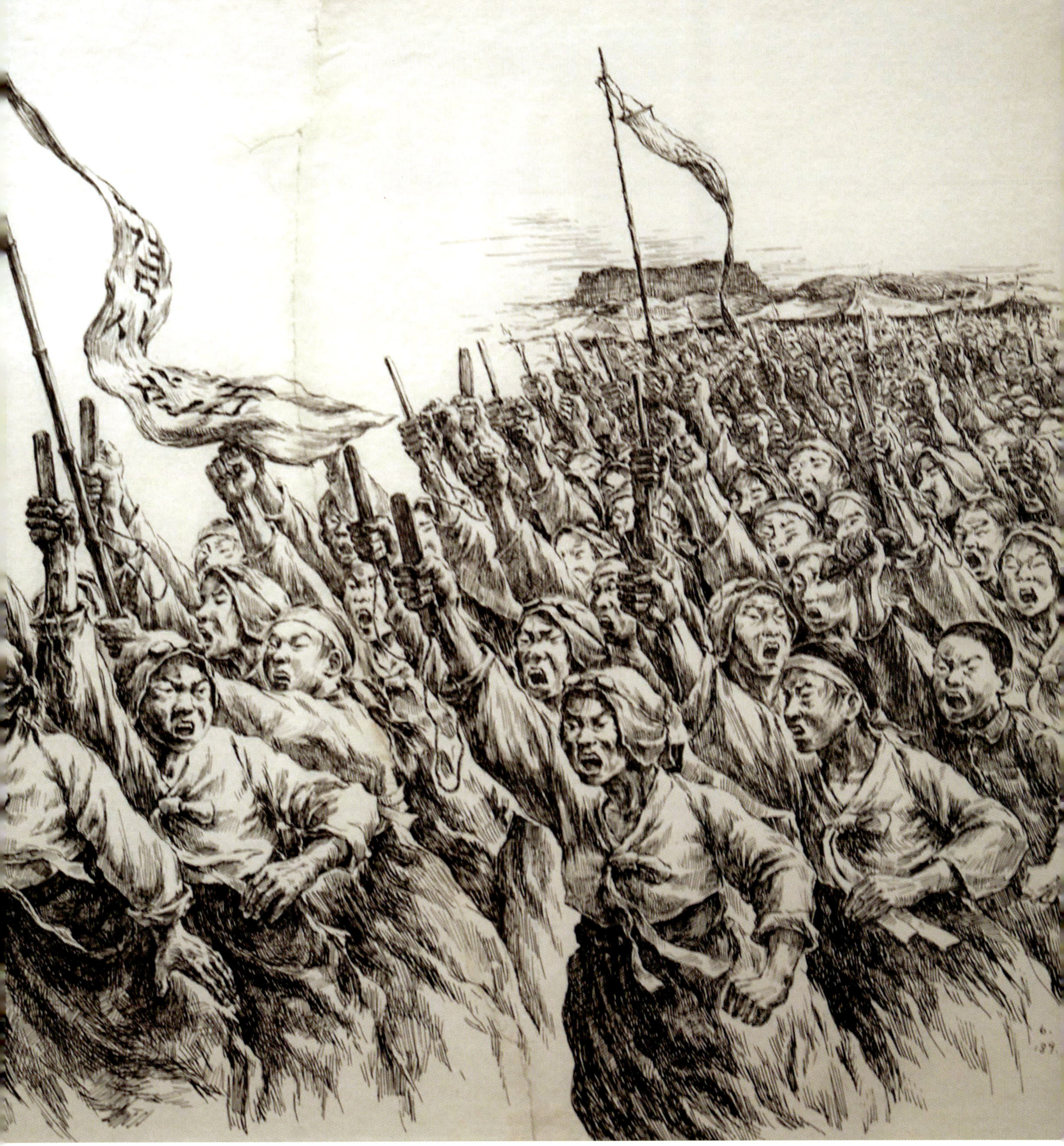

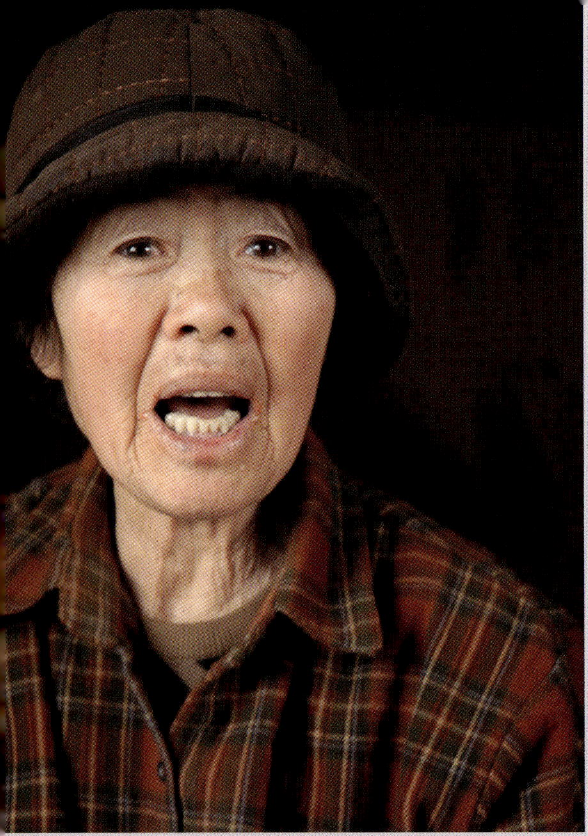

KANG DEOK-KYO

b.1935 | Hallim-eup

Memories of the Massacre

I have a lot of memories of the April 3 Massacre. I was around 15 at that time. My brother was 19 and attended high school, but he was killed as a student. Because of that tragedy, my family has lived a miserable life. My mother grieved for my brother—an only son—who passed away much too early. Now many years have passed. Sometimes, I forget about it because I have my own children, and they make me happy. But I still feel pain whenever I think about that time.

We were scared of policemen. They threatened and took away the villagers whom they suspected of being insurgent sympathizers. People from the mountains often came down and instigated the young people in the villages. If the police learned of this, they came after those rumored to have met with the insurgents. They were killed without any investigation.

Many innocent youths were killed by the police. The insurgents from the mountain tried to recruit smart youths and students. They were too young to have better judgment. Someone probably informed the police about them. My brother and two other students were killed on the playground of a middle school. He was taken out of the classroom while he was studying. He was very smart—a good student. What I remember is that we had a happy childhood together. Before he died, my family didn't have any difficulties or worries. After his death, we had so much pain in our hearts. My family suffered terrible grief. I didn't go out or go to school after that. His gravesite is in Suwon, where I was born, in a cemetery nearby.

As a *Haenyeo*

In Jeju, in the seashore villages, almost all of the women used to be *haenyeo*. My mother and mother-in-law were *haenyeo*, too. There are about 50 woman divers in our village. The eldest is 85 and lives down the road. Because she's still healthy and still dives, a lot of TV and newspaper reporters often come here to interview her. Most of us are in our 60s. There are a few in their 40s, but none who are younger than that.

I first learned how to dive in my teens—around 16 or 17. I got married when I was 19 years old and began diving little by little. I feel so happy in the water, even when I don't catch many things. These days, I catch turban shells. While I'm in the water, I don't feel any pain. I just concentrate on the catch.

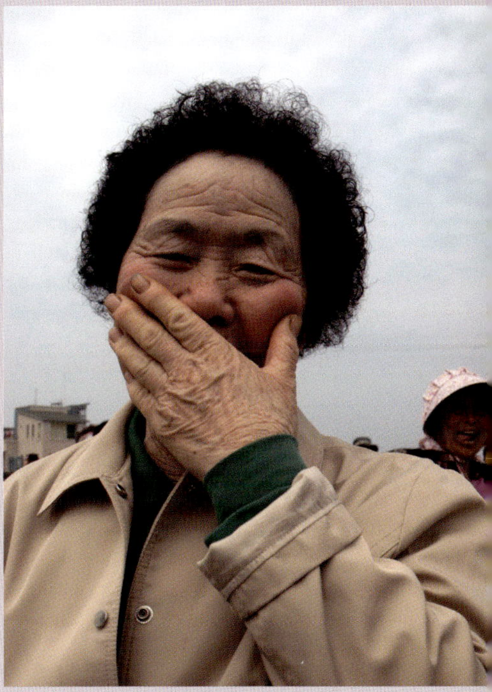

When I see abalone under the water, it makes me so excited. It's not easy to catch these days. We see them only once in a while. Fishermen still scatter abalone seeds in the sea, but more than half of them die before they grow. We can only see their shells in the water. The sea is getting warmer and saltier. It's not as clean as it used to be. Everything is changing under the water.

And yet I often dream about being in the sea and catching many things. I'll probably stop diving within five years. The older we get, the harder it is to stay underwater. Also, once I'm out of the water I feel exhausted and don't feel like doing anything. When we were younger, the sea was always our playground. These days, young children don't even know how to swim.

(March, 2009)

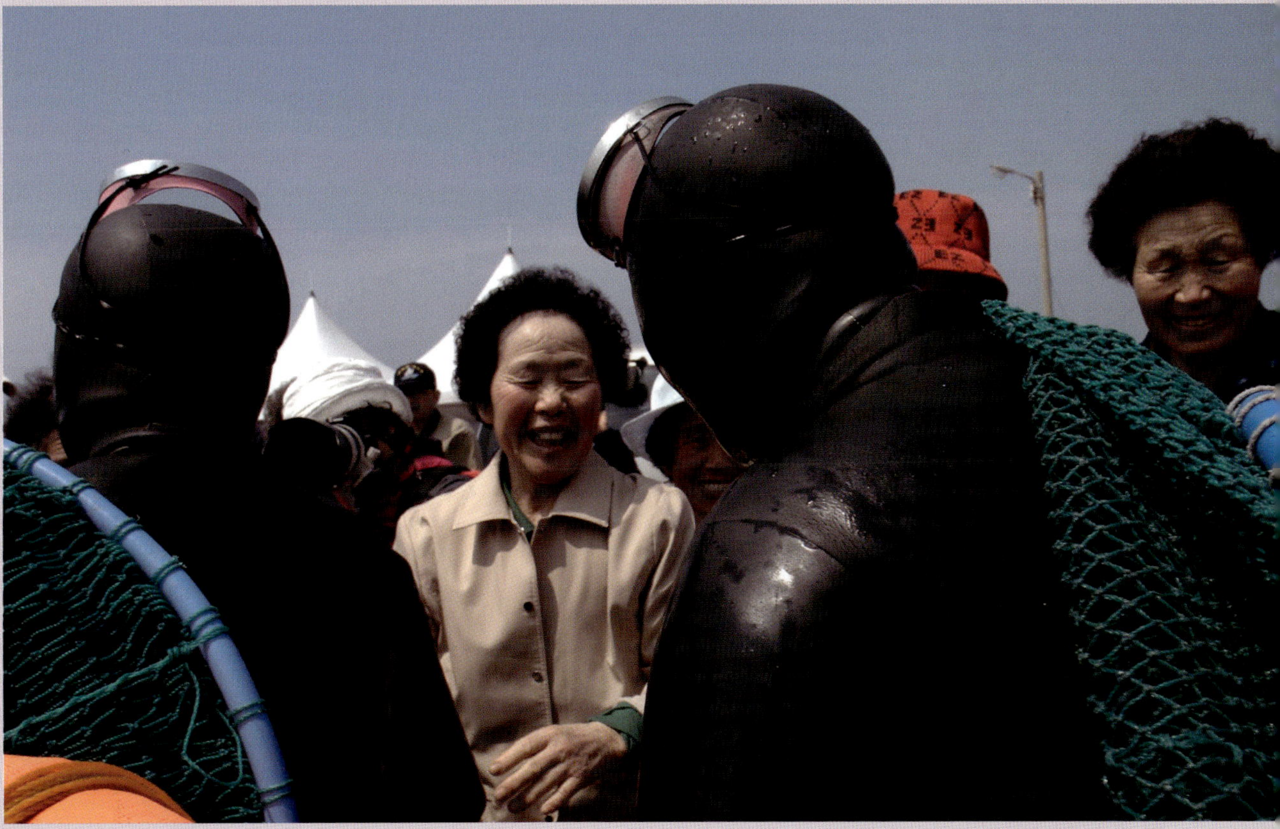

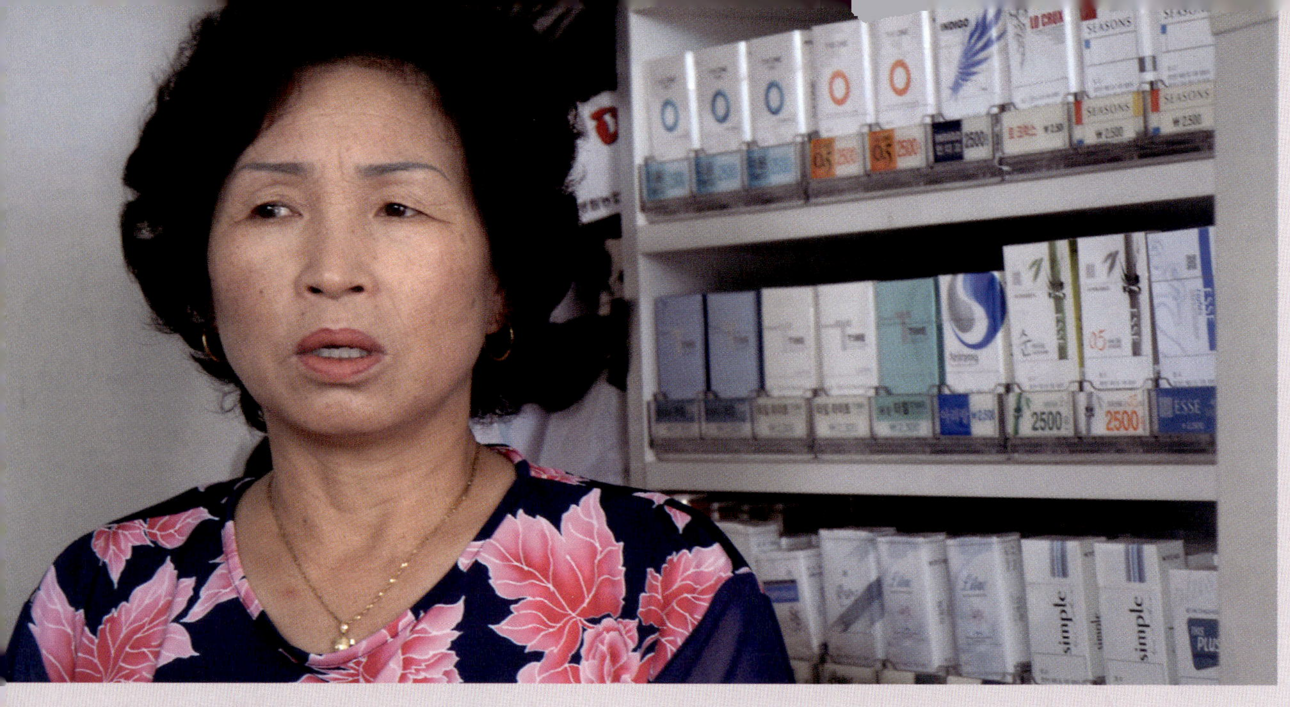

HONG SUK-JA

b.1953 | Bukchon-ri

Even though I am 56 years old, I haven't worked as a *haenyeo* for nearly 30 years. The reason is that I survived a terrible accident in 1982 on the road where we used to sell our sea products. While I was walking along the road looking out for cars, a truck that should've been in the opposite lane hit me from behind, crossing the centerline. The driver was drunk. After he ran me over the first time, I was still conscious and tried to raise my head. I remember seeing the engine of the truck and the driver's legs stepping down. But as soon as he saw me, he returned to the truck and re-started the engine, and he ran over me two more times. Later, another driver passing by rushed me to the hospital in Jeju-si. I remained unconscious for three days. The doctors thought I wouldn't survive. But my father was insistent upon sending me to a bigger hospital in Seoul for one last chance. We occupied nine seats on the airplane: me, an accompanying doctor, a nurse, my father, and my uncle.

As soon as we arrived in Seoul, I was transported to Severance Hospital. It was there that I had surgery. The doctors say a miracle saved me because I suffered broken bones all over my body except for my head and hands. My recovery took a long time, as I was transferred back to Jeju-si and stayed in the hospital for six years. That's why I worked indoors at a minimart in Gimnyeong. Now that I don't do that anymore, I stay in my house here in Bukchon. Neighbors often drop by and reminisce about the painful past.

Painful Remains of the Past

Our village suffered great losses during the April 3 Massacre in 1948, more than other villages. My mother—also a *haenyeo*—and her older sister managed to escape the atrocities. They hid under a pile of seaweed before escaping by boat to the mainland.

Their younger sister, who was expected to deliver a baby any day, didn't want to leave. She had been waiting for her husband's return home. When he didn't arrive as expected, she went looking for him at the farm. According to some witnesses, my aunt was shot by the police. The baby came out, cried for a while, and then died with my aunt. Many people who witnessed these brutalities hid in the pine forests. They couldn't stop the killings. Otherwise, they, too, would've been shot. My aunt and cousin's graves are located together in a cemetery. Whenever my other cousin visits his mother's grave, he brings water because he was told that she cried out for water as she was dying. I've heard so many gruesome stories about the massacre.

Some villagers killed family members of the police and the soldiers. But they also killed each other if they suspected someone of collaborating. One of my relatives who served as a police officer was killed with a bamboo spear. Nearly every family in our village has relatives who either killed or were killed.

Many of our villagers are believed to have been insurgents. They hid in the mountains and would come down to the village at night. That's why the police and soldiers often targeted individuals indiscriminately. The insurgents often forced civilians to inform on each other. If they didn't, they were killed. People were forced to betray those they knew in order to save their own lives.

Many of the atrocities took place at the playground of an elementary school. The soldiers would come to the village, set the houses on fire, and force the people to stand in rows at the school while they were being shot on suspicion of being communists. Luckily, some villagers were able to escape by hiding in the mountains.

Many of the old *halmeoni* who survived the massacre used to visit me and recall their memories. One woman told me that four of her sons were killed, and she prepared no less than 17 *jesa* in one year because most of her family members were killed. Most of the older survivors have since passed away. But I know one who lives in a nursing home because of her dementia. Recently, a memorial tower has been erected in memory of those whose lives were sacrificed during the massacre.

(March, 2008)

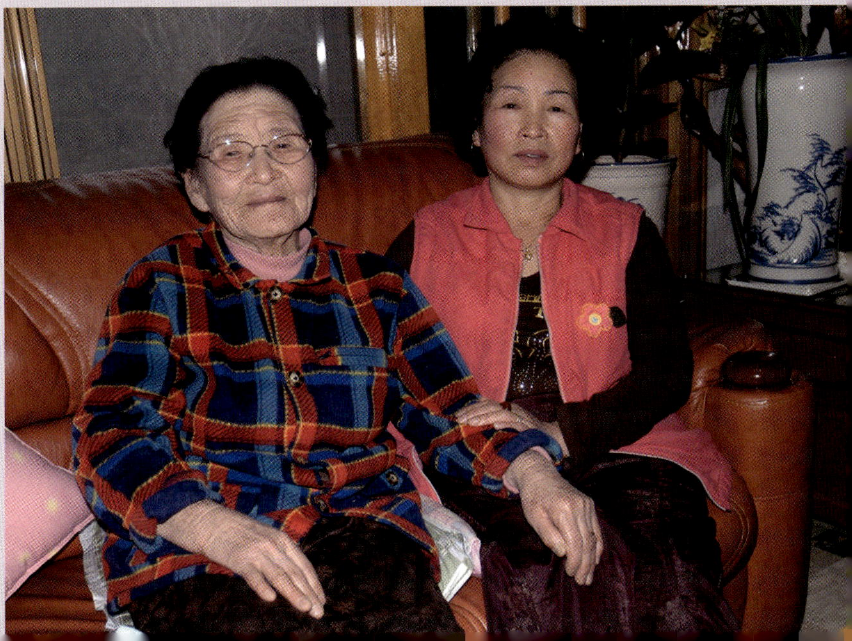

Hong Suk-ja and her mother—a former *haenyeo*—at their home in Bukchon-ri.

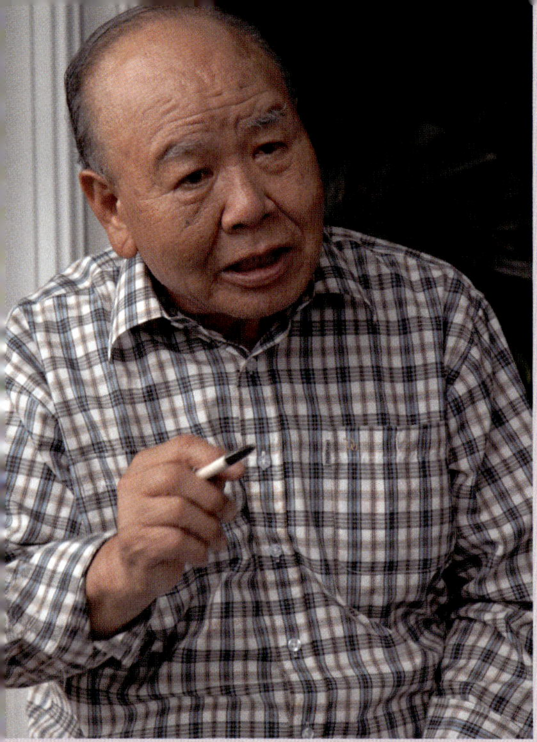

KO CHANG-HO

b.1932 | Onpyeong-ri

My mother was a *haenyeo* for 70 years. Although she had never learned how to read and write, she went to Cheongjin, North Korea, to work as a *haenyeo*. My family had lots of debts. Many women would work outside Jeju for between one and three years. They left their children to be taken care of by sisters and brothers—some as young as eight years old. They couldn't even dream of going to school. By age eight, so many of the young girls had begun to go to the sea and work near the seashore. Then they learned how to dive and eventually to work in the deep ocean.

In Jeju, each village has its own designated area for diving. During the time the *haenyeo* were gathering seaweed to rebuild the village's school after it burned down, I helped to designate one area as "sea for the school." The *miyeok* (brown seaweed) caught at that spot would contribute toward the reconstruction. Every *haenyeo*, every child, every man in the village participated in collecting seaweed. The money saved within one year subsidized the construction of one classroom.

At first, they saved and collected money to build only two classrooms. But they thought they needed at least five rooms, which included an office and more space for all of the children. So the *haenyeo* also established a school support association. People were asked to donate whatever they could, such as wood, bamboo, and other items. People came to help make mud bricks. When it became apparent that they also needed a roof, some villagers from a refinery in Incheon assisted. Once the school was built, we still needed desks and chairs. With the help of reporters, we obtained a list of Koreans living in Japan. We sent letters asking for their help. One by one, they replied to our appeals and sent us money to buy the necessary supplies. Our school became the best in Jeju.

Unfortunately, many poor students still couldn't afford to attend higher-level schools. So once again the *haenyeo* addressed this problem, by establishing a scholarship club. Most of the active members were widows. I suggested setting up a monument in honor of these *haenyeo*. Only a few of them are still alive. I meet them on occasion.

The economic situation in Korea was terrible then. You see, at that time Jeju people weren't allowed to go to the mainland because of the April 3 Massacre and the Korean War. During the massacre, a lot of young men were killed. Because of the Korean War, most young men had to join the army. So women had to do all the things that the men normally did. That's when they became involved in everything.

Under martial law at that time, we were told that if someone could understand Hangeul, they could obtain a seal that would allow them to work on the mainland. So during the early mornings—before the women went out to dive—around 60 *haenyeo* attended a Hangeul language class. Since I was hired as a teacher there in 1954, I taught the class.

The women were able to read and write basic Hangeul within two months. They studied very hard and received high marks, understanding the necessity of being literate. Most of the women studying were in their 40s. If they were alive today, they would be over 100 years old. These days, after receiving a higher education the younger generation can make easier money by working on the mainland. Even in Onpyeong, approximately 60 percent of our youth study on the mainland and never return to live here on the island.

Japanese Exploitation

Koreans and Jeju *haenyeo* were terribly exploited during Japanese occupation (1910–1945). There were many *haenyeo* who worked in Japan, China, and Russia during that time. They dived for sea urchins, which were valued as a side dish for Japanese soldiers. So the Japanese authorities forced the *haenyeo* to work far away catching them. But they never earned the money they deserved.

Another example of exploitation was the manufacture of buttons. At that time, buttons were made of seashells, such as abalone. But we didn't have the machines to make the buttonholes. After the buttons were manufactured, they were sent to Japan to have the holes cut. But when they were returned, they were priced more than 10 times their initial cost.

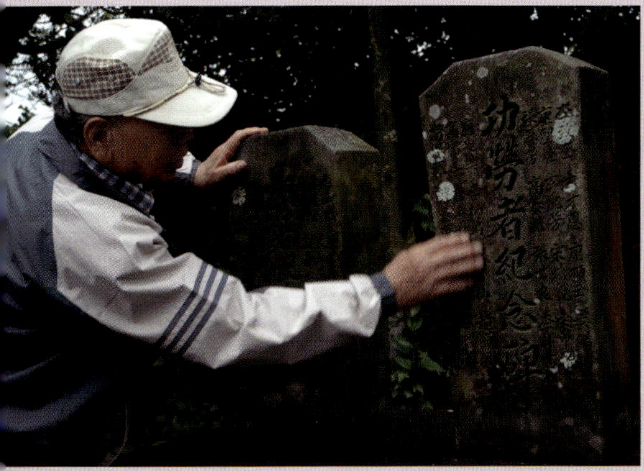

During the Pacific War, there were some battles waged off Jeju Island. Sometimes, corpses floated to the seashore. The Japanese authorities ordered the *haenyeo* to get rid of them. The Japanese authorities also built several processing facilities for the seafood and collected it at dirt-cheap prices from the *haenyeo*. Eventually, the women began to realize how much they were being exploited by the Japanese authorities.

(June, 2007)

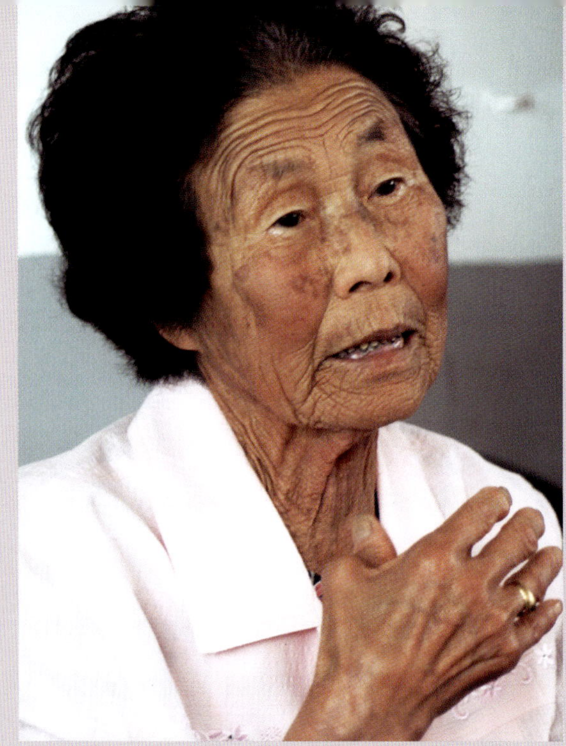
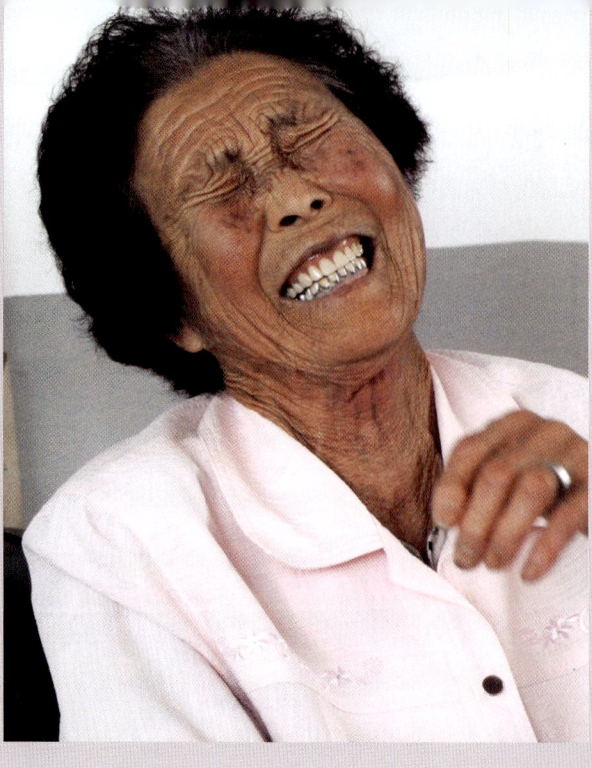

CHO SUN-WOL

b.1926 | Onpyeong-ri

I got married when I was 18 years old. But my husband and I didn't live together for a long time because I worked on the mainland as a *haenyeo*. I also went to Japan during the Japanese occupation. After the Pacific War ended in 1945, I returned to Onpyeong.

I was elected as a *haenyeo* leader in the 50s when I was 29 years old. It was a long time ago. I'm not sure I can recall the memories well enough. All of the leaders at that time have passed away except for me. Off and on, I served for a total of 10 years. When I finished my last term, I was awarded a brass bowl—the kind used for *jesa*—as a token of appreciation. I am very proud of it.

While serving as leader, we began to rebuild a school that had burned down in our village. The accidental fire had been caused by pine cones that were used for heating. Unfortunately, a fire broke out and destroyed the school. We installed electricity by selling the *miyeok* we caught. At first, we were going to reserve half of our catch for the school, but then we decided to contribute all of it for our children's education.

The *miyeok* around Onpyeong was considered to be the best quality. We could sell it to the army base at a higher price than in other areas. We didn't have an *eochongye* (diving association) until 1963, so we managed everything from diving to drying and selling. Fortunately, though, by harvesting and selling the dried *miyeok* we were able to buy foxtail millet. With thirteen pounds of dried seaweed, for example, we could get five gallons of millet and use it for our suppers. Young *haenyeo* used to tell me to stop working. I would tell them that I would dive until I am 85 years old. That made them laugh a lot.

(June, 2007)

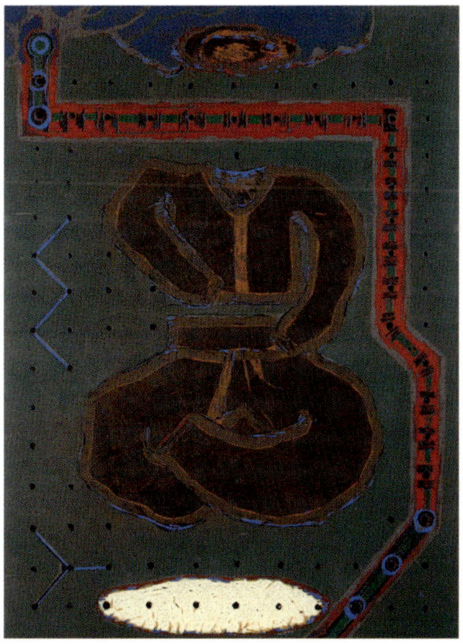

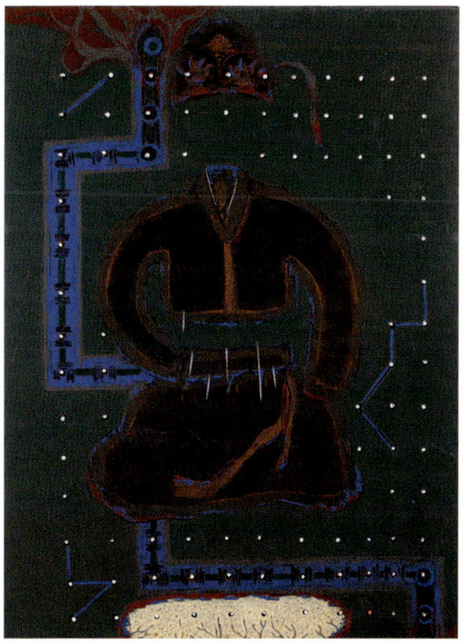

Umbilical Cord of Mythology 1 & 2 60x90 cm | woodcut | 2008

Samseung & Gusamseung Halmang

Known as the goddess of childbirth, Samseung Halmang is believed to have been a maiden from a country associated with long life. During earlier mythical times, legend has it that she competed in a flower contest with a maiden from the ocean and won. Thus, she was appointed the goddess of childbirth, whereas the other maiden left for the world, becoming Gusamseung, the goddess of death, who would take children's souls out of the hand of Samseung Halmang. Grief and loss have always been present as companions in the dangerous lives of the *haenyeo*.

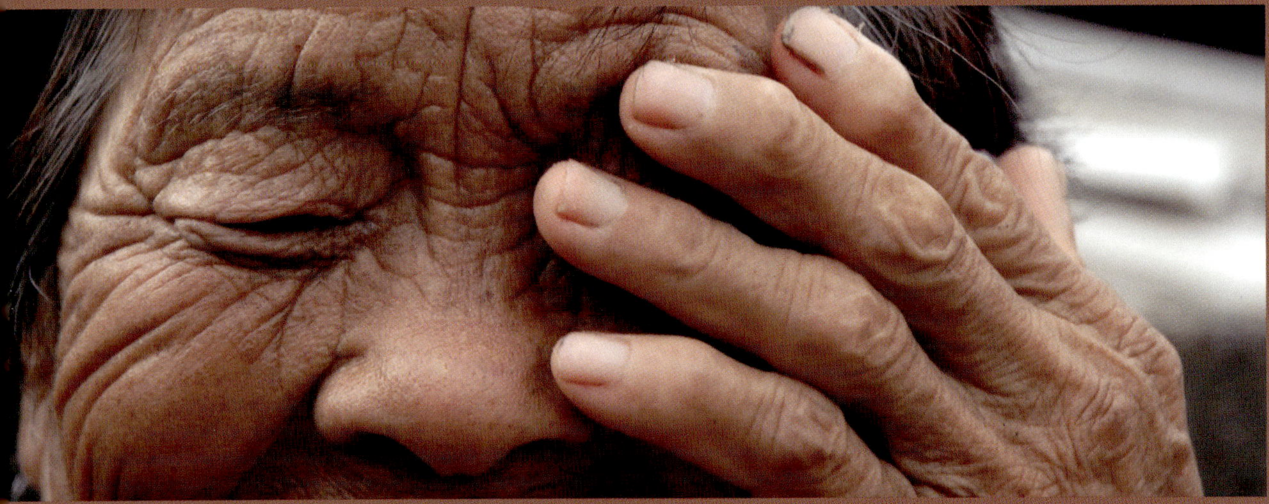

Chapter Four

LIFTED BY THE WIND & TIDES OF AGING

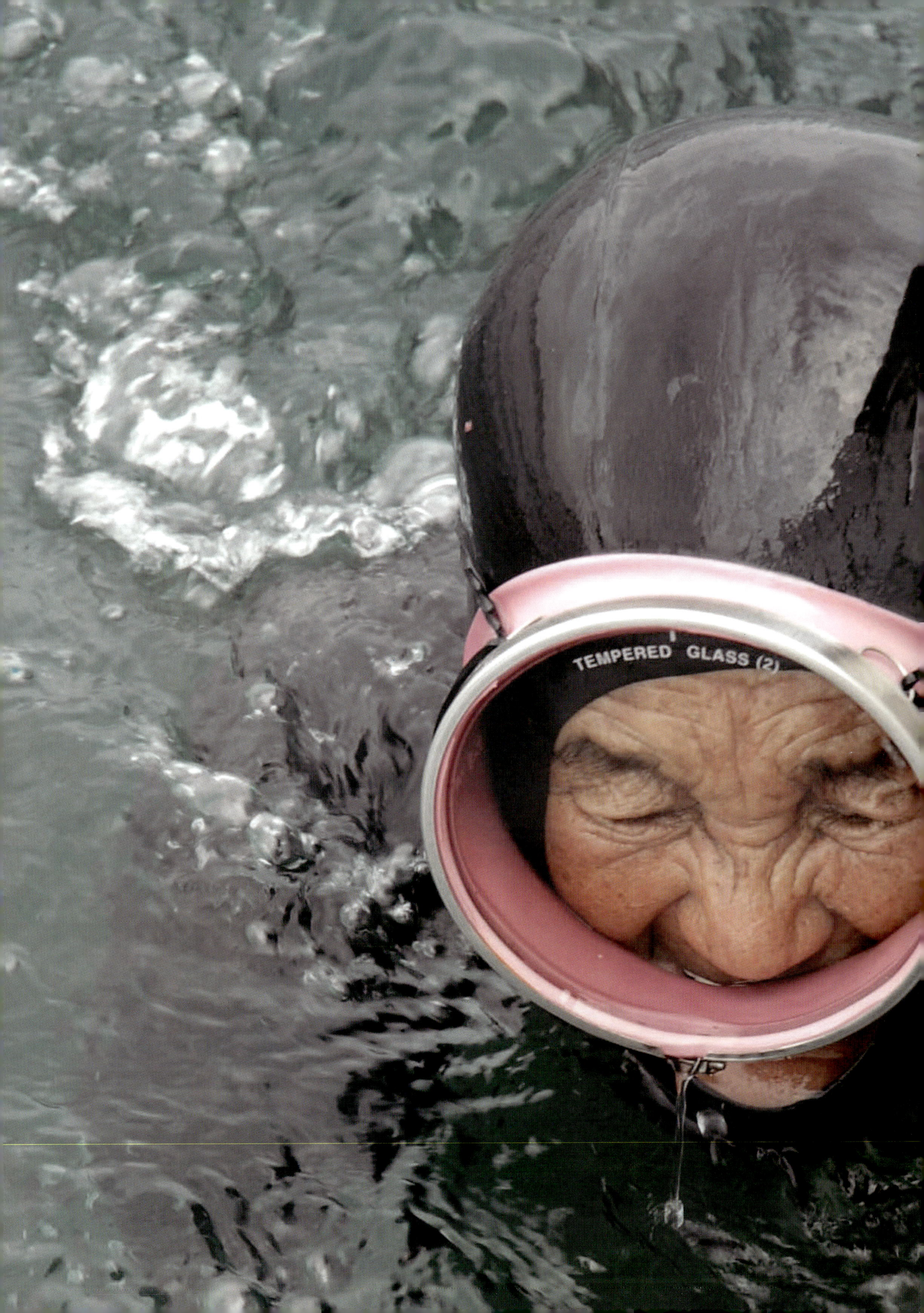
TEMPERED GLASS (2)

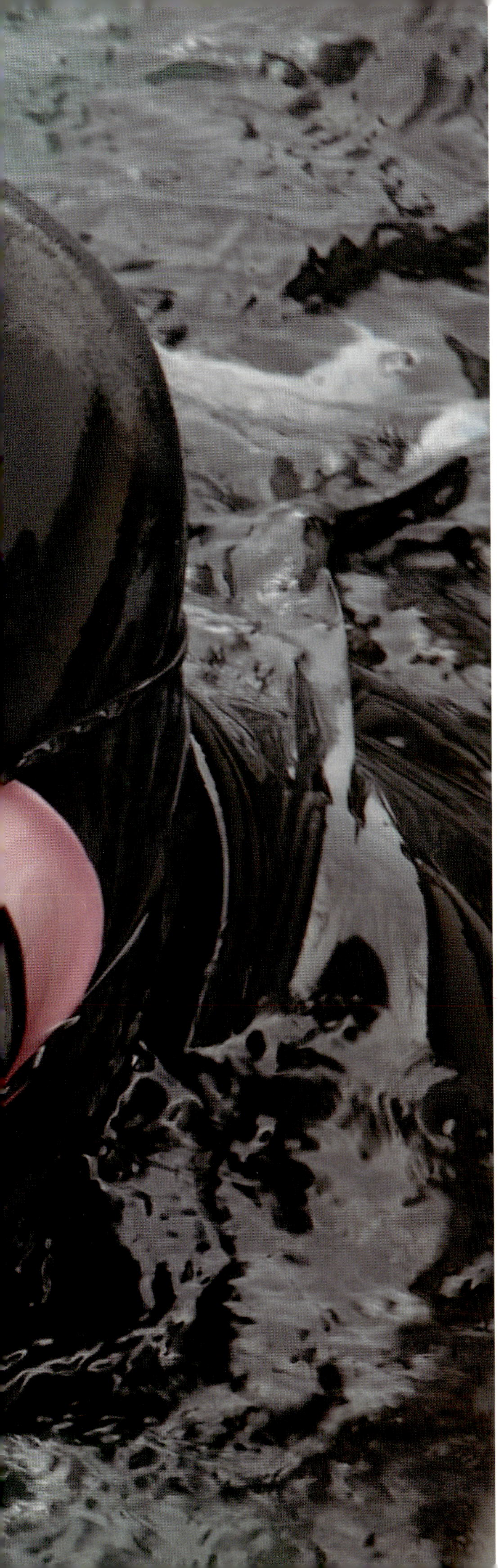

From the first house I occupied in Gimnyeong-ri, I could walk to the sea in 30 seconds. The open area, which was covered with grass and a concrete walkway, included a wooden pavilion known as a *jeongja*. If it wasn't too windy, elderly women would often meet there to escape their indoor loneliness and the boredom of watching TV. Men and children also enjoyed the space with no walls. Most of all, they came for the fresh air, cool breeze, and casual communion.

One afternoon, I decided to join them. "*Annyeong-haseyo*?" I said, while shyly stepping on to the platform barefooted. "How are you?" There were no other words exchanged. It was actually reassuring to be so graciously ignored. What I remember about that moment was the sheer pleasure and purity of being in their presence. My own grandmother had passed away in 1967. This was the first time in decades that I had physically sat and lain down among Korean grannies. (I had been living in Vietnam at the time, where over 60 percent of the population was born after 1975.) The women's wrinkled faces and bended bodies expressed their own wisdom and familiarity.

As the days went by, I would walk to the pavilion on a regular basis, each time hoping that my somewhat foreign face would become more familiar. Clearly, I was the one who was self-conscious. On a few occasions, Youngsook and I brought along my camera and tape recorder—hoping that the grannies would be willing to be interviewed. Usually, she would talk to them first in *Hangungmal* (Korean) and explain why I had come to Jeju: "This writer has come to Jeju to learn more about the *haenyeo*'s life. Would you mind if she asked you some questions?" Whenever possible, I would speak enough Korean to break the ice and apologize for not speaking fluently. I could rattle off one sentence flawlessly. The translation: "A long time ago, I learned Korean from my grandmother, but I can't speak it very well. I'm sorry." Very often, the fluency of my apology elicited their undeserved praise: "Oh, you speak well."

Their candor often surprised me. For example, Hwang Chun-seok admitted, "I married a useless man. He had four wives, in addition to me. In the past, men often had more than one wife. They weren't legal wives, just concubines."

Most of the woman divers over 60 years old had not been educated beyond elementary school. Very often, they didn't know how to read and write. Said Hwang, "In those days, people didn't think girls needed to be educated. I just wanted to earn money and pick berries in the field."

I found it reassuring that the elderly never stop dreaming. If only the young would stop to listen.

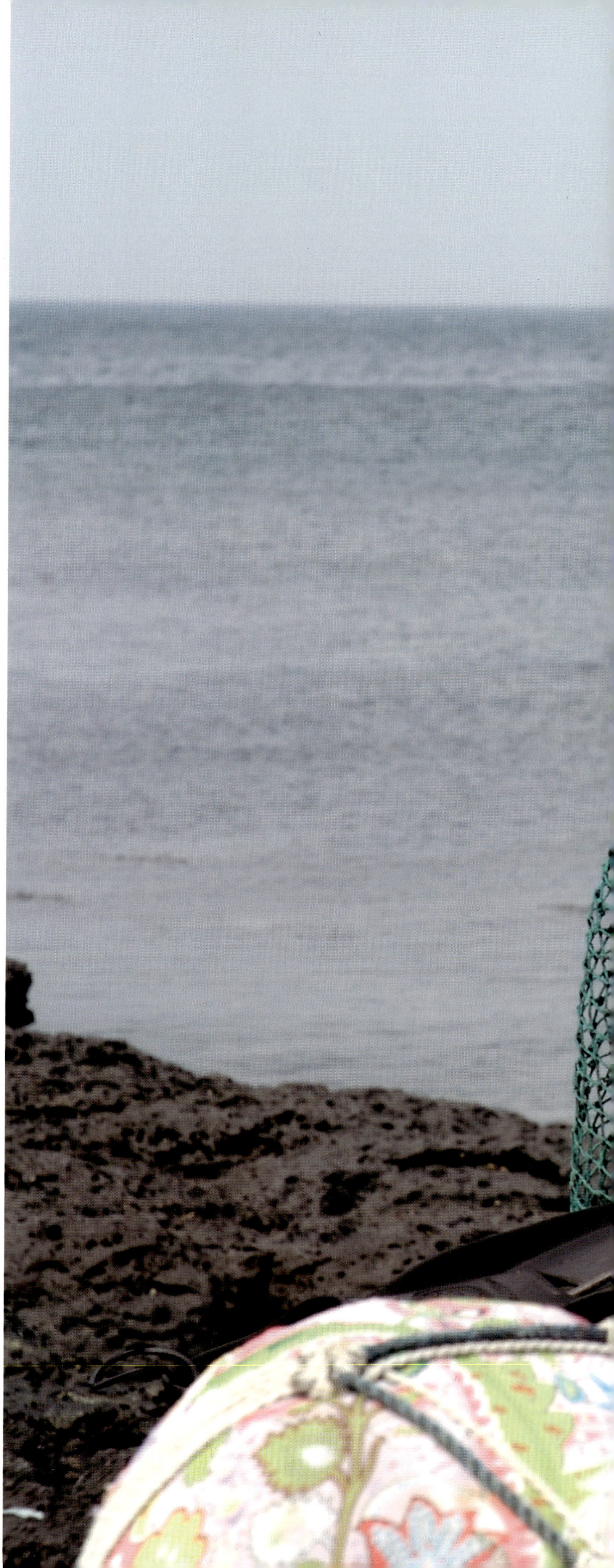

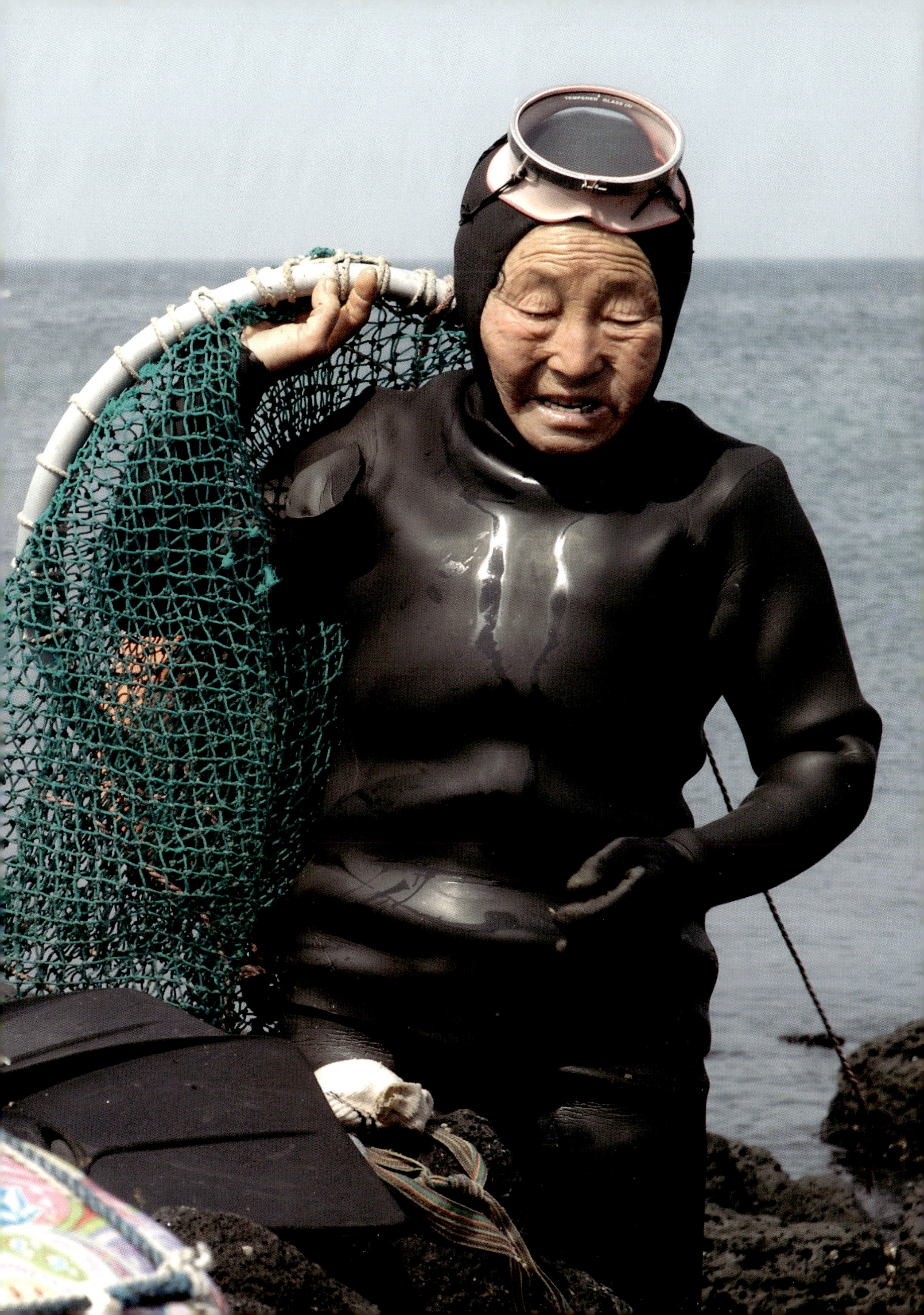
TEMPERED GLASS

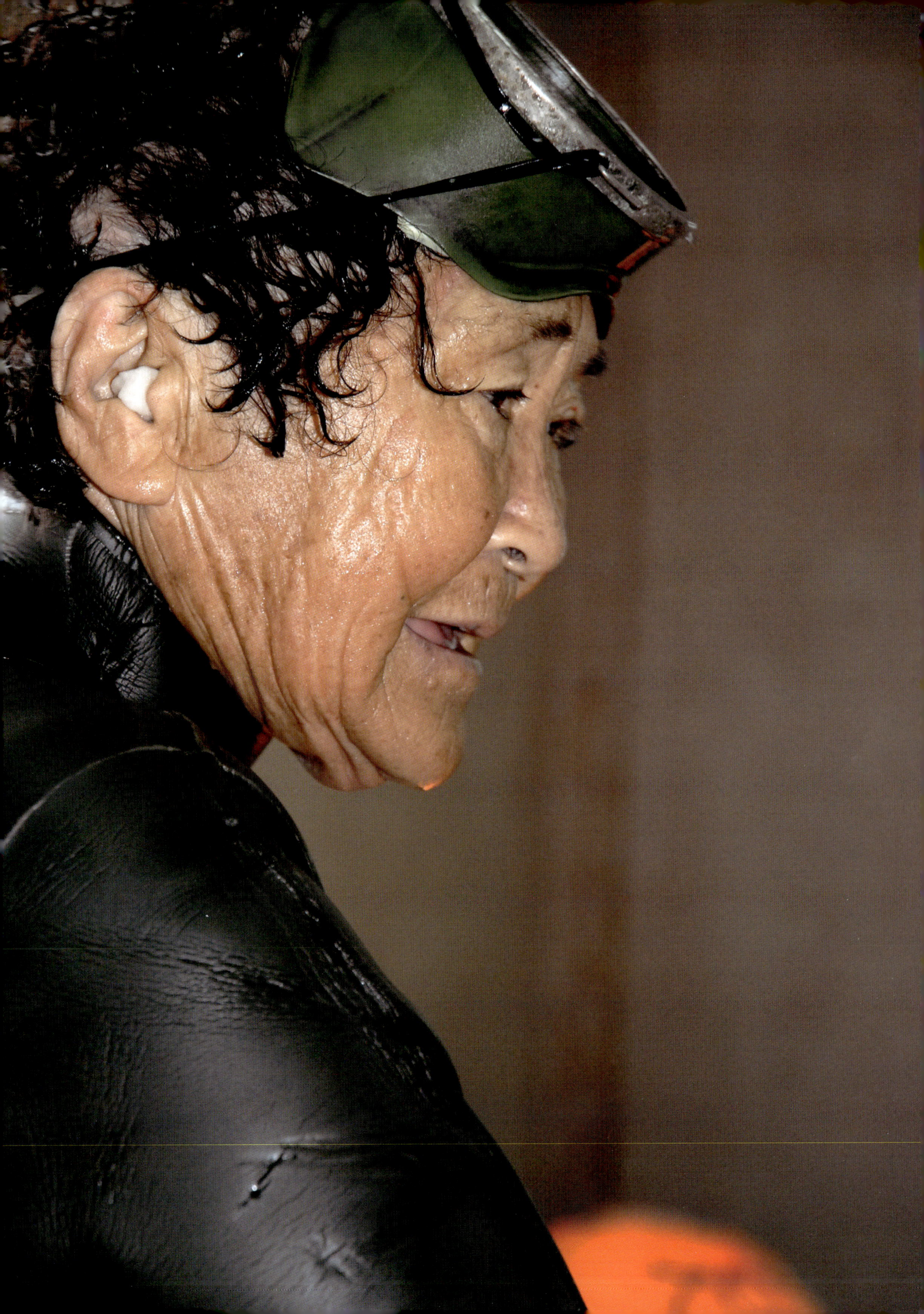

GRANNIES OF THE SEA

Each wrinkle on a *haenyeo*'s face is like the ridges and folds she encounters under the sea—a time capsule of Mother Nature's terrestrial and marine life creations.

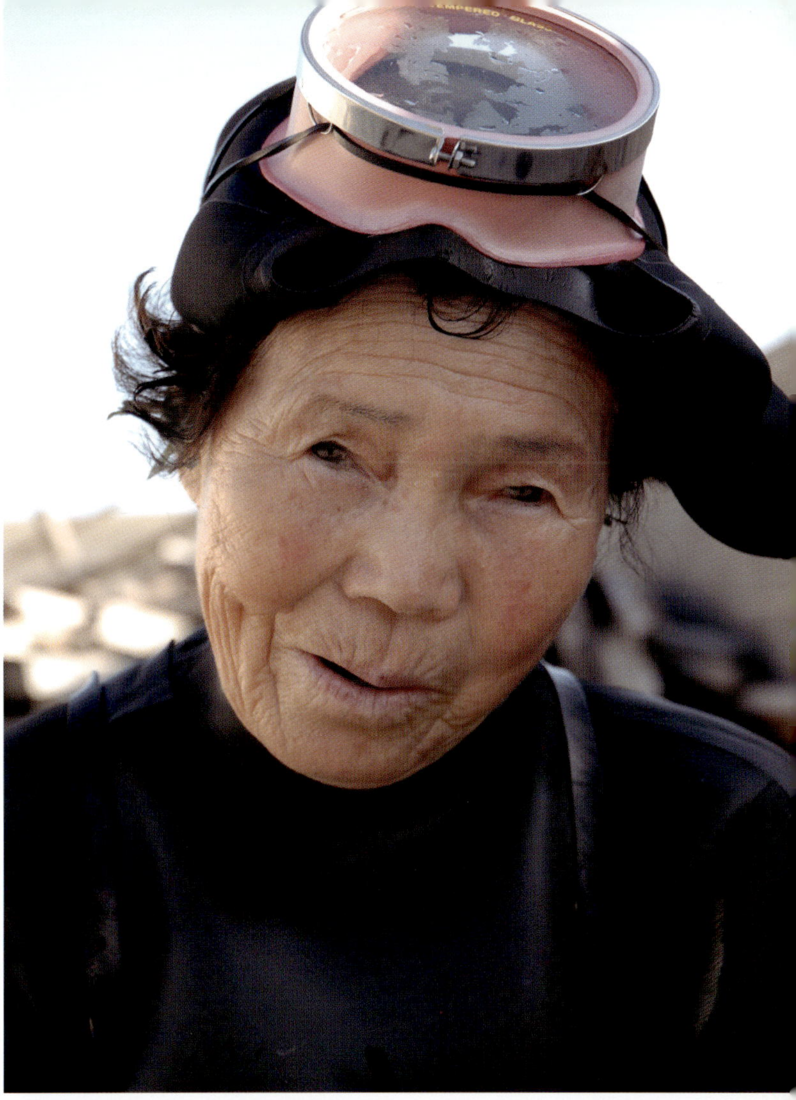

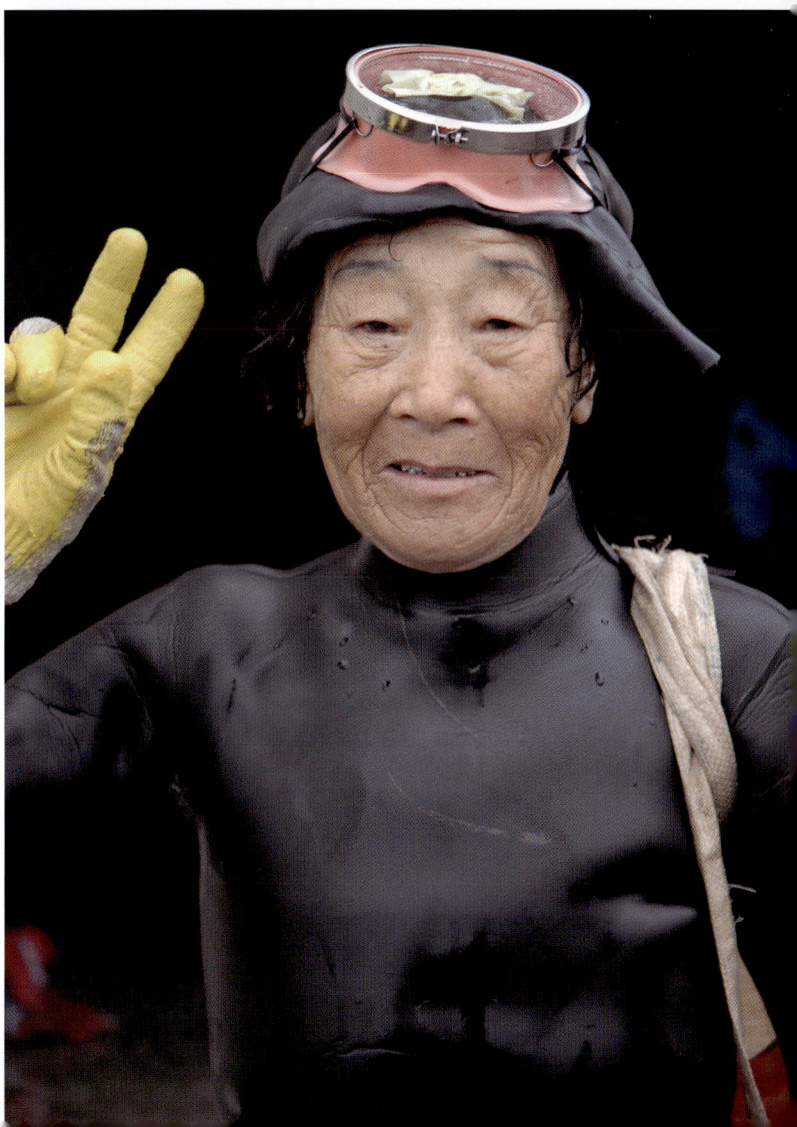

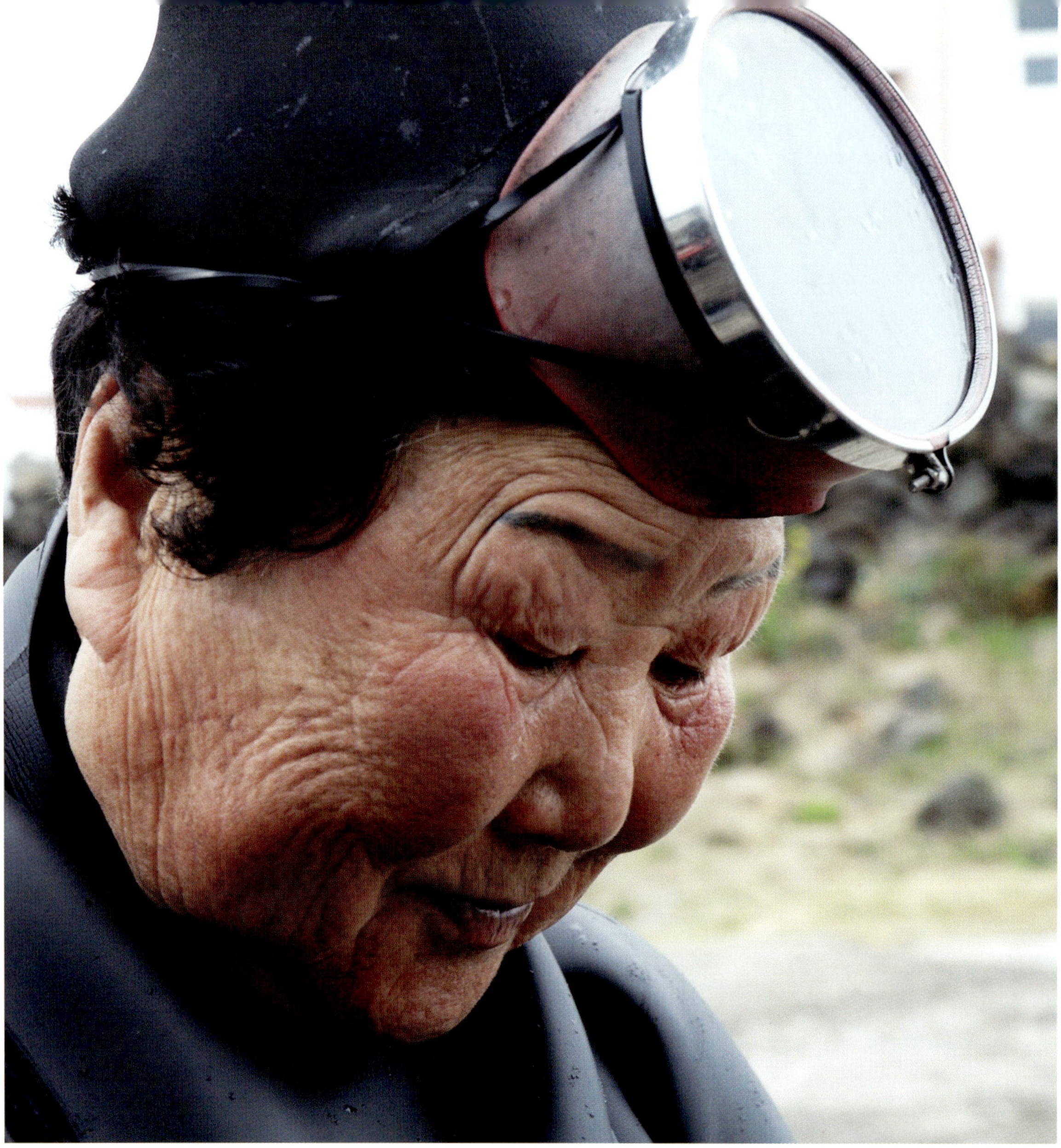

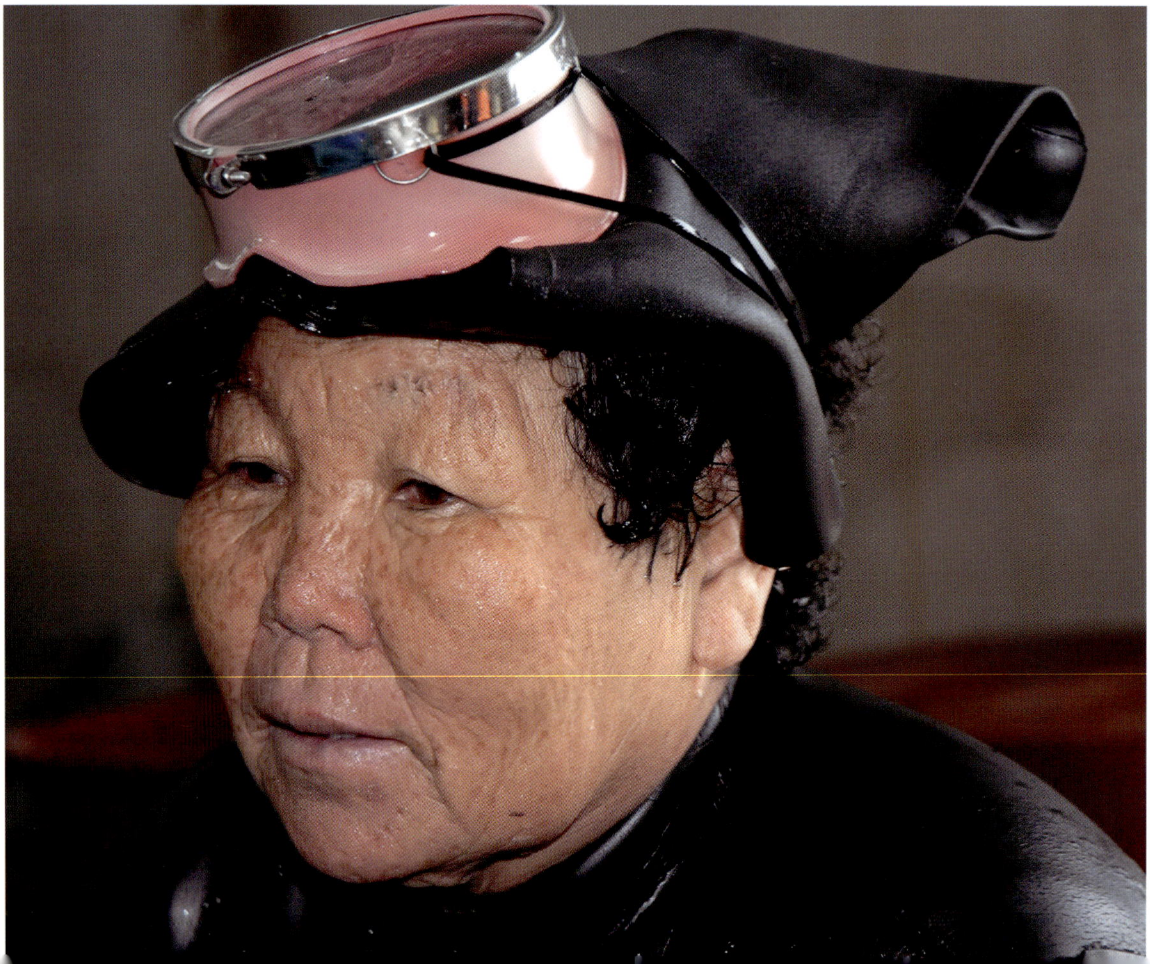

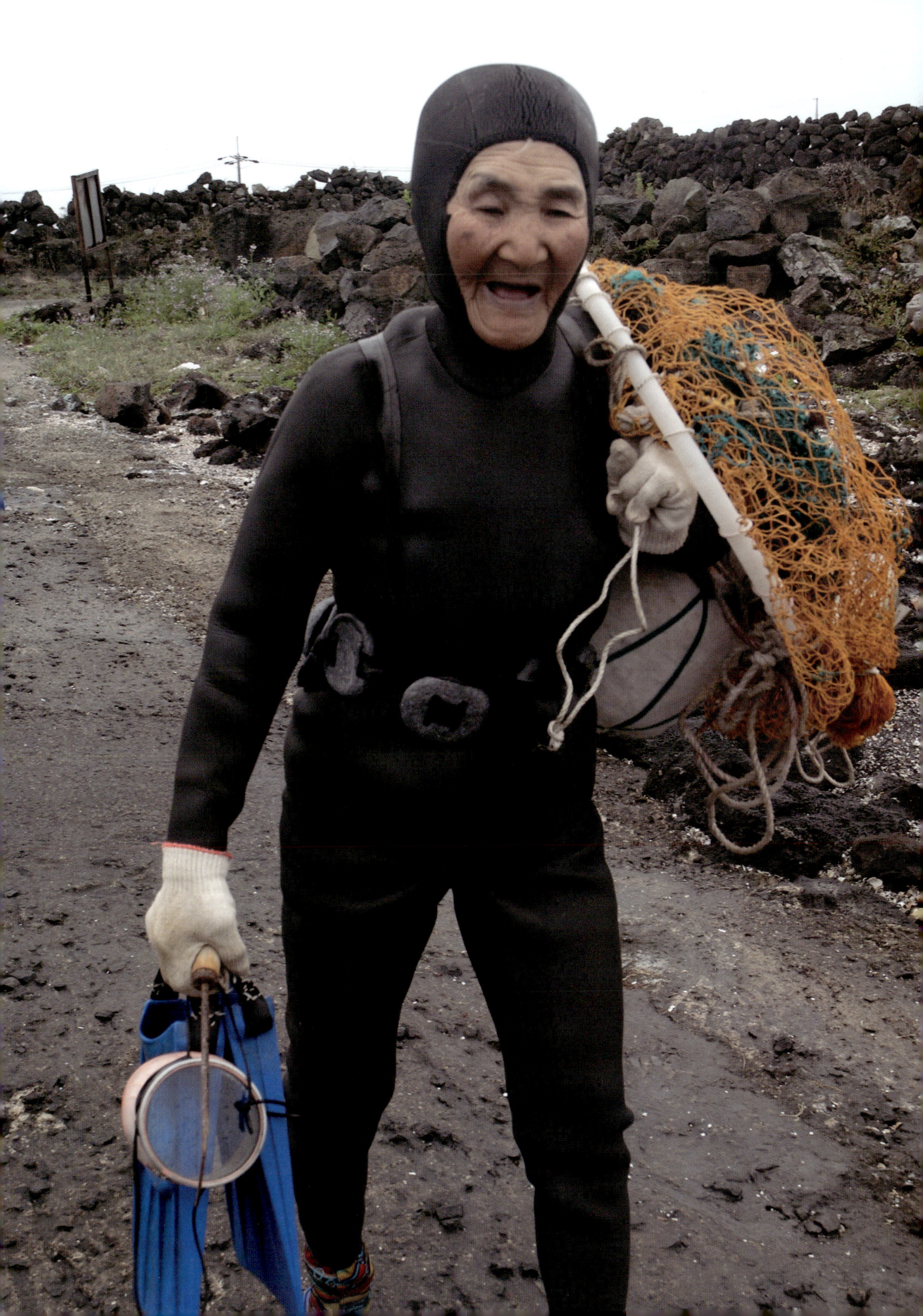

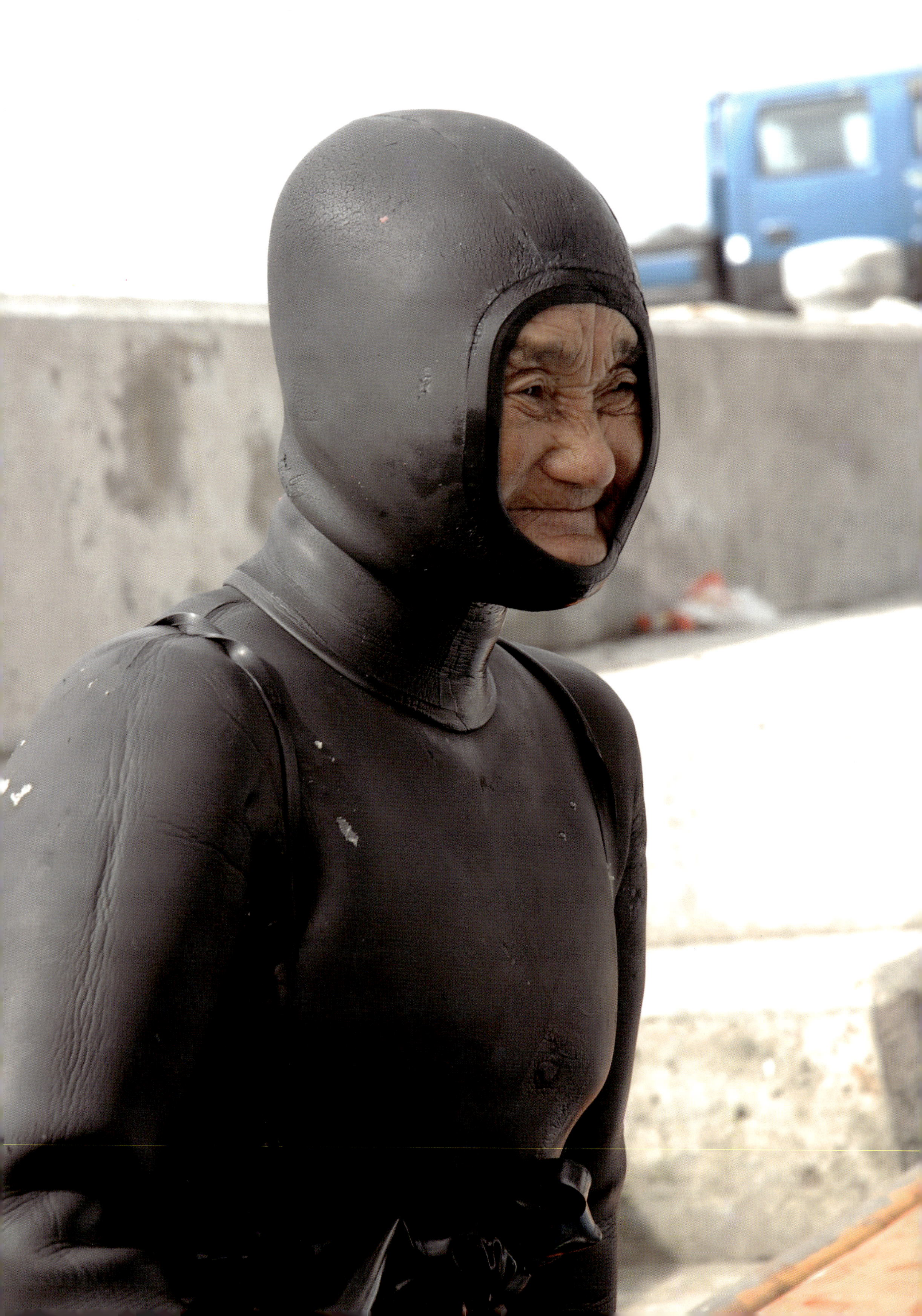

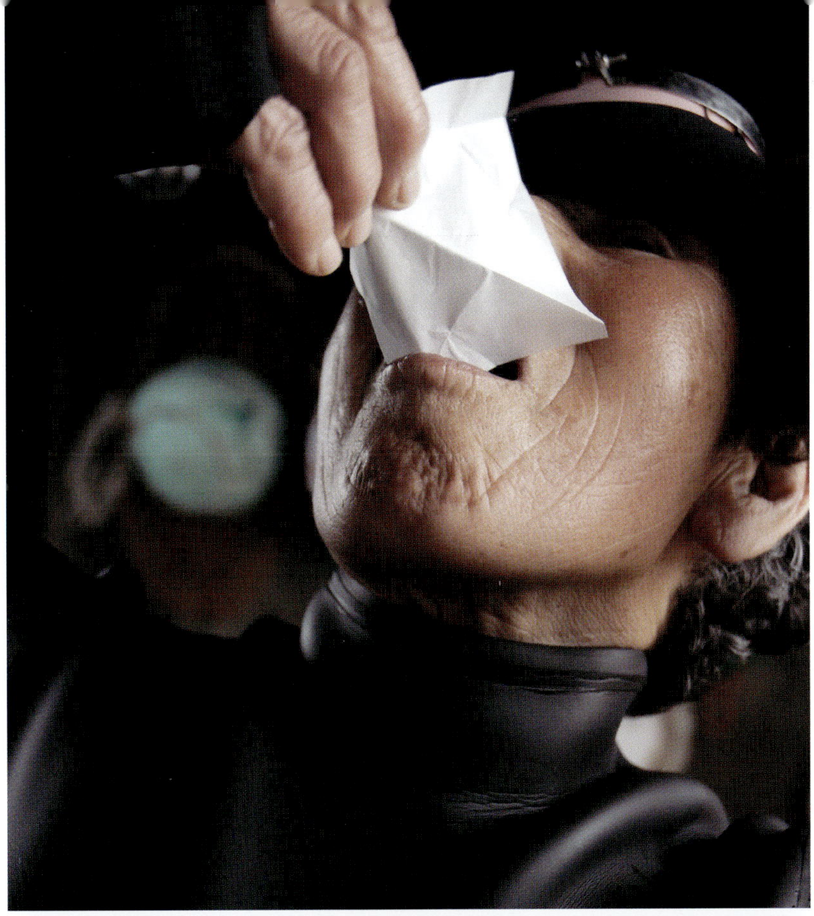

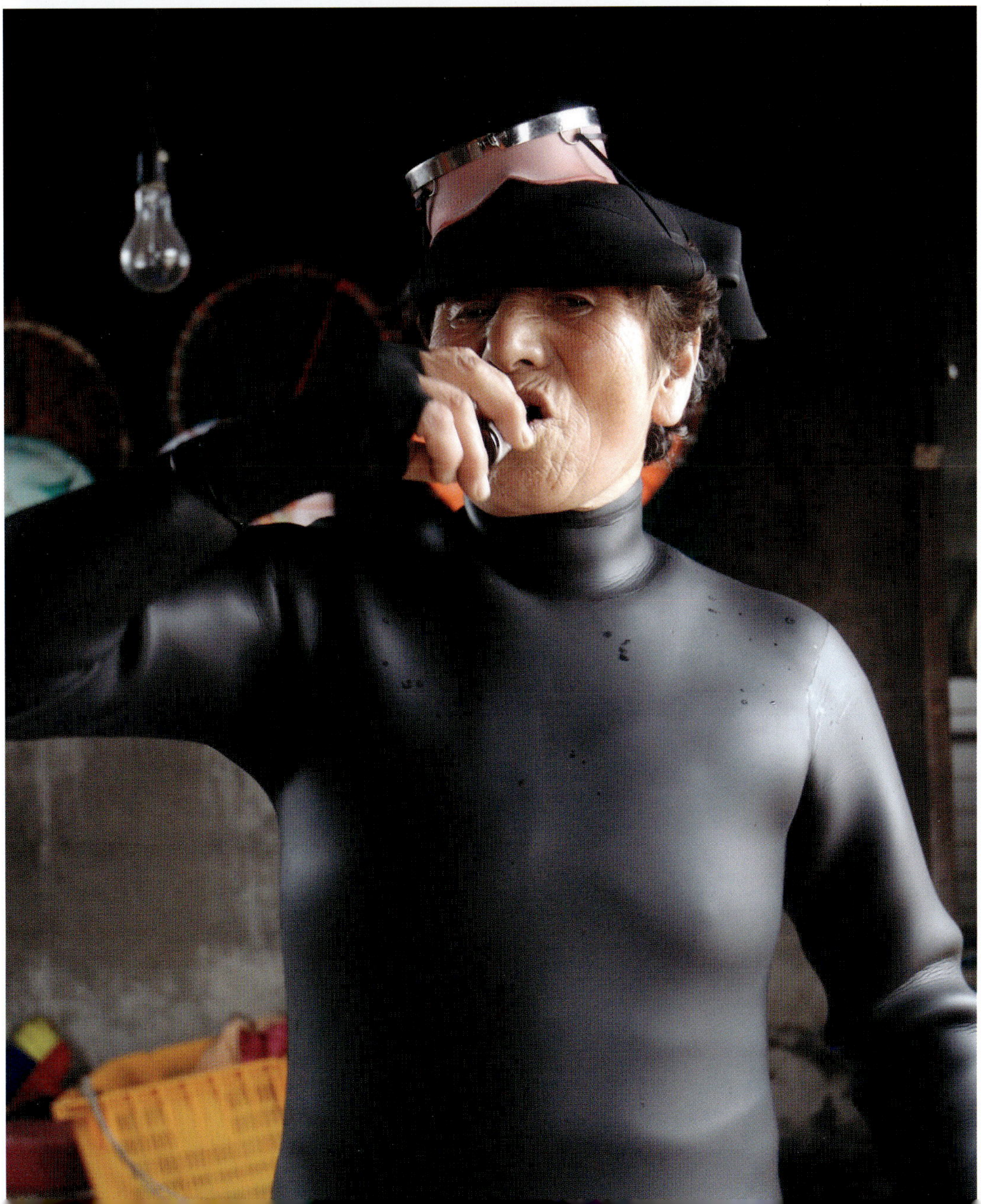

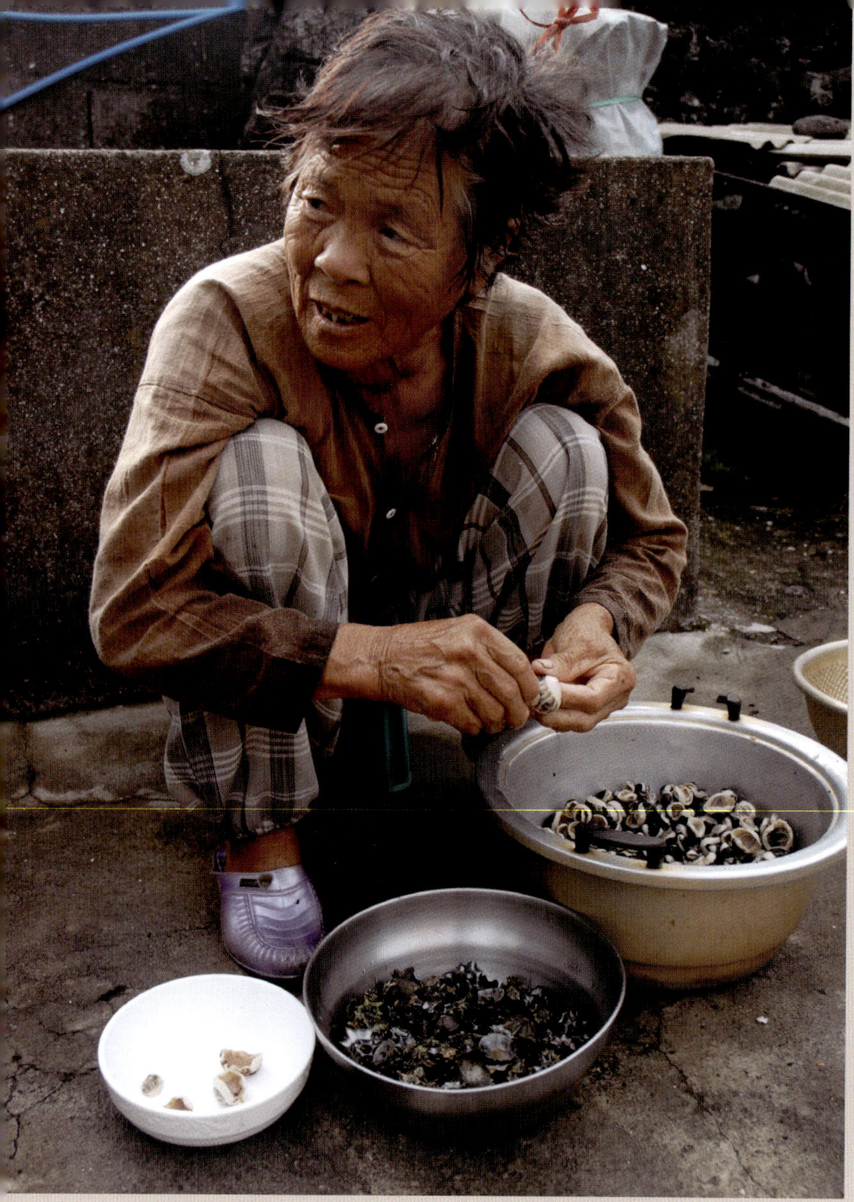

KIM DO-WON

b.1923 I Jongdal-ri

As a child, I didn't have anything in my life that made me happy. We had a hard life under the Japanese occupation. There weren't any houses in front of my house. It was like a swamp. Sometimes, the water from the sea came up to the front of our house. There used to be a salt field nearby when I was young. Now it's gone, and the area has been changed into a rice field. I wish it had remained like the olden days.

Marriage and Family

I got married when I was 19 years old, but I couldn't have a baby. At age 24, I got a divorce from my first husband. Then I met my second husband on the mainland, and we had three children together.

I had my first child at age 28. When delivering my first child, I began my labor at five in the afternoon, but my son wasn't delivered until eight the next morning, at my mother's house. Although many *haenyeo* go diving again right after childbirth, I was lucky. My children were born in the winter when we weren't diving.

My second husband and I didn't live together very long because he already had a wife. I came back to Jeju and lived here. You see, after divorcing my first husband because I couldn't have babies, I wanted children so badly that it didn't matter who the man was. That's why it was okay that he was already married. My first husband also remarried with another woman and had children. They are also living in this village now.

I don't see my children very often. My first daughter lives in Jeju-si. My younger daughter lives in Masan in Gyeongsang Province. I usually go to Jeju-si on New Year's

Day, Chuseok, or *jesa*. My children ask me to visit, but I don't feel like going. I just want to be free by myself. Besides, my children have to work even when I visit them. I don't want to stay at their empty house. My son, who used to be a police officer, retired last year. I don't know what he's doing now. He's 59 years old.

Becoming a *Haenyeo*

I learned swimming by myself when I was really young. Both my mother and grandmother were also *haenyeo*. I've lived here in Jongdal all my life except when I went to the mainland in Jeolla and Gyeongsang Provinces to do *haenyeo* work. I even went to Wonsan in North Korea, and also Japan for a short time.

Someone came to every village to recruit *haenyeo*. So those of us who wanted to go let the person know. All of the *haenyeo* who went with me to the mainland have already died. I'm the only one still alive, and I'm still doing *haenyeo* work at my age.

When I went for the first time, I didn't feel excited or scared. It was just something I had to do. I went at age 18 with my two sisters. There were three of us *haenyeo* in our village. We met a lot of women from Hallim on the boat. We boarded a boat in Seongsan Port and went to Busan. I didn't get seasick. Housing was arranged for us on the mainland.

In those days, we wore cotton suits that we made ourselves. We didn't even use lead weights. Just our thin cotton suits, a towel wrapped around our head, small metal goggles, and our bare feet. I could dive 8 to 10 *bal* (one *bal* is equivalent to the distance from one fingertip to the other with arms outstretched).

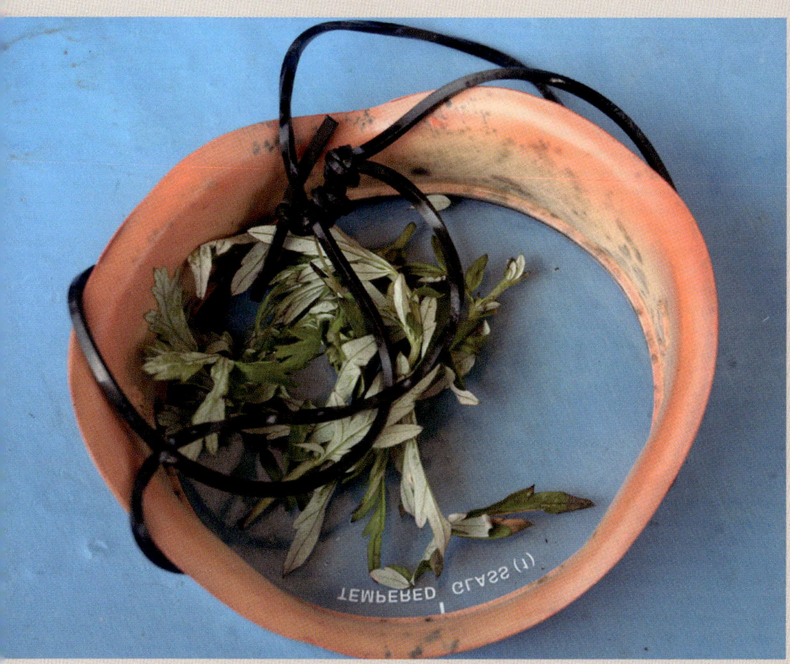

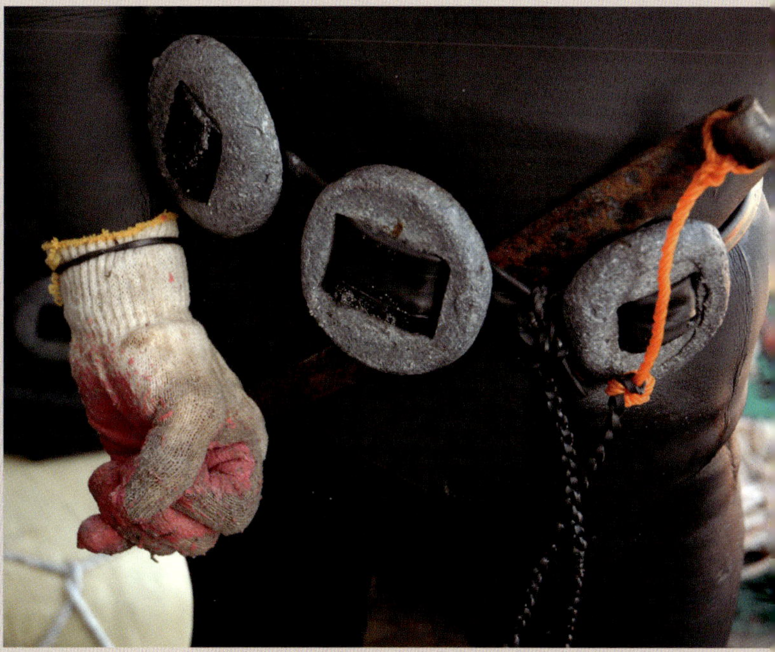

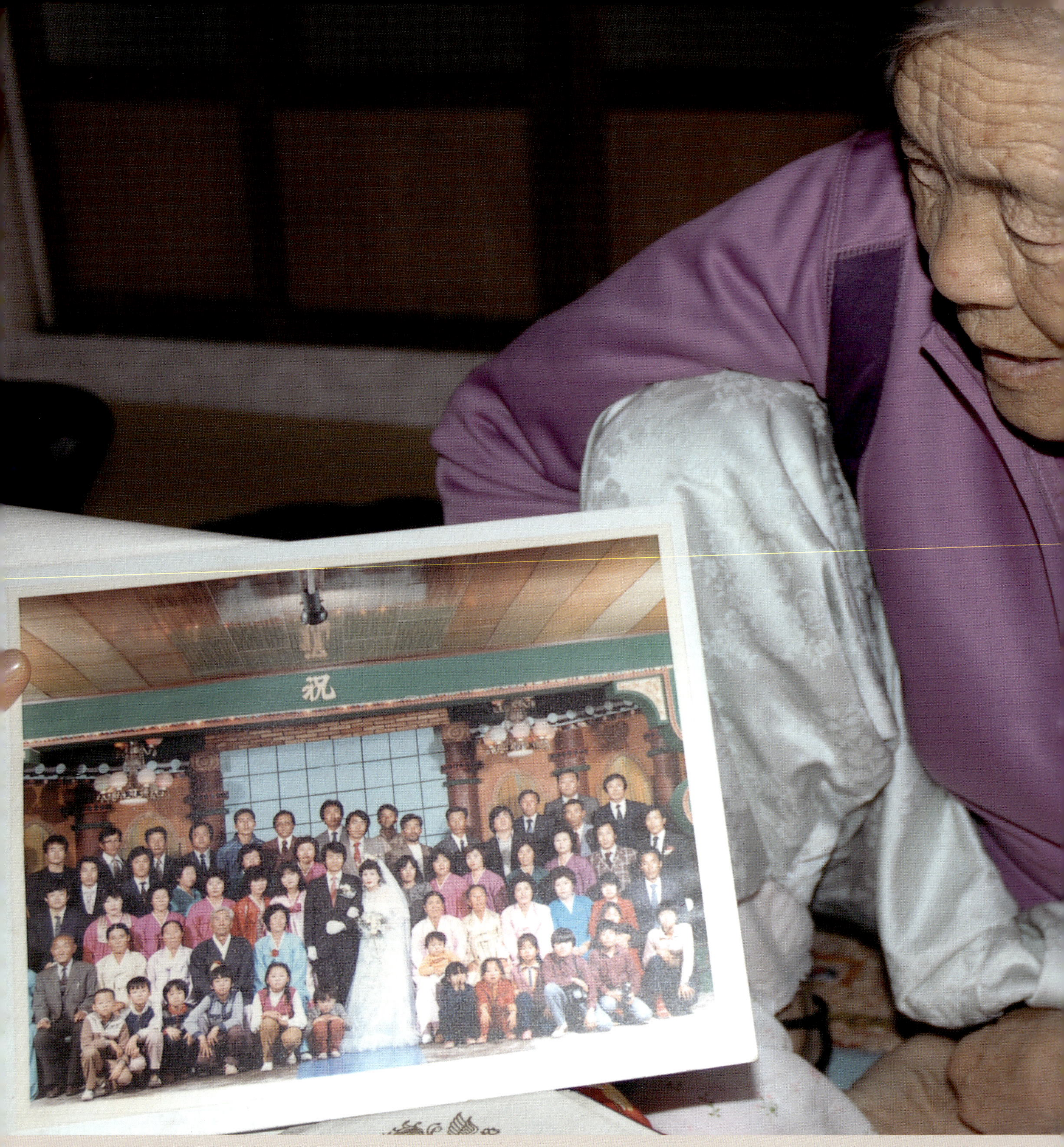

We dived for agar-agar and were paid by the amount we caught. After collecting the agar-agar, we dried it, and the boat owner measured and stored it. A company representative came to buy it for sale in Busan. The company paid the owner of the boat for the catch, and he paid us according to the amount we had collected.

When I was in Japan, near Gogoshima, I dived for sea urchins. But I couldn't make much money there, so I stayed for only two months. I could earn more money diving

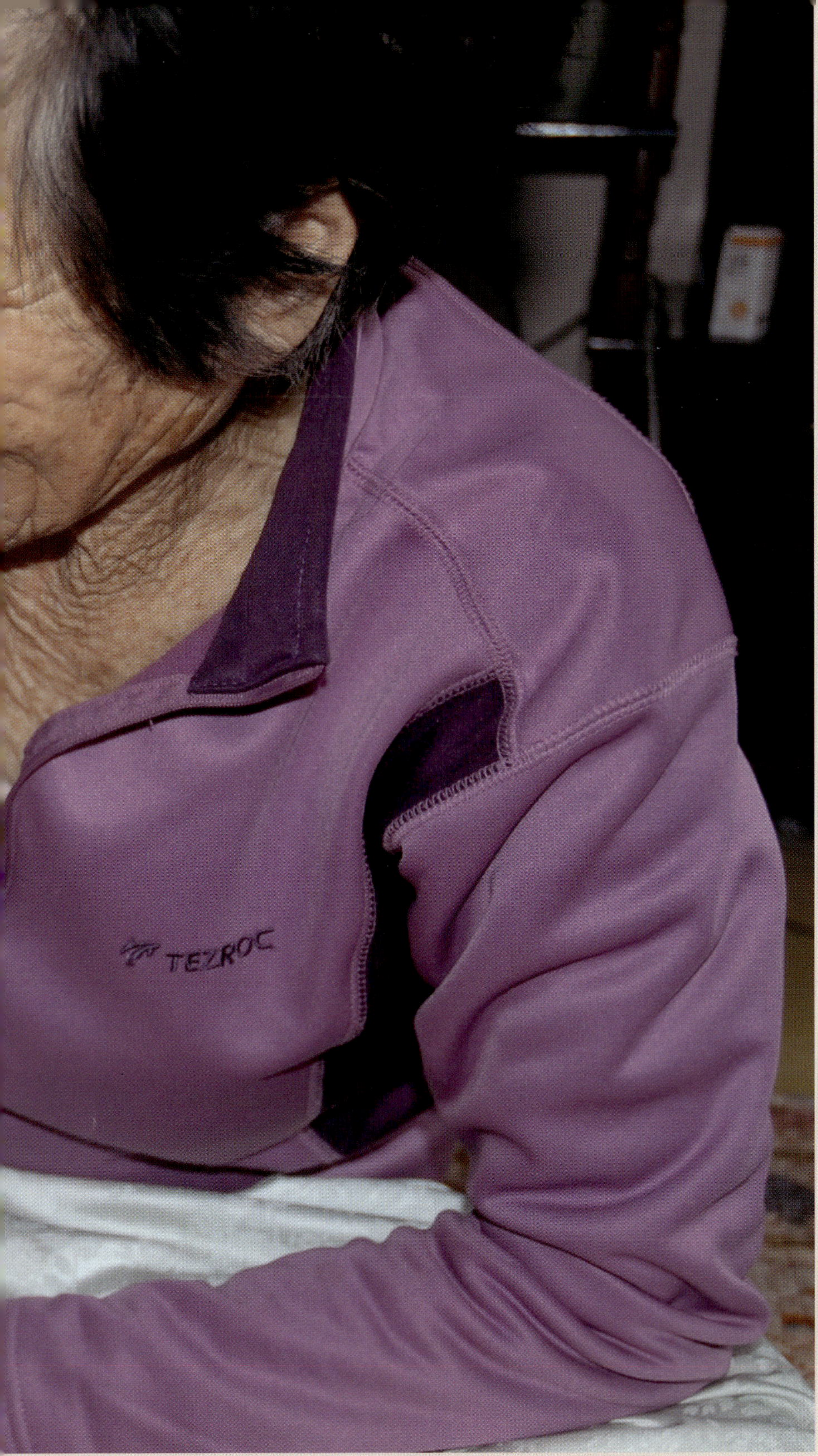

on mainland Korea. I usually left in March and came back in August. I went back and forth to the mainland every year for 20 years. I could earn around 50,000 won over five months. It was not much money even 50 years ago. I spent it on living expenses and taking care of my children's needs for the year. It made me feel good.

I've enjoyed doing *haenyeo* work. If I didn't, I couldn't have stood it. I stopped diving on the mainland at age 45 in order to take care of my children, who were in school at that time.

I don't do farming, except to help my sister on occasion. I make my living doing *haenyeo* work. I don't know how much I will get this year. Last year, I made 1.5 million won ($1,300) gathering agar-agar. It's not a lot, but I just manage to live. I don't do anything other than collecting agar-agar because I am too old. I work together with other *haenyeo* in the shallow waters. I used to be one of the best *haenyeo* when I was younger. But now I only do as much as I can.

I'll probably quit next year. My sister doesn't want me to go to the sea because it's dangerous for me at my age. But I like working in the sea much better than working on a farm. I'll have to see. My health changes year by year.

I think the youngest *haenyeo* in our village is in her 40s. Generally, younger women don't want to do *haenyeo* work because it is too hard.

Health

I am the oldest *haenyeo* here. People say I am very healthy and strong. I don't have any health problems, and I rarely go to see a doctor. Only when I catch a cold. Once a year, the staff from a big hospital in Jeju-si visits our village in a big vehicle to examine the elderly and young people for free. But not for dental problems. If they do find some health problems during the check-up, they'll send us a notice to get further examinations. I never got any notices until now.

I have poor eyesight because of my age and don't wear glasses except when I use a sewing machine. I used to have beautiful hands when I was younger. But now, with my arthritis, my fingers are getting deformed. That's why I use the machine to make floor cushions, pouches for farming, wrapping cloths, and other simple items.

My children also don't want me to go to the sea. But I want to go because I'm bored and have nothing to do. I do it as a kind of exercise. I often have a hard time sleeping because of insomnia. When I work during the day, I sleep better. These days, I don't work, and so I lie idly around. I went to the drugstore to buy sleeping pills, but they didn't sell them to me because they can be dangerous if they're abused.

Sometimes, I wish I could die when I'm healthy. I don't want to cause trouble to my son by living too long or having dementia. But I heard I've passed the time to get dementia. They say those younger than 80 are more likely to have it.

If I want to relax, I get together with other *haenyeo* under the big tree and chat. Or I watch TV dramas. I cry a lot watching soap operas that show couples who are divorced because they can't have babies. They are like my life story.

(August, 2007 & March, 2008)

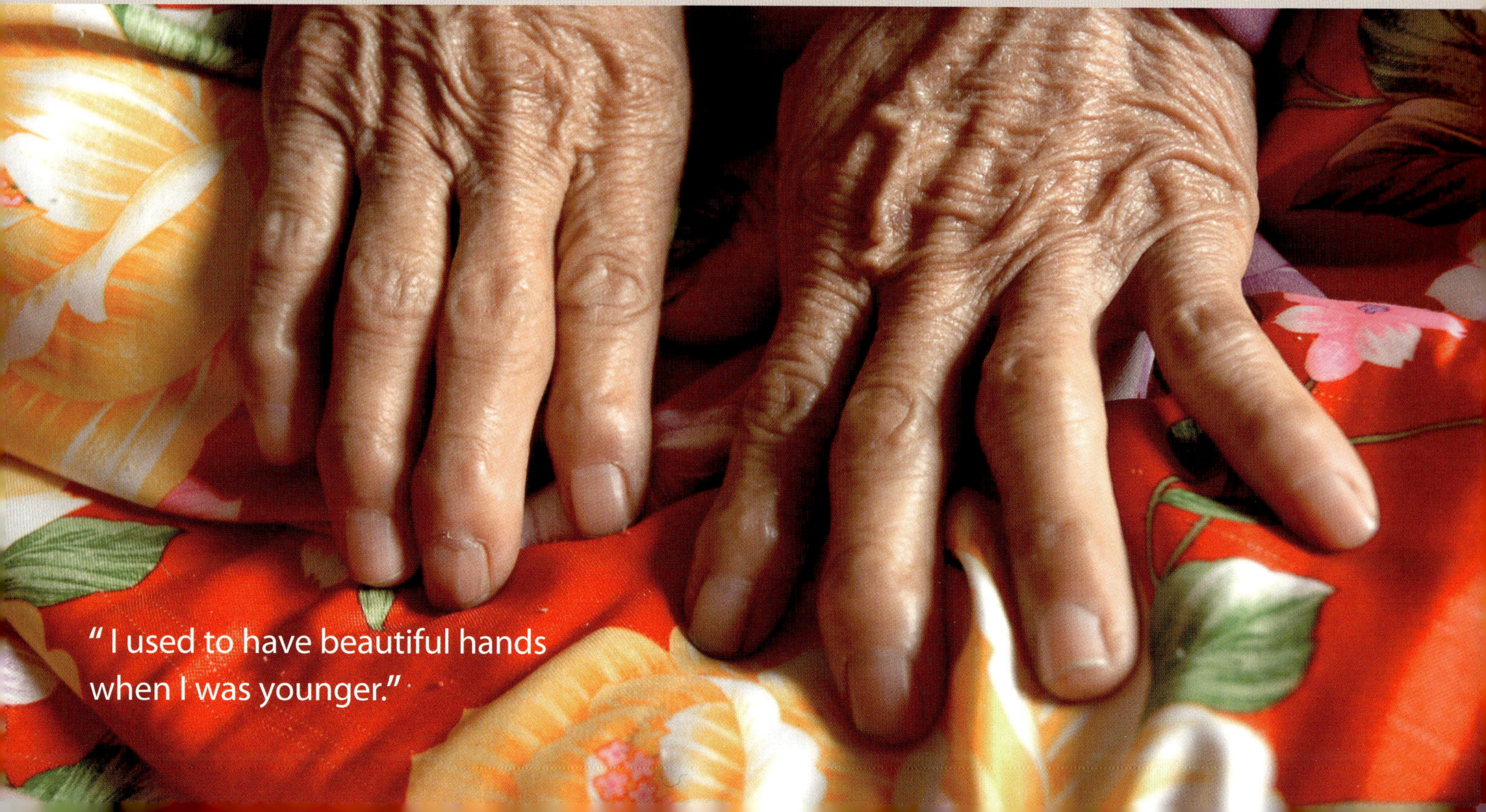

" I used to have beautiful hands when I was younger."

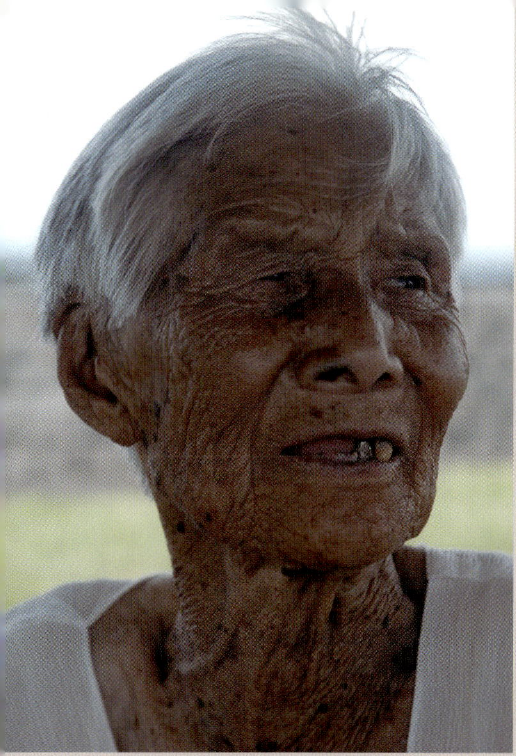

OH YEONG-GEUM

b.1915 | Gimnyeong-ri

I learned to be a *haenyeo* all by myself when I began going to the sea when I was 15 years old. My mother and grandmother were not *haenyeo*. I have never been to the mainland or the deep sea to do *haenyeo* work. I've only dived around here, and only in shallow waters.

I have four daughters and two sons. Two daughters live in Jeju-si, and two live here in Gimnyeong. Two of them are *haenyeo*. But the second one who lives here doesn't do *haenyeo* work because she is afraid of the deep water. I live with my first son's family here in Gimnyeong. He is 59 years old, a barber. My second son lives in Seoul.

I stopped doing *haenyeo* work when I was 75 years old, in the late 80s. In past times, we never had rubber suits like the woman divers wear today. During the winter, we were really freezing when we came out of the sea. We shivered terribly and even gritted our teeth a lot because it was so cold. These days, the *haenyeo* wear rubber suits, so they can stay in the water for five hours straight without feeling too cold.

When I was a *haenyeo*, we made a fire with dry grass or wood brought from home around a *bulteok* (fire pit). These days, the *haenyeo* can take a hot shower after diving. They're so lucky. When we washed out our cotton suits and laid them on the rocks to dry, they quickly became frozen. It was such a miserable and tough time back then. One of my daughters went to the mainland to do *haenyeo* work. No one has had a harder life than her.

Before I got married, I took a boat for two days to Japan. Many people got seasick and vomited and couldn't eat anything. But I was okay. I worked at a zipper factory. While there, I met my husband through a matchmaker at age 20. He is originally from Gimnyeong. I am from Sehwa. After we were married, and I got pregnant, we came back to Gimnyeong when the Pacific War broke out.

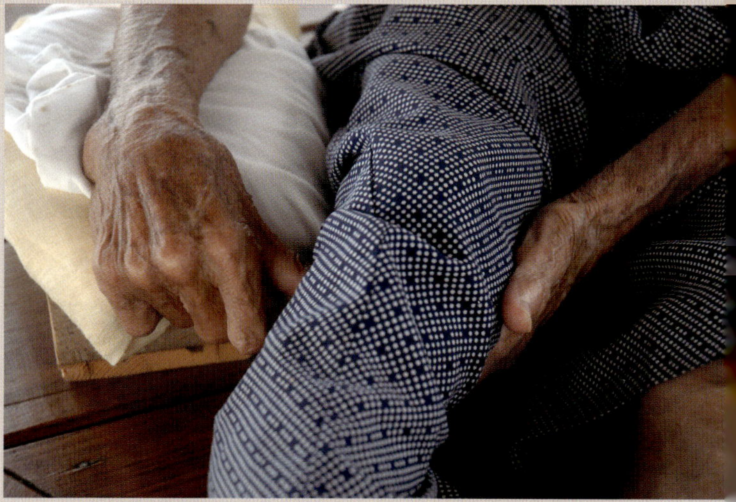

Throughout the years, I did farming and *haenyeo* work. I never went to school. Because my husband was an only son, he didn't do much of anything. While farming, I grew barley and foxtail millet.

(August, 2007)

HWANG CHUN-SEOK

b.1928 | Gimnyeong-ri

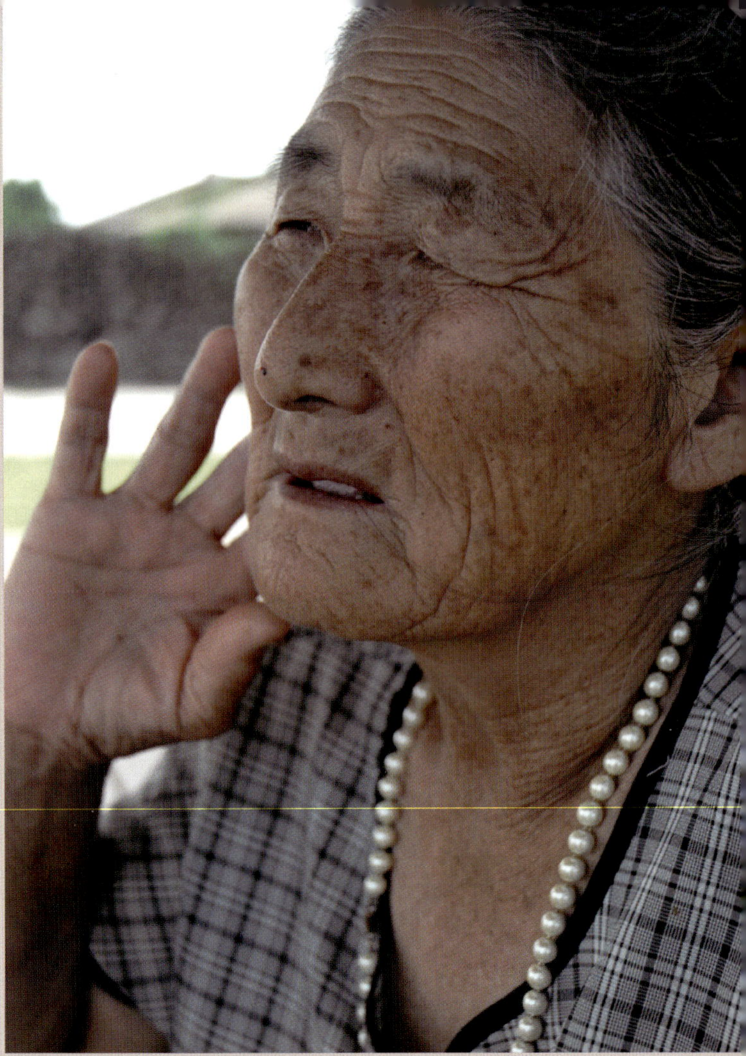

Once on a cold day—a long time ago—I was coming out of the water after catching *miyeok*. My breasts were exposed as I rushed to the *bulteok* in order to warm up before the fire went out. The other *haenyeo* had already come in, but I stayed out later in order to catch more sea products. Everyone around the fire was laughing out loud. They said an American had taken a picture of me! Even now, it makes me laugh whenever I think about it. I wanted to catch him, but he had already disappeared.

I used to do *haenyeo* work when I was younger, but I quit when I was 47 years old because I didn't like the work very much. I learned to swim when I was 8 or 10 years old, as most of us did at that time.

My family was very poor. When I got married, I discovered that my in-laws were also very poor. They didn't have any land to grow grain. We rented land for farming. Because we didn't have enough food, we often cooked rice with seaweed. Sometimes, we cooked the seaweed with flour. I've had many difficulties making a living, and I have never had anything luxurious. I never spend the money I earn, and I wear clothes and items given to me by my neighbors.

I married a useless man. He had four wives, in addition to me. In the past, men often had more than one wife. They weren't legal wives, just concubines. I think it's because there were always more women than men on Jeju Island. Of course, I felt miserable. It was OK that he had other women. What I didn't like was that he was really hard on me. Whenever I saved a little money, he pressured me to give it to him. I didn't feel like making money for that, so I stopped working as a *haenyeo*.

My husband died when he was 57. We weren't living together at the time. So I had to raise my children alone. I've done everything for them. I have two sons and two daughters. My first daughter lives in Busan and runs a raw fish seafood restaurant. My second daughter and second son live in Jeju-si. My first son lives here in Gimnyeong.

In those earlier years, if we didn't have anything to eat we caught and boiled small sea crabs. After I grew up, I worked hard to get enough food because I hated the memory of hunger. Even now, I never idle away my time. Whenever I am asked to give a helping hand, I do it. Otherwise, I like to come to the *jeongja* to meet my neighbors and chat. It's cooler than in my house. When I'm at home, I watch TV, but when I get sick and tired of it I like to come outside.

Childhood and Education

As a child, I didn't attend school. I could never think about it because we were too poor. Years later, our village community center offered classes in reading and writing. I went to the night classes for a while, but I didn't think it would be necessary. In those days, people didn't think girls needed to be educated. I just wanted to earn money and pick berries in the field. If I had been interested in learning, perhaps I would know more. I can only read the names in a phone directory and write my name. Now, I'd like to learn more, but there's nobody to teach me anything. What would be the point?

Health and Happiness

I wear this necklace to reduce the pain in my body. I tripped at the farm and hurt my head and legs. I suffer from neuralgia. There's nothing that I really want or need. I have enough money to support myself and buy whatever food I want to eat. Sometimes, my first son, who is a fisherman, brings me fish. Otherwise, there's only one thing that would make me happy. That's if my only grandson got married. He's 32 years old. I want to have a great-grandson. My only regret is that I couldn't study when I was younger. I always tell my children to study, whatever they do.

(August, 2007)

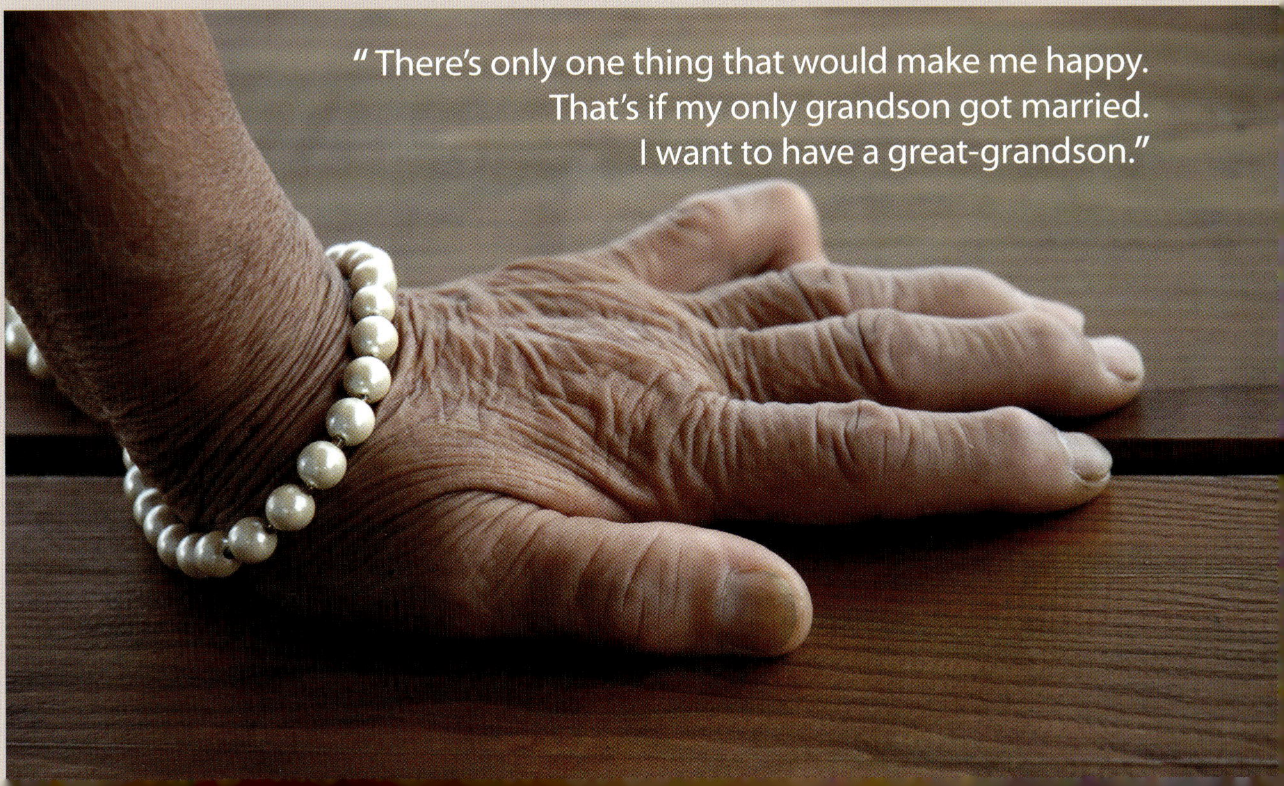

KANG DO-SAENG

b.1913 | Gimnyeong-ri

I am one of the *haenyeo* who went overseas to Japan to dive. We *haenyeo* would dive into the deep sea with our waists attached to a rope from a boat. We would dive for *umi* (agar-agar), which grows in the deep sea. Those of us who worked in Japan made more money than if we had stayed in Jeju. I stayed in Japan for about five months at a time, came home, and then went back again. Meanwhile, my husband took care of our children. My father-in-law's younger brother lived in Japan and was a boatman there. So I worked with him. He rowed the boat while I dived. I was not scared at all of diving in the deep sea.

I don't know what made me such a good diver. People who couldn't hold their breath for a long time couldn't work in the deep sea. But I could hold my breath for a very long time, so I could dive pretty deep. The deeper I dived under the sea, the more *umi* I gathered. At that time, I took a basket with me while diving, collected the *umi*, and then sent a signal to the boatman that it was full. Then he pulled me and my catch up using a pulley. It was enjoyable then because with all that *umi* I could make a lot of money.

Now I don't miss being in the sea, because I am too old. The last time I worked in the

Many *haenyeo* in Gimnyeong regard Kang as one of the best deep sea divers.

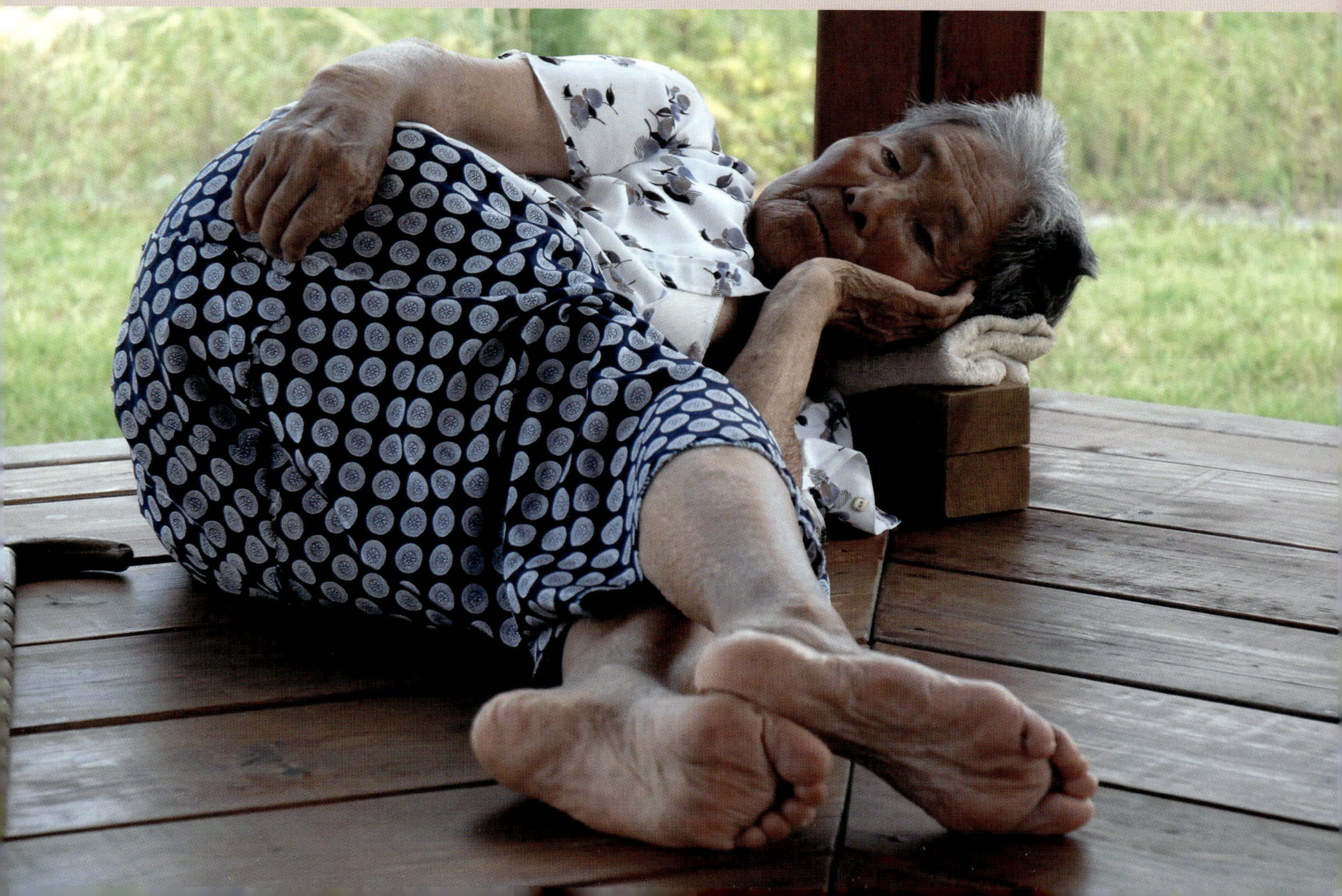

water was in my 60s. When I was younger, I was elected as a *haenyeo hoejang* (leader) and served for 10 years. I organized *haenyeo* activities.

I have nine children. I had four with my husband, and there are five others who were actually fathered with my husband's "second" wife. Both of us were born in Gimnyeong and met through a matchmaker, like many other couples. My first son lives in North Korea. When he was living in Japan, he went there and never returned. I have his pictures. Now he's 72 years old. When reunification comes to Korea, I'll be able to see him.

These days, I live alone. When I'm bored, I come out here to the pavilion. Otherwise, I watch TV. I don't use a cell phone because I don't know how to press the buttons and numbers. I can just use an old-fashioned one.

I'm healthy and rarely go to see a doctor. I can eat whatever I want. Sometimes I feel like eating pork. But when I used to dive, I did take *noeseon* in order to prevent headaches while diving. Even now I still take it whenever I have a headache. I don't remember exactly when I started using it—probably when I began diving in my teens.

One of the other things I also did years ago was serving as a midwife for fellow *haenyeo* and my neighbors. In former times, many *haenyeo* went back to work immediately after delivering their baby, sometimes at sea. Babies born at sea were referred to as "*baeseonie*" (boat babies).

(August, 2007)

" These days, I live alone. When I'm bored, I come out here to the pavilion."

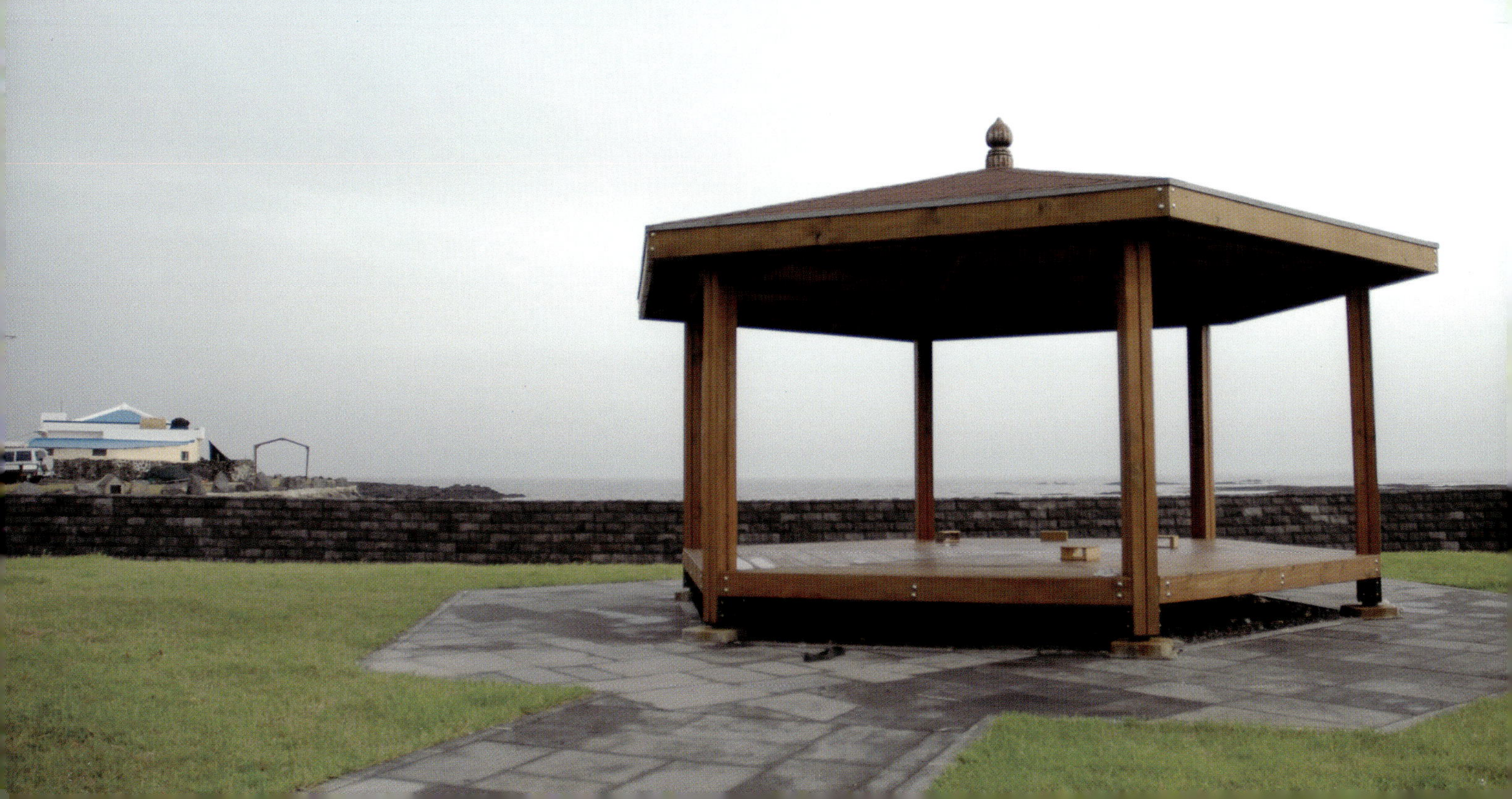

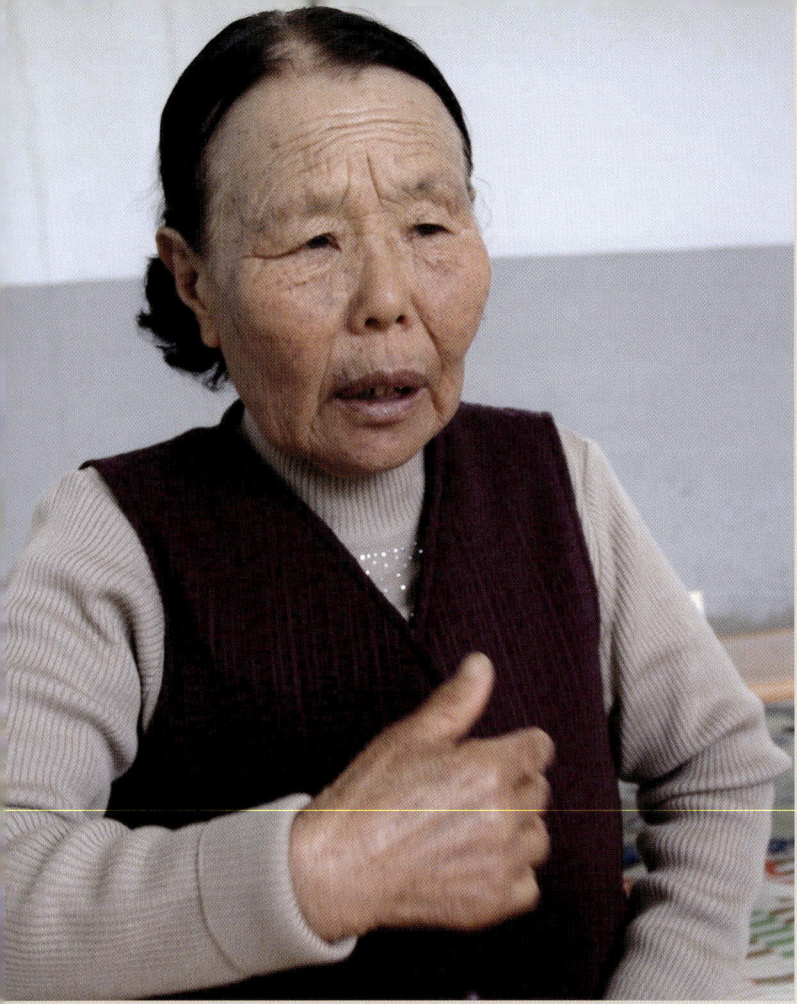

KO CHUN-HWA

b.1927 | Gimnyeong-ri

I was born in Japan and grew up here in Gimnyeong. Both of my grandmothers and my mother were *haenyeo*. My mother was one of the best *haenyeo* when she was young. My family came back to Gimnyeong when I was three years old. But we went to Japan again when I was 13 and finally returned to Jeju when I was 20, a year before Korea's independence from Japan.

I learned to dive when I was 21 and went to Geoje Island on the southern coast of Gyeongsang Province. After I returned from Geoje Island at age 25, I got married, and I had my first child a year later. Coincidentally, my husband and I were both born on the same day. We're both 82 years old. I'm not pretty, but my sons are quite handsome and tall because they look like him.

We have two sons and two daughters. I raised them and provided for their higher education by working as a *haenyeo*. All of them graduated from college. I even sent my first son to graduate school, even though I began with empty hands. While my children were growing up, we needed a lot of money. So when I wasn't diving, I helped on my neighbors' farms and made extra money for my children's school expenses. After they grew up, I could buy farmland and build a house. I don't envy anyone. Everything I gained was from working as a *haenyeo*.

Our earnings differ depending on how much each diver catches. On average, I earn a daily wage of 35,000 won ($30). Sometimes, I'll earn more than that. Young divers (there are about 12 in Gimnyeong) can earn 100,000 won ($86) per day. Gimnyeong used to have more than 100 *haenyeo* in our village. Now there are approximately eighty.

I don't think I can continue diving for many more years. I've already earned as much as I need for the rest of my life. I won't rely on my sons later. I think we should prepare for our old age. Maybe I'll dive for another two years. After that, I'll continue with some farming, relax, and enjoy my nine grandsons and three granddaughters. Fortunately, they already have a deep appreciation of the *haenyeo*'s life. I've never been ashamed of being a *haenyeo*. Before, when I worked on the mainland, people looked down on us. Now we're acknowledged by many people.

People are always curious how my husband and I remain so healthy. There's nothing special about my being healthy. I am naturally healthy. I've never caught a cold. But one thing is that I don't eat too much—mainly a lot of vegetables. For breakfast, I eat a little rice and vegetables with *doenjang* (fermented bean paste). I drink some coffee, but never any soft drinks. I eat nearly the same food for lunch and dinner. My weight is the same as what it was when I married. There's no one like me in Gimnyeong. I hardly ever see a doctor.

" I don't envy anyone. Everything I gained was from working as a *haenyeo*."

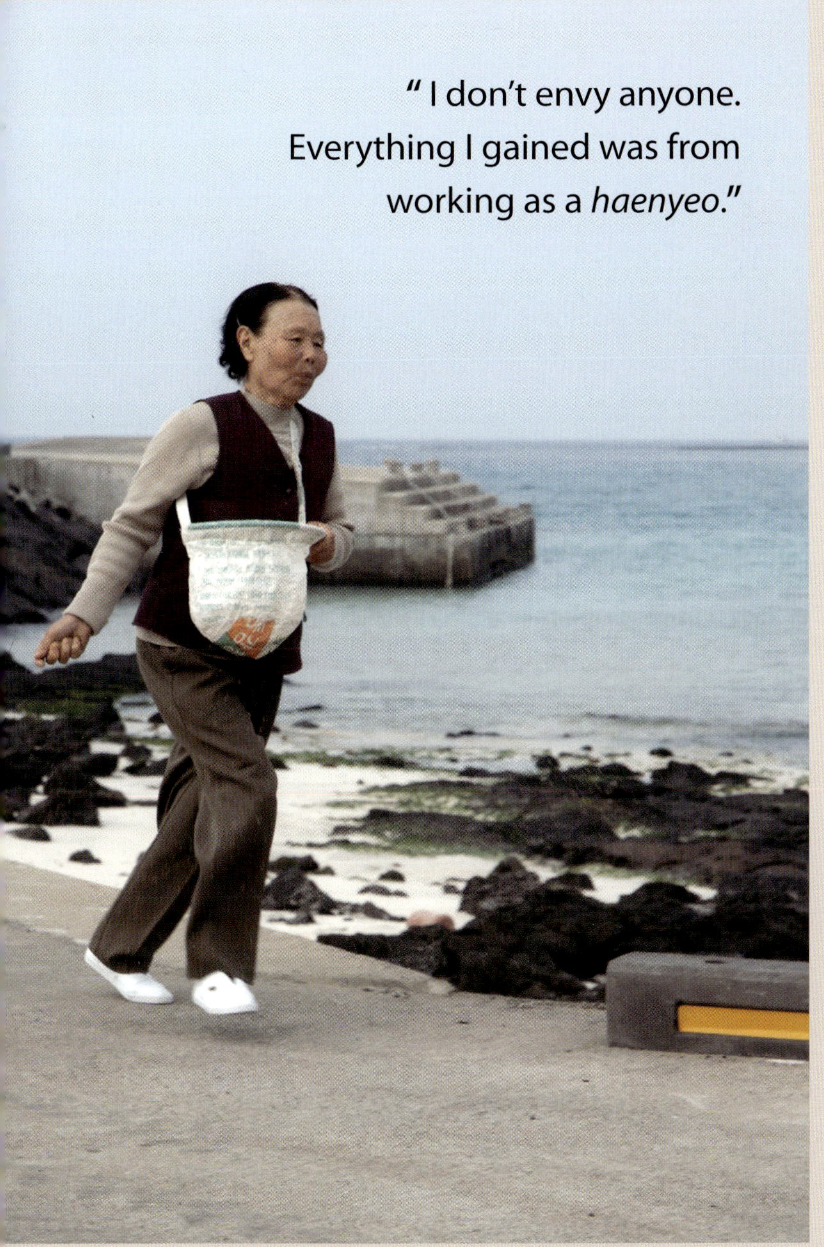

I think I'm very determined. I've been interested in making money only because I lived a poor life, and I try to stay out of debt. Whenever there's something to do, my neighbors ask for my help. Usually, people my age are seldom asked to work. But I'm an exception. A few days ago, I picked tangerines and planted cabbage.

In former days, we women couldn't help but work as *haenyeo*. There wasn't any other way for us to earn a living. If we didn't know how to dive, we couldn't even get married. Everybody thought that those who didn't know how to dive were fools.

I was never afraid of water because I used to swim in the river while I was in Japan. I still dive, and I usually go to the deep sea by boat. Sometimes I stay near the coast. I used to dive as deep as 10 meters, but now I dive only five meters. When younger, I could hold my breath for more than two minutes, but now I can only do it for around one minute, which is still longer than others the same age can do.

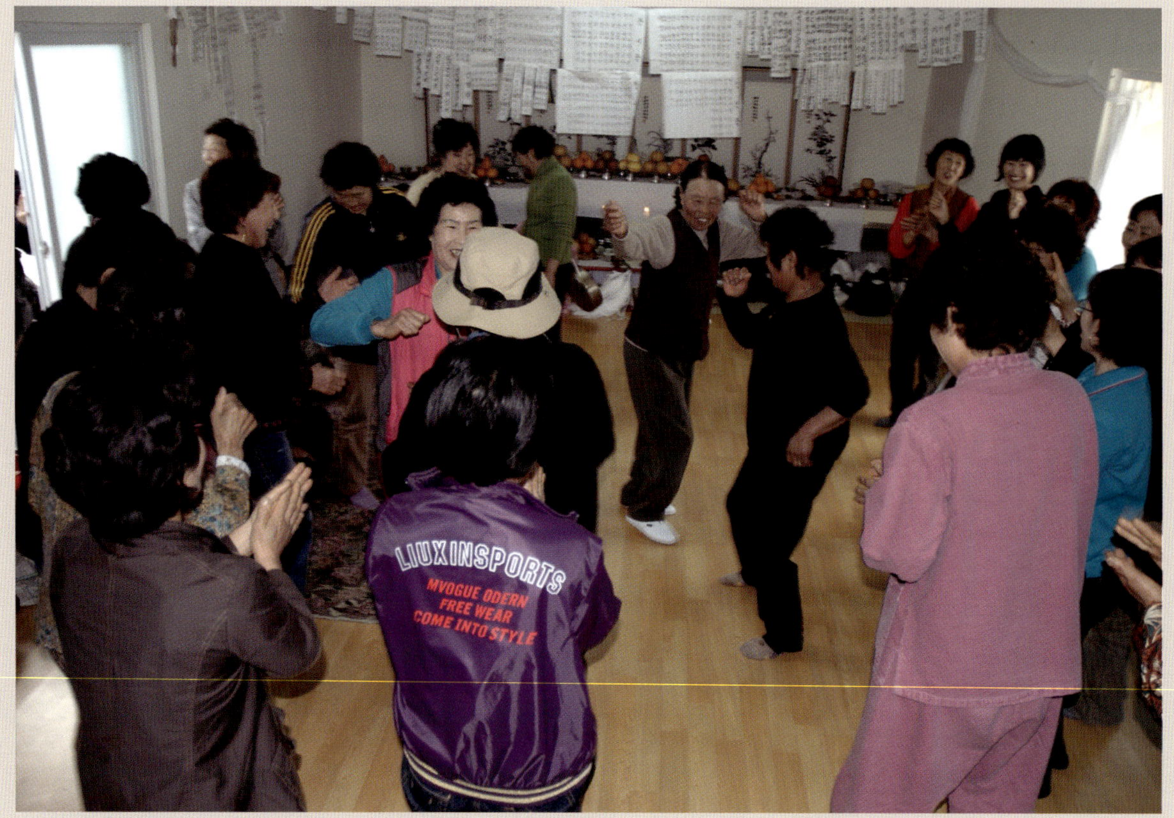

Joyful *haenyeo* from Gimnyeong-ri dance after a special *gut* to ask the gods for their protection and an abundant harvest.

We *haenyeo* dive as a group, following orders from our *eochongye* leader. I've never been in an accident. There are people who have died while diving. You can lose your breath while struggling to get back to the surface in time. That's why we can't help believing in shamanism. Because of the strong energy in our shamanist shrines, and thanks to the *gut*, we have fewer accidents. In fact, we haven't had any diving accidents at all lately.

During the April 3 Massacre in 1948, Gimnyeong didn't experience the same degree of losses as other villages on the island. People from the mountain villages came down here to avoid the massacre because they believed the energy from our shrines would protect them.

(April, 2008)

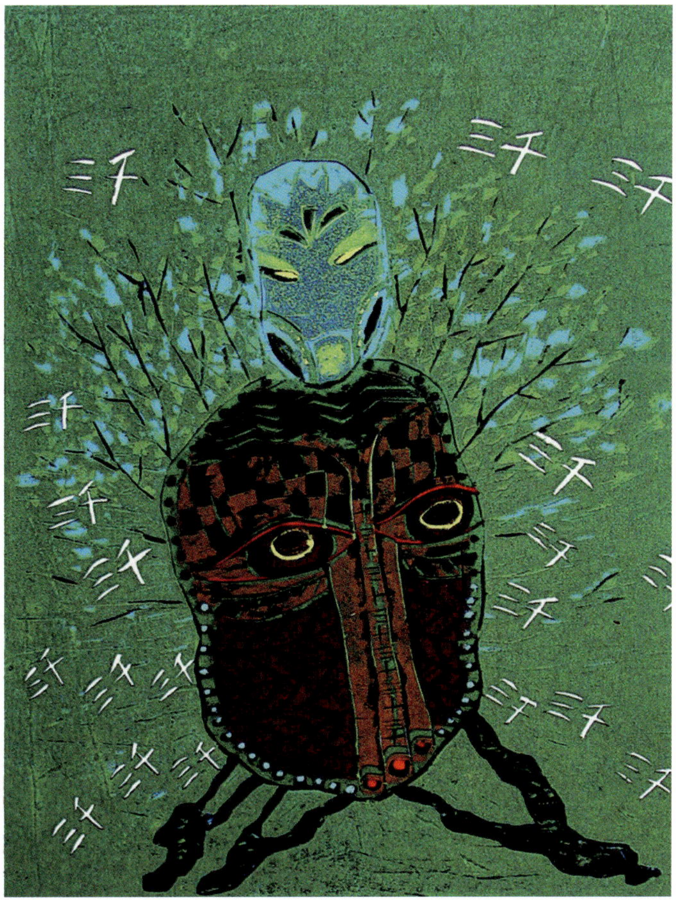

The God of Longevity 40x60 cm | woodcut | 2005

Samani

(God of Longevity)

The god Samani was a poor man who bought a musket in order to hunt and sell animals at the market. One night, a woman in a white dress came to him in a dream. She claimed to be the spirit of an animal that had been killed by a musket like Samani's. She asked that he collect and take good care of her skull and bones. The next morning, Samani went to the spot designated in his dream and dutifully collected the remains. Upon returning home, he paid his respects by preparing food and drink for the skull and bones. From that day on, he became a very rich hunter. One night, three messengers from the Underworld came to visit him. The messengers told him that his 33 years on earth were over. They had come to escort him to the Underworld. Samani treated them all night to delicious food and drink. In return for his generosity, the messengers extended Samani's life to 3,000 years. By the time Samani reached the age of 3,000, he again treated the messengers with more food and drink. This time, he was allowed to live to the age of 40,000 years. That's why he has been given the name Samani, which means "40,000-year-old-man."

Like many other characters in Jeju shamanist myths, the god Samani represents a human being's basic desire for longevity and wealth. This god myth also reflects Jeju islanders' belief in skull worship, which was prevalent in ancient times.

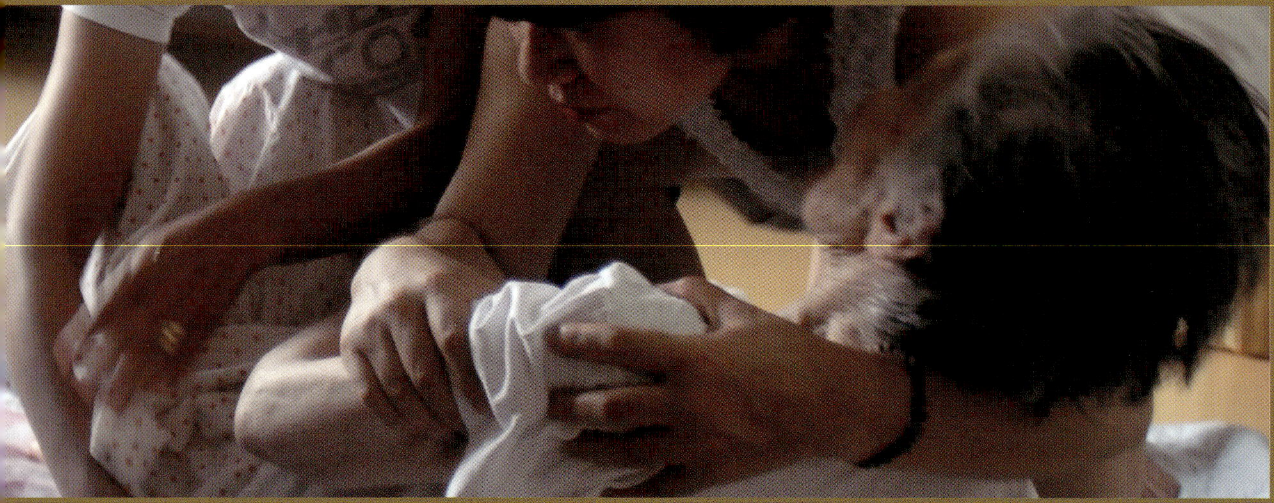

Chapter Five

LIFTED BY THE WIND & TIDES OF COMPASSION

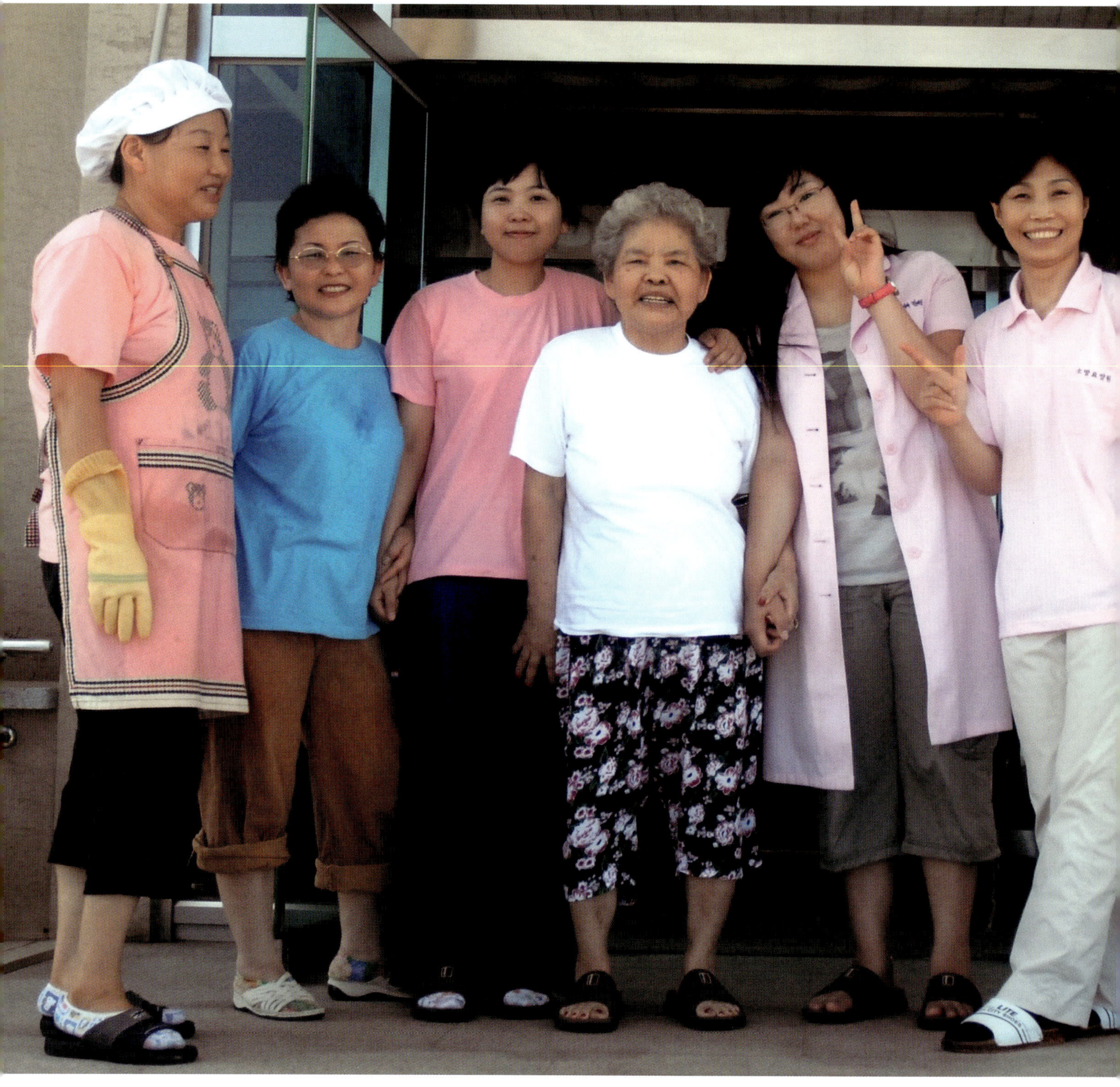

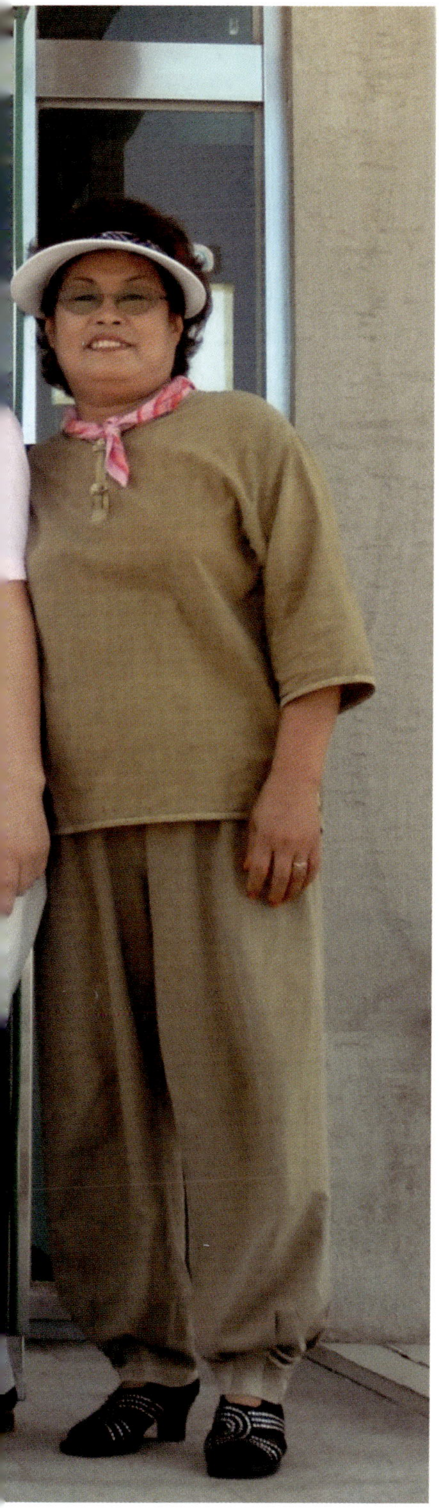

For much of my adult life, I've always tried to devote some portion of my time to volunteer work. Within the past eight years, I've continued to offer online grief support for bereaved parents who've also survived the death of a child. One of my friends—in her 70s—saves a few thousand dollars each year in order to spend three months volunteering in places such as India, Israel, Cambodia, Vietnam, and various countries in Africa. "This is my contribution to world peace," she says.

Interestingly, I came across a recent study, funded by the National Institute on Aging in the US. It was co-authored by researchers at the University of California, Los Angeles. Their results suggested that volunteering may prevent the elderly from becoming frail. The correlation is probably a farfetched consideration for the *haenyeo*. Regardless of any health benefits, I discovered that the *haenyeo* have selflessly volunteered in various ways to benefit the ocean environment, the elderly, and the abused.

One morning, I went out on a boat with 25 *haenyeo*. They didn't dive for abalone or octopuses that spring day. Each woman leapt into the waters to remove plastic bins, fish traps, glass bottles, rubber hoses, ropes, candy wrappers, and cigarette butts. They retrieved anything that indicated human disregard for the sea and marine life. While the women scoured the sea, the boat driver quickly returned to the pier to pick up no fewer than a dozen more elderly *haenyeo*—all dressed in baggy pants and cotton bonnets. Their job was to haul the findings into the boat and help sort them out. While a few of them leaned over the boat to grab the litter handed over to them, other grannies let the sun rays beam on their faces as they waited for their turn to assist.

On another occasion, I visited a local residence and health facility for Alzheimer's patients. A group of "younger" *haenyeo* from Gimnyeong (in their 40s and 50s) volunteered to feed and bathe the elderly—many of whom were former *haenyeo* and their friends' relatives. I remember

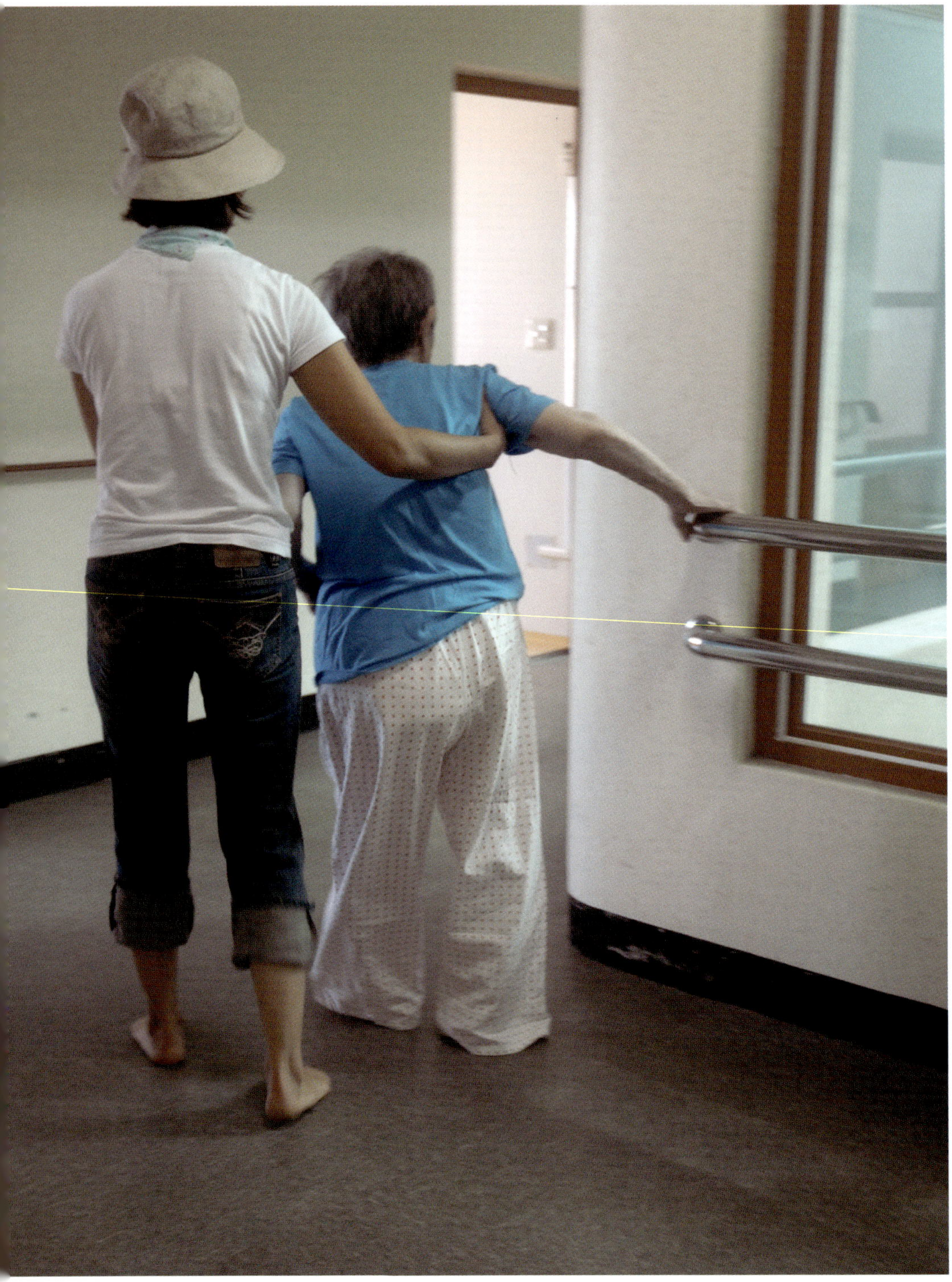

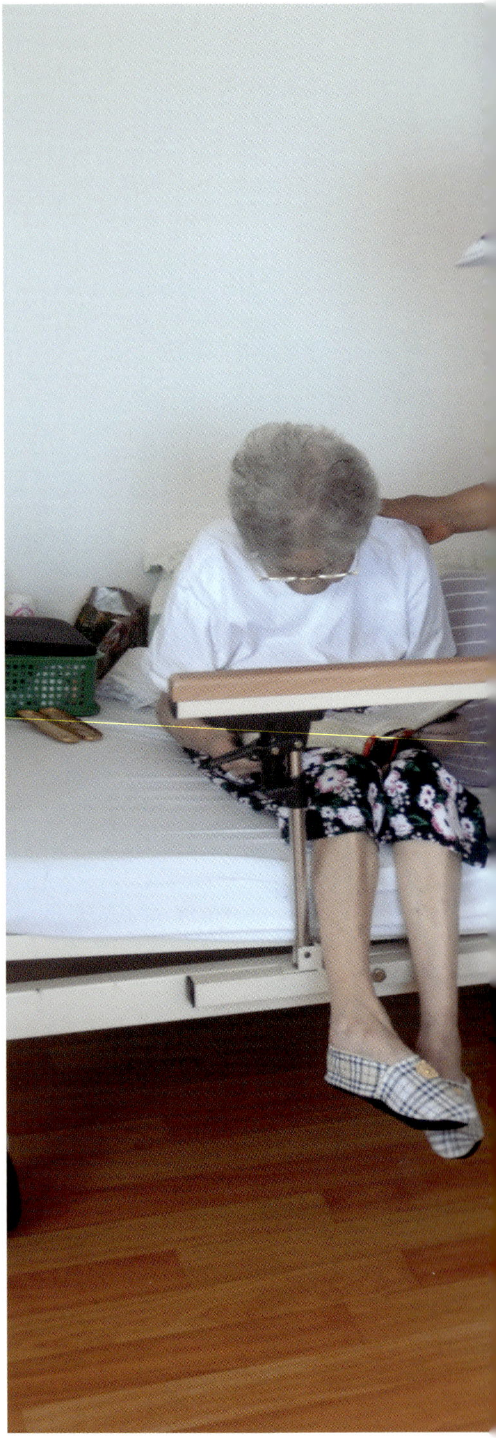

the big grin on one woman's face as she showed off her hand-painted red nails—a simple pleasure provided by the volunteers.

The most illuminating volunteer activity I learned about was the work of Han Ki-ok. A former *haenyeo*, she served as managing director of the Hanbit Foundation between 1997 and 2006. In 1999, she and her daughter founded the Hanbit Shelter for Women—a residence and service organization that assists survivors of domestic violence and abuse. "I've had a tough life as a woman, too, being widowed at 37 and having to

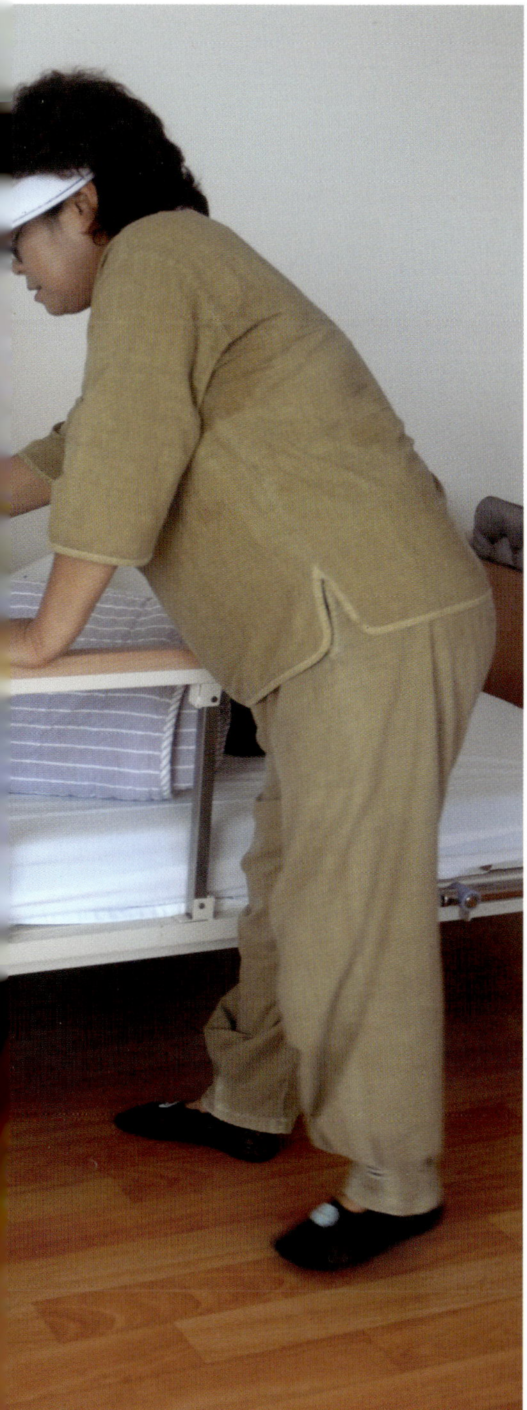

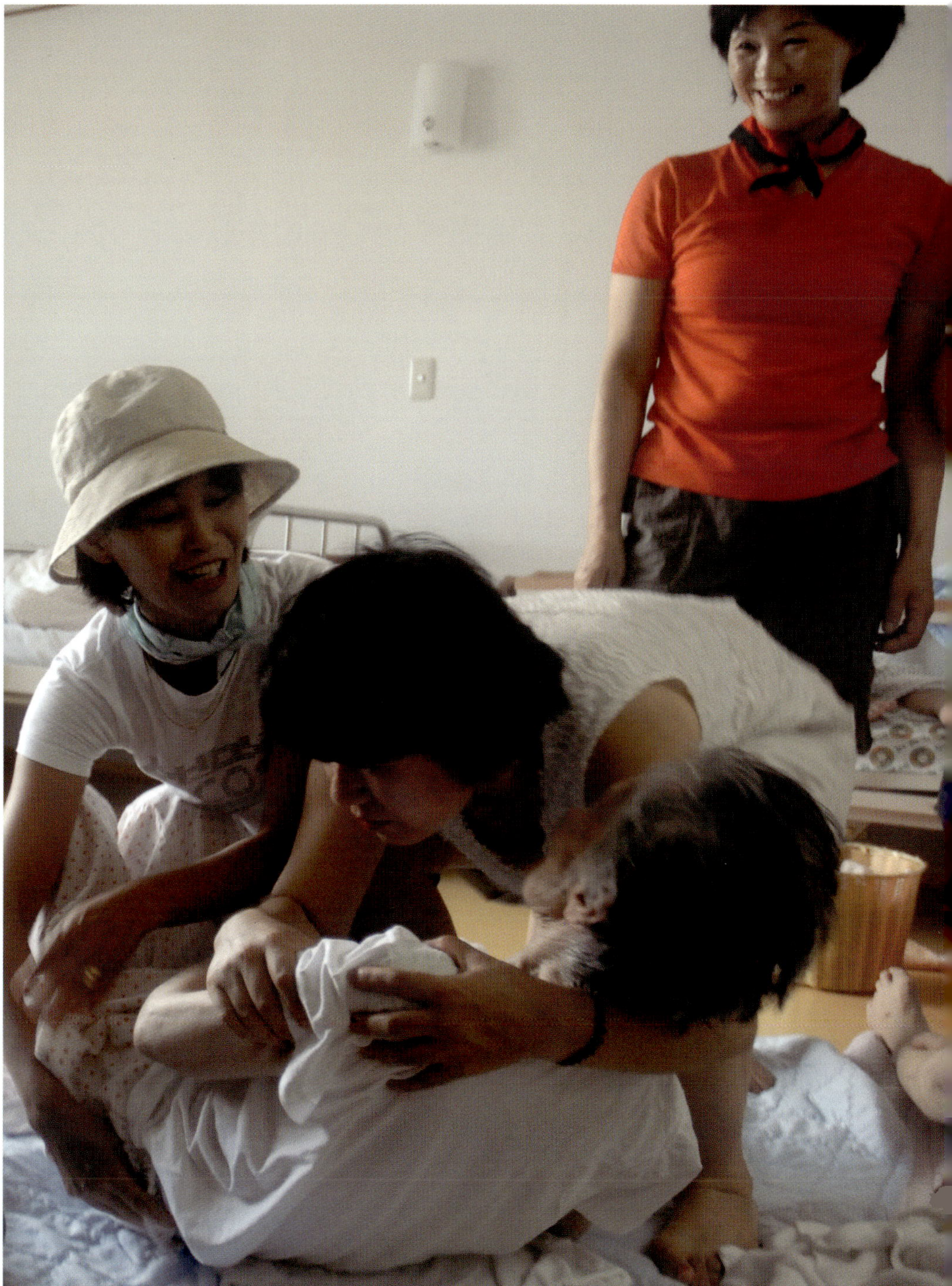

raise children on my own. I'd like to help women who are lonely and suffering," she said during an interview conducted by her daughter, Kang Mi-kyoung, director of the shelter.

One is reminded that even though Jeju exalts the prowess of its women—especially the shamans and sea divers—there are still thousands of other women who remain vulnerable to the exploitative aspects of a Korean patriarchal society.

Caring for the Elderly

A group of working *haenyeo* from Gimnyeong-ri volunteers regularly to bathe and feed other *haenyeo* who suffer from Alzheimer's disease and dementia.

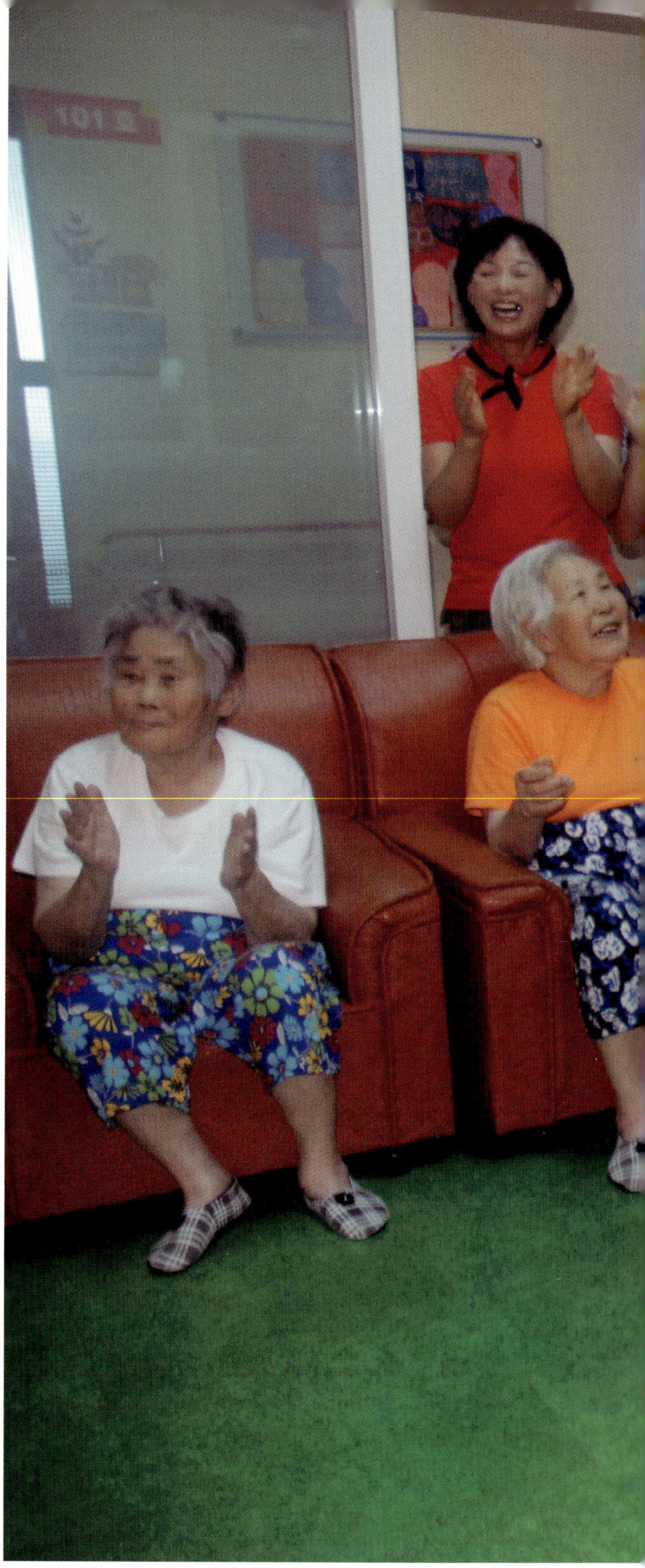

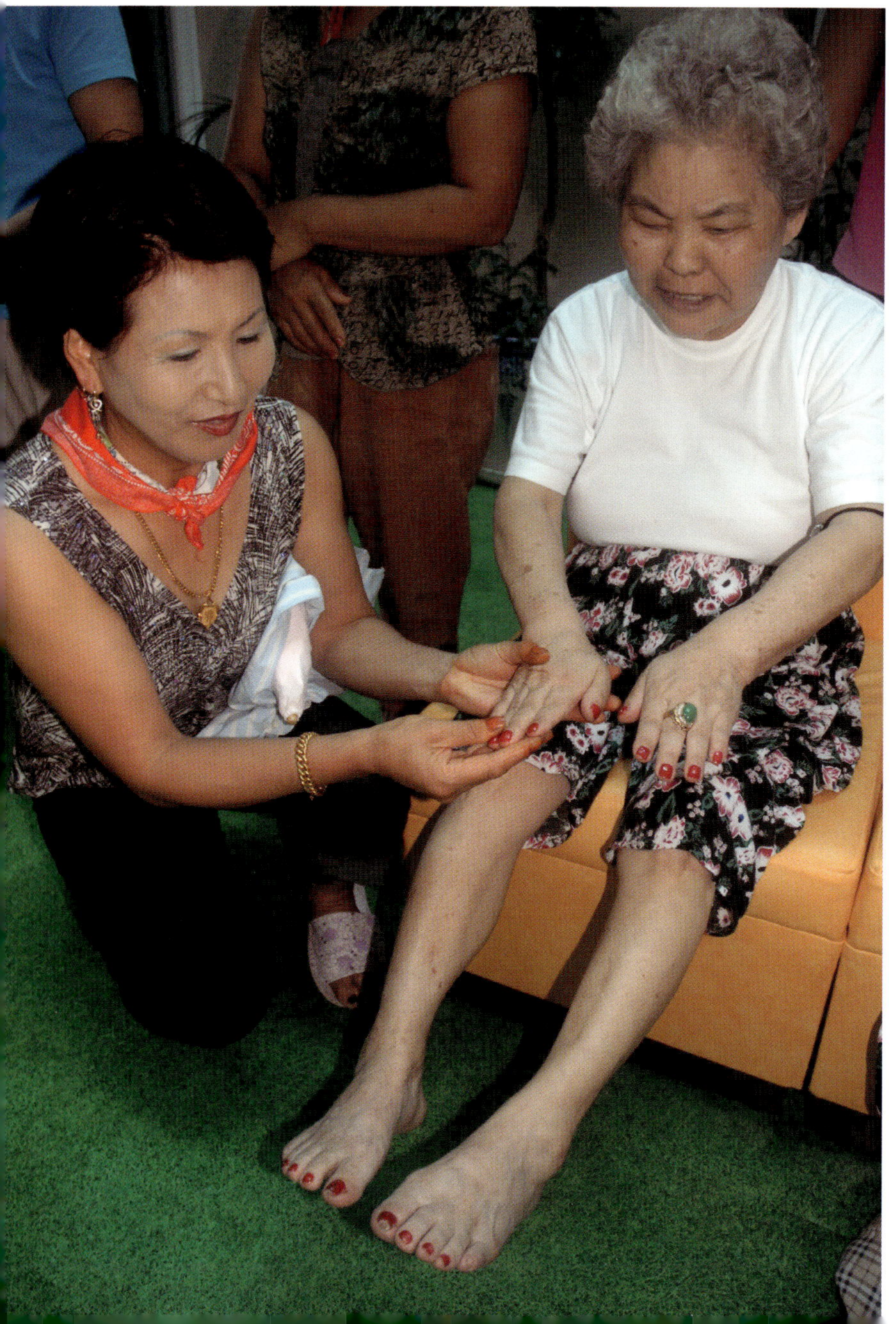

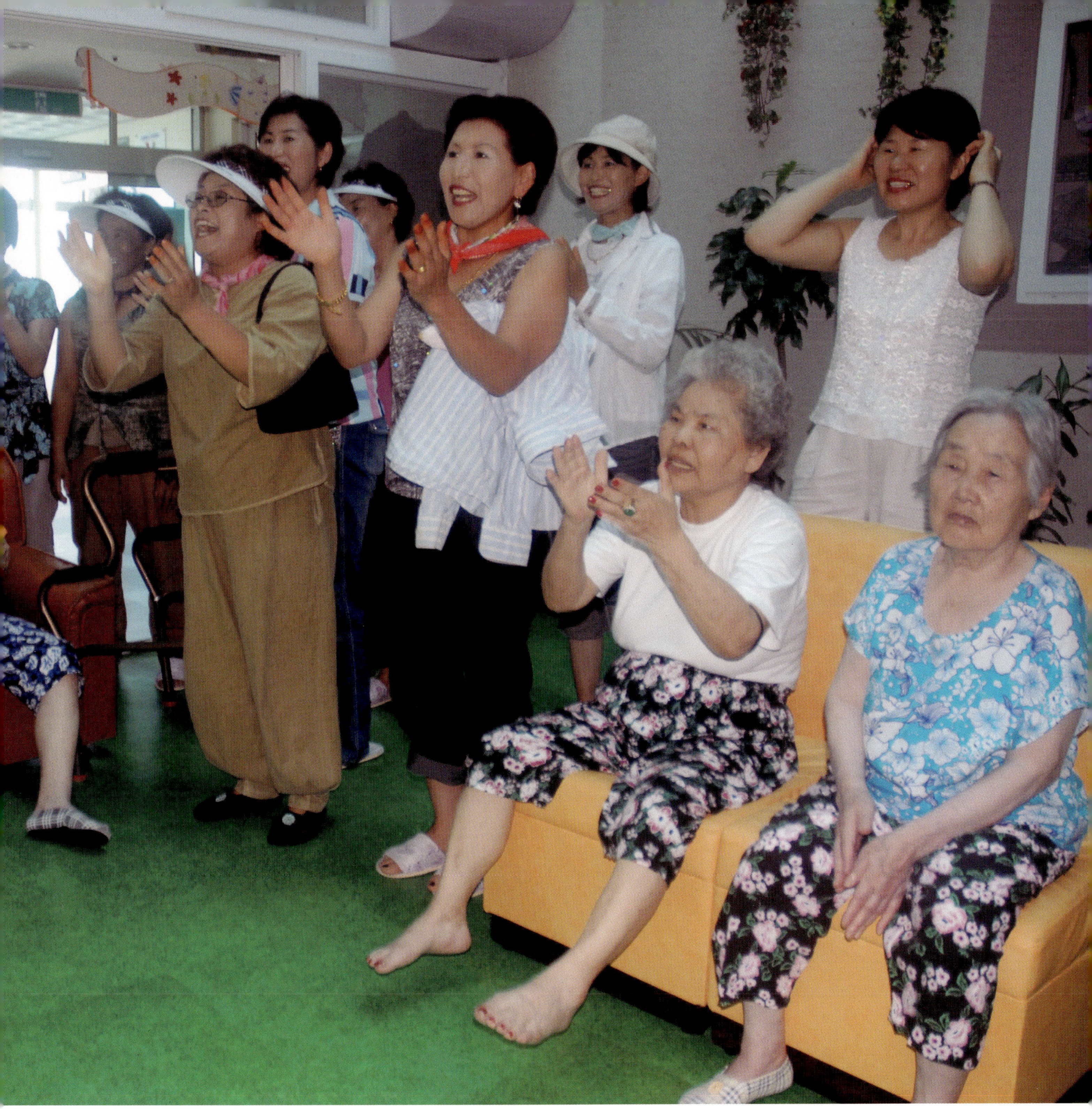

" Among all the households in our village,
we know who needs the most help.
Everyone knows everyone."

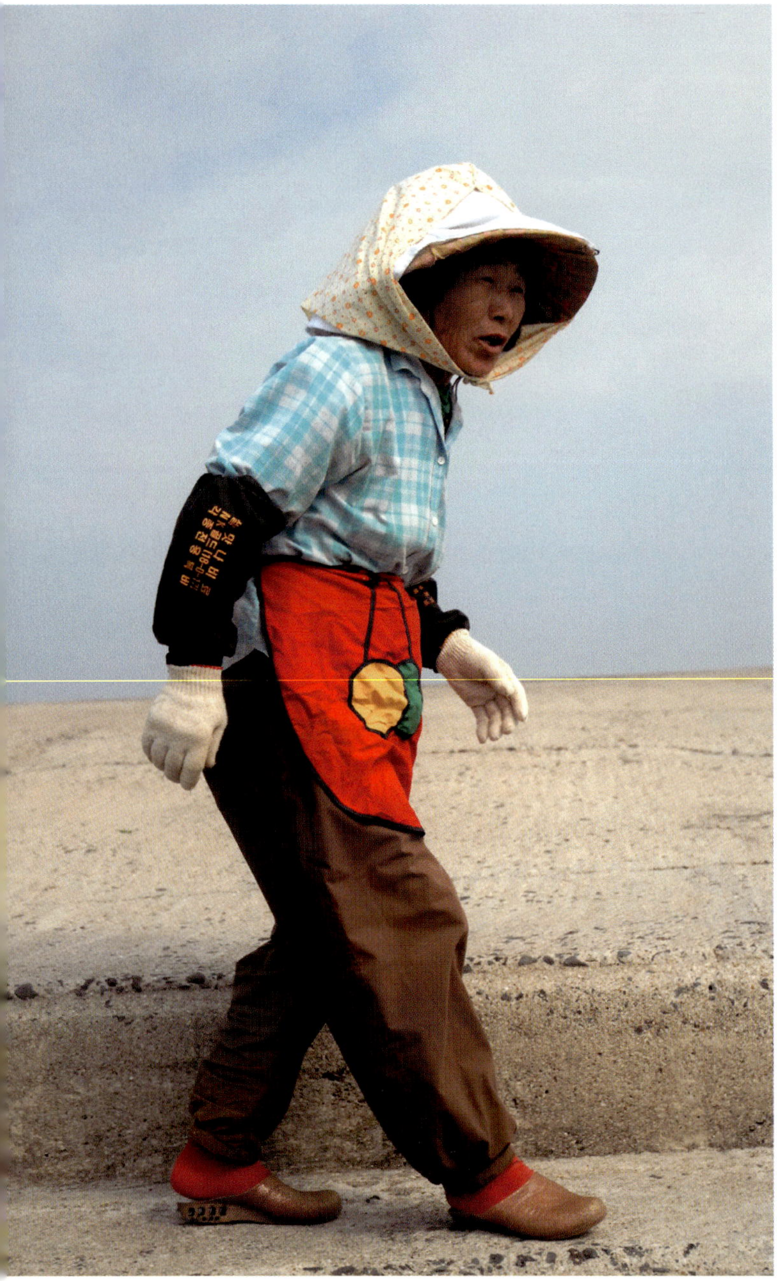

Caring for the Sea

Each month, working *haenyeo* dive into the sea to clean up the ocean environment. Retired divers remain on board to haul and sort the garbage, such as oyster traps, bits of Styrofoam, plastic bottles, and other garbage.

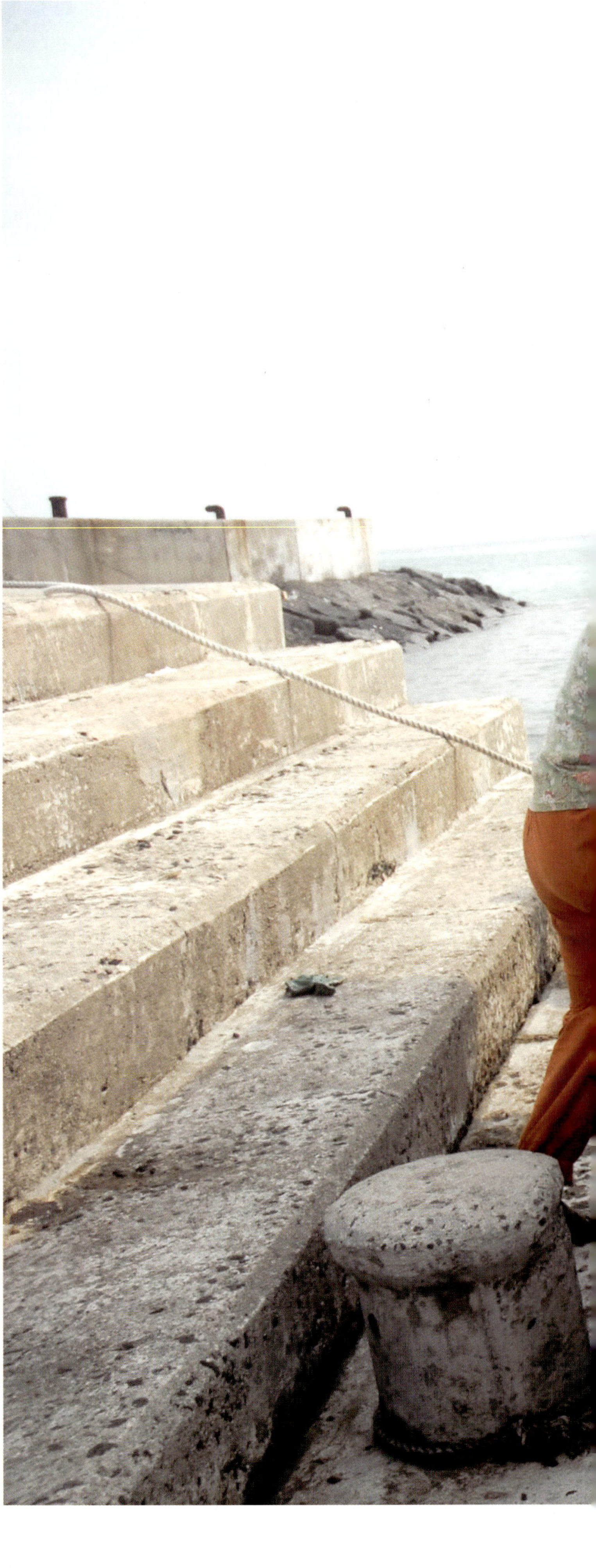

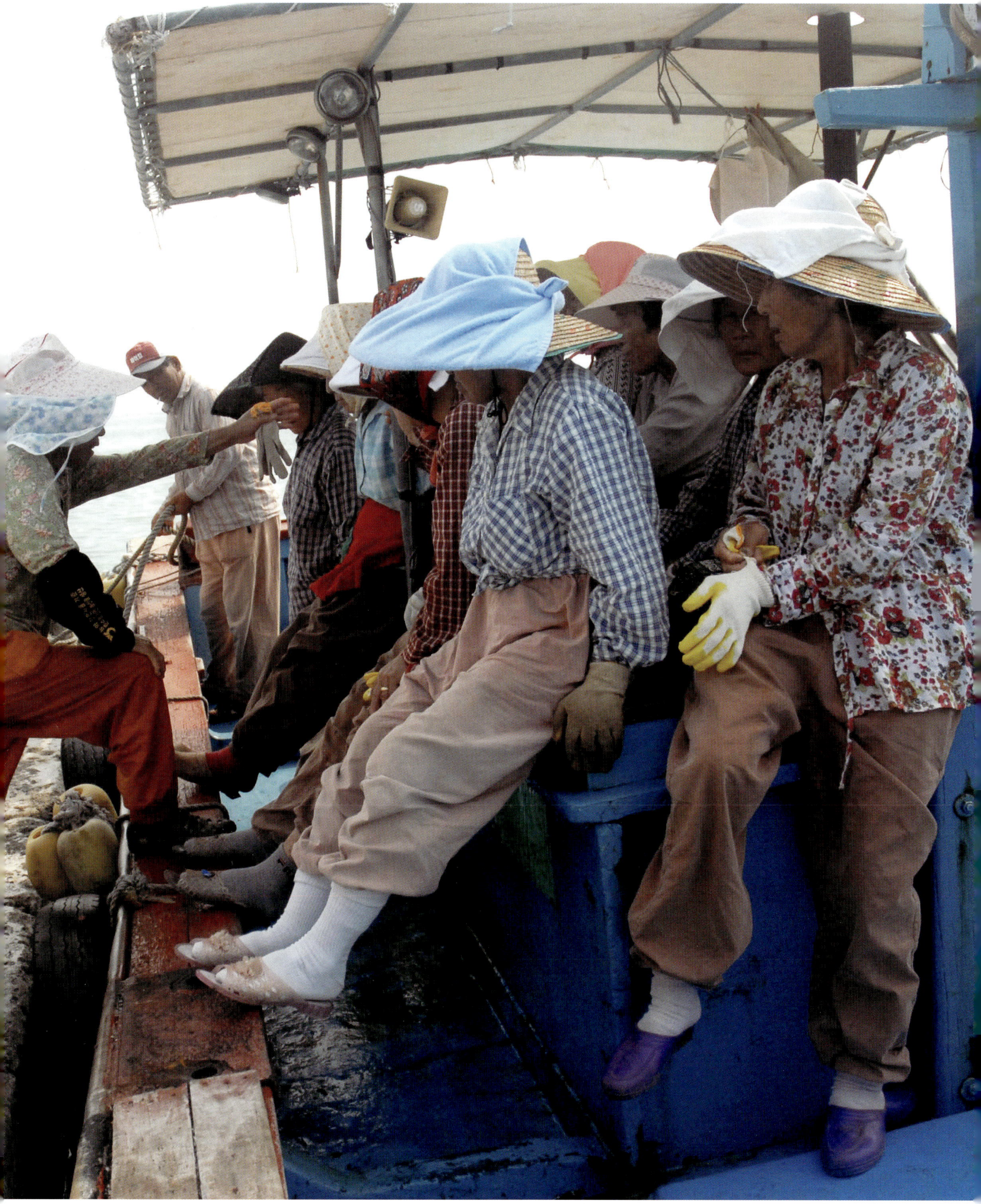

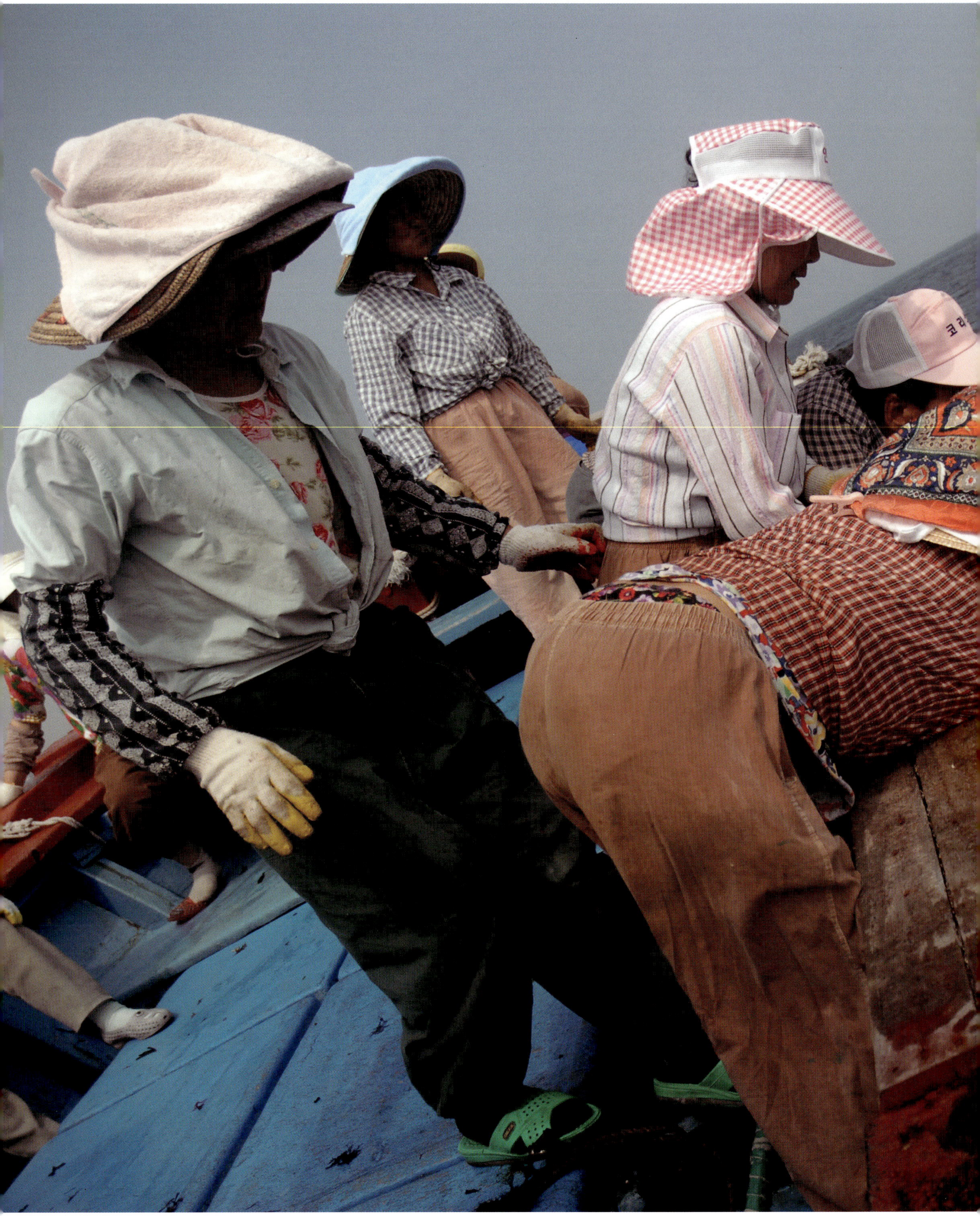

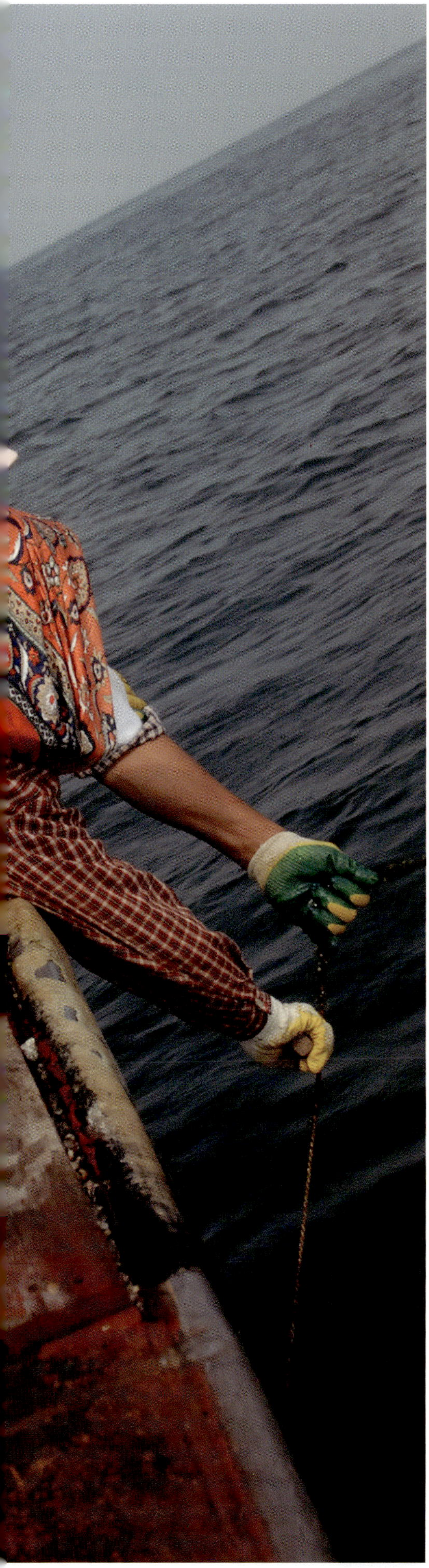

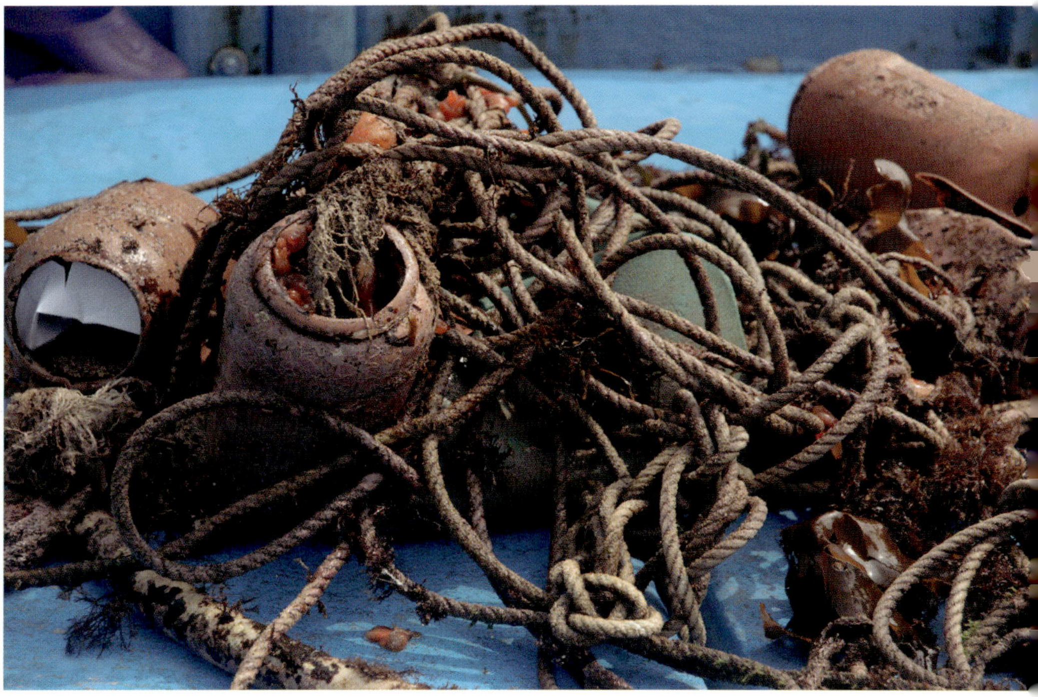

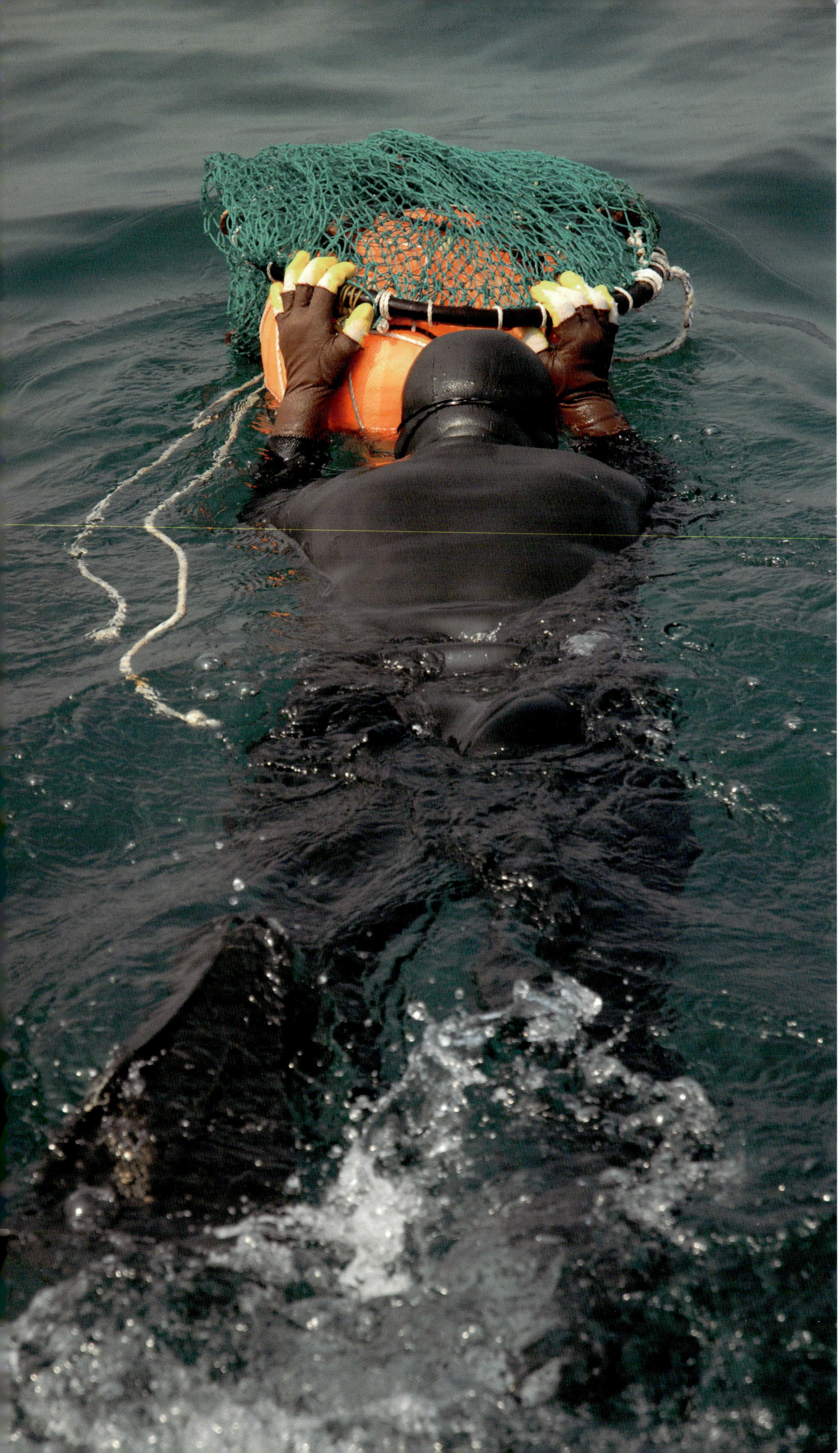

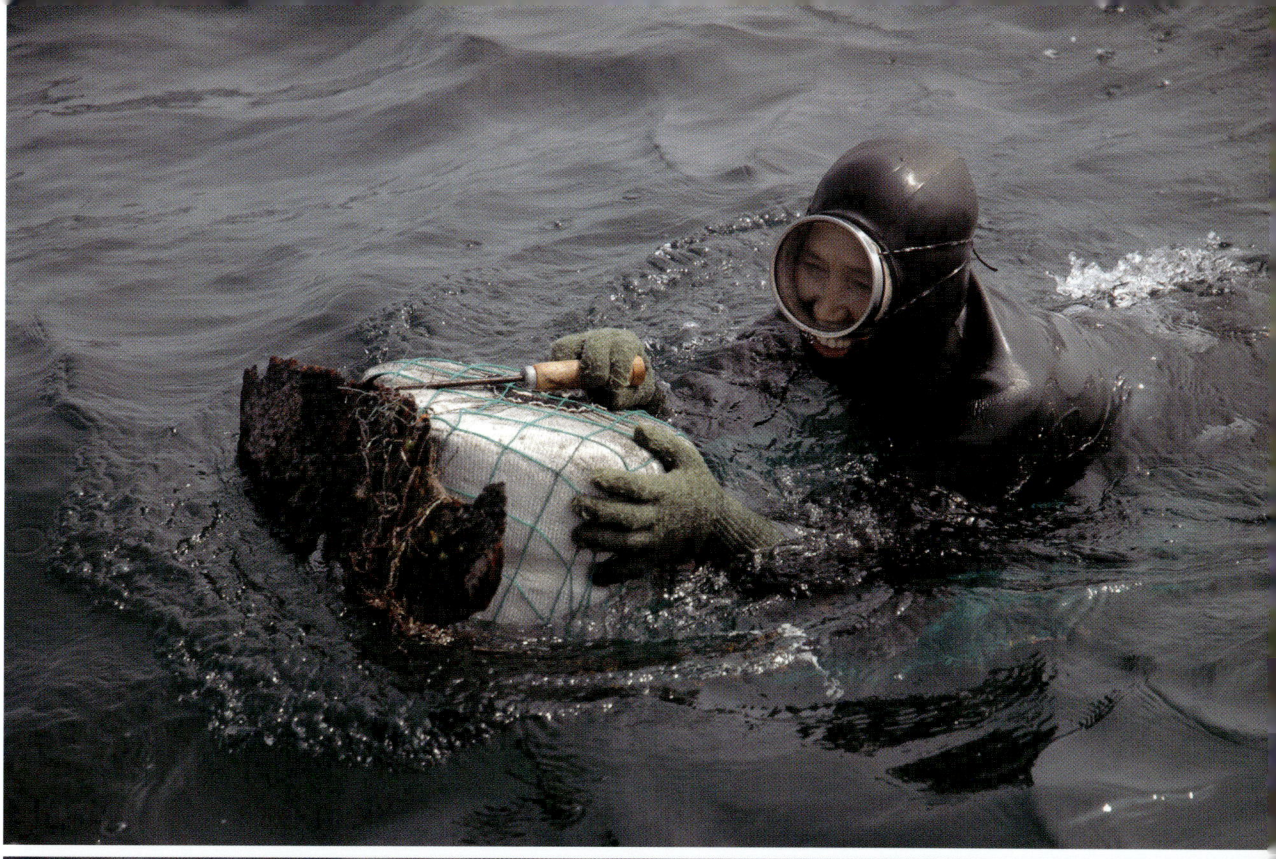

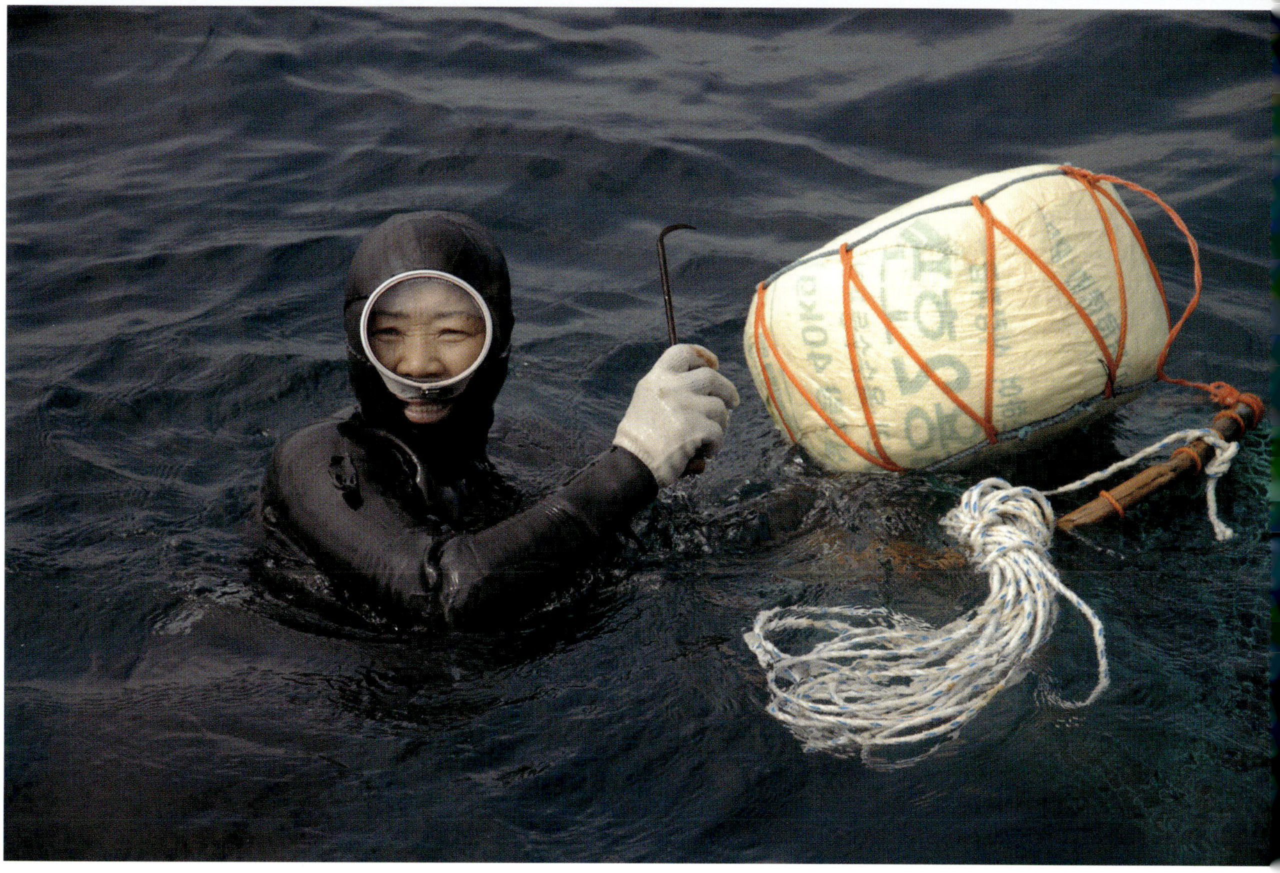

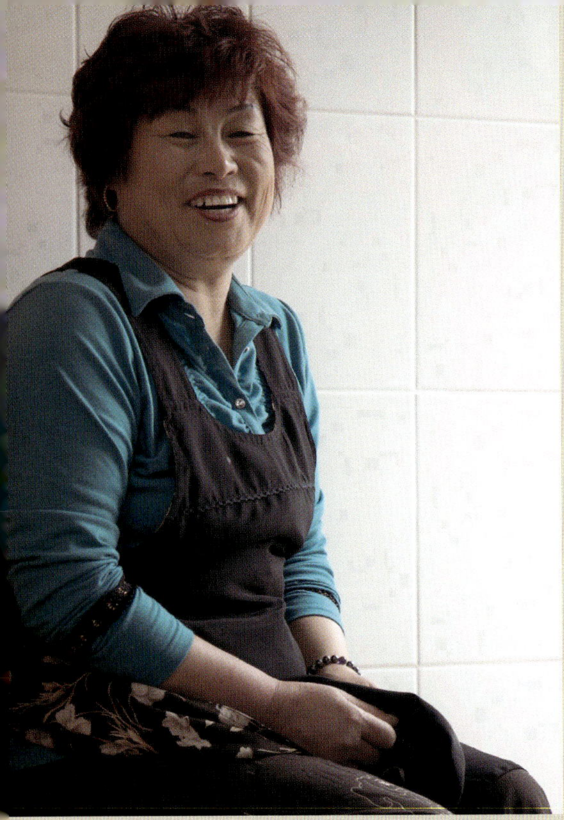

JUNG DONG-SEON

b.1955 I Gimnyeong-ri

These days, people are comfortable with senior facilities. The staff takes good care of the patients. They're given regular baths, and they don't smell. The rooms are clean. What's troubling is when the elderly remain at home but aren't taken care of very well.

The reason families admit their elderly relatives here is because they can't take care of them at home, especially if the relative suffers from dementia. Someone has to be at home all the time, which is difficult because of people's jobs. At this center, there are approximately 50 senior citizens. Most of the women here are former *haenyeo*. There are more elderly people currently on the waiting list. A lot of Jeju villages have their own centers. The one here in Gimnyeong is good, and the people who live here are not only from our village but from others as well. It was built in 2007.

Our group in Gimnyeong doesn't only volunteer to bathe and feed the elderly here. We have also been talking to the credit union about cultivating a nearby field in order to grow plants. I think it's a rewarding experience doing volunteer work at the center.

We have 18 members. There are eight neighborhoods in Gimnyeong. We first help young students who are living with their grandparents. Then we help the elderly in the facility for those with dementia. There are approximately 52 people who are afflicted. Most of them are over 70 years old. These days, it is becoming more common to send the elderly to nursing homes.

Sometimes we cook the seafood we've caught and serve it to the staff and patients. We help with cleaning at the hospital and assist the patients in taking a bath. Many of them can't move freely, so they need our assistance.

Among all the households in our village, we know who needs the most help. Everyone knows everyone. We want to help all those in need, but we don't have enough money. We feel sorry about that. There are many volunteer groups here, but ours is the only one in which we work ourselves and earn money to help others.

(August, 2007)

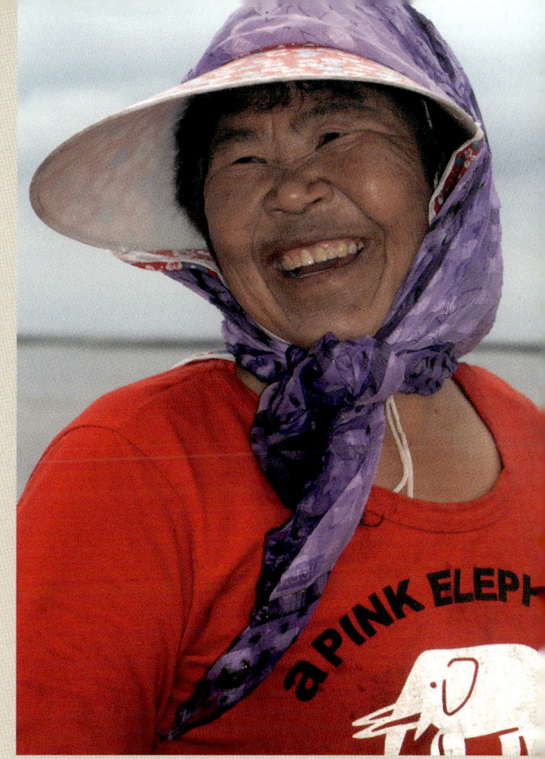

WOO BONG-SEON

b.1944 | Jongdal-ri

My father passed away when I was three years old. My mother became a widow in her 40s. She had five children and often went off to the mainland to make money. We waited and waited for her to return. She didn't do any farming—only diving—because she was from Udo. We were poor and hungry all the time. She wasn't able to make enough money by diving because *haenyeo* could only stay in the water for an hour in their cotton suits. Because my mother was often away, I cooked for my brothers and did all of the housework. They went to school, and I worked on a small farm nearby. I began working on a farm at the age of seven. I stopped attending school in the fourth grade and started learning to be a *haenyeo* at age 13.

(June, 2007)

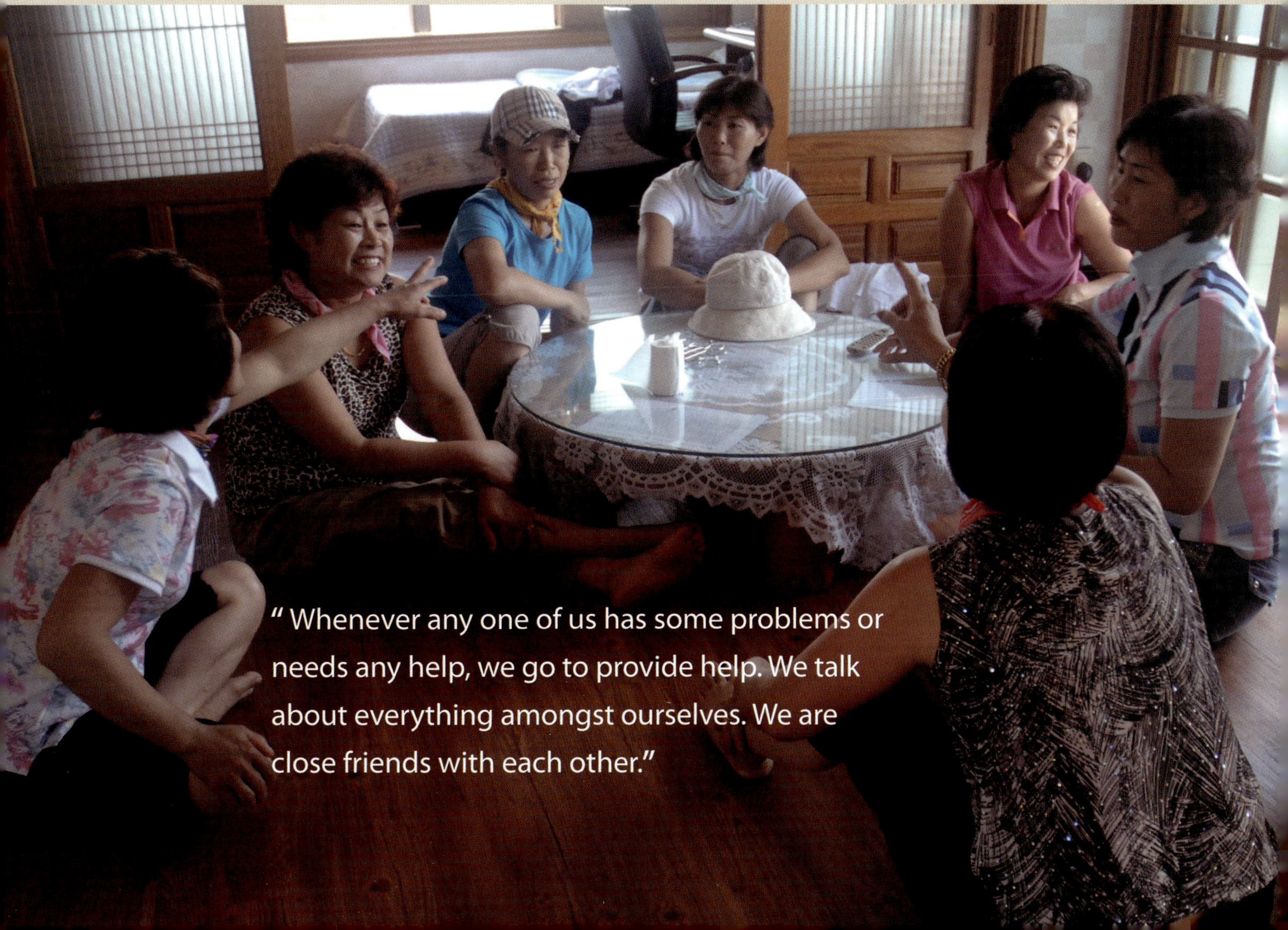

" Whenever any one of us has some problems or needs any help, we go to provide help. We talk about everything amongst ourselves. We are close friends with each other."

The following interview with former *haenyeo* Han Ki-ok, 73, was conducted in April 2010 by her daughter, Kang Mi-kyoung, director of the Hanbit Shelter for Women on Jeju Island. Han Ki-ok volunteers at the center—another example of how the *haenyeo* participate in community activities. She was born in a small, beautiful village with a long history on the beach of Bomok-dong in Seogwipo. A widow, Han Ki-ok has three daughters and one son.

HAN KI-OK

b.1938 I Seogwipo-si

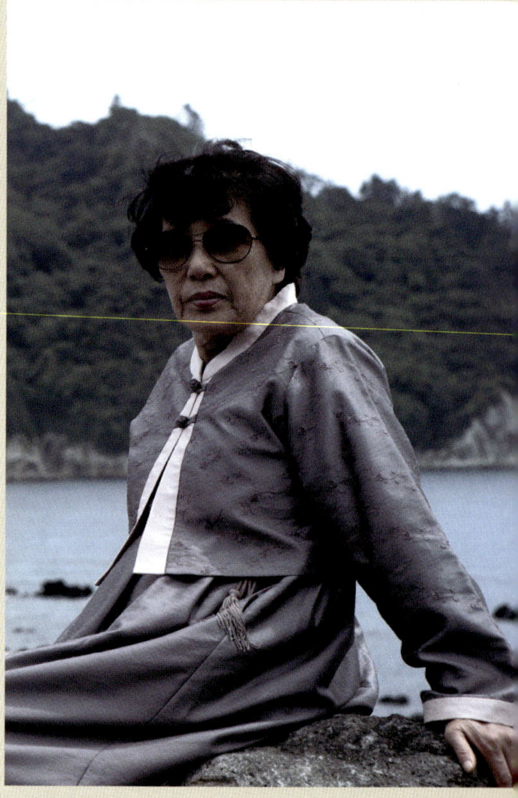

How many generations of* haenyeo *have there been in your family?

My mother, maternal grandmother, and maternal great-grandmother. I can't be certain which generation it is, but from what I can recall there have been three.

***How many years did you work as a* haenyeo?**

From age 16 to 20, I dived off Bomok-dong in Seogwipo and Uljin, North Gyeongsang Province. After marrying, I dived near Jeongbang Falls in Seogwipo from age 20 to 23.

***What did you do after retiring as a* haenyeo?**

After being widowed at age 37, I did whatever work I could in order to clothe, feed, and educate my four children on my own. I worked in a rice shop, sold fish, farmed bananas and mandarin oranges, and worked at a fishery and electronics shop.

What led you to develop an interest in domestic violence against women?

After being widowed at an early age, I had a very difficult time raising my children by myself. I naturally came to develop an interest in the psychological difficulties women confront, including domestic violence and sexual assault, as well as their economic problems and difficulties with their children's education.

What prompted you to assist in the foundation of your daughter's shelter?

Mi-kyoung, my eldest daughter, developed a feminist consciousness as her own married life gave her a sense of the difficulties women encounter. Over the course of 10 years working as a women's counselor for Seogwipo City Hall, she provided counseling to women who suffered from domestic violence and sexual assault, sex workers, and single mothers, and she came to sense a kind of social need for a place that would care for women and allow them to rest, but such a place did not exist in Jeju. I heard that and said, "I've had a tough life as a woman, too, widowed at 37 and having to raise children on my own. I'd like to help out women who are lonely and suffering." So I took my savings, and we used it to establish the social welfare corporation Hanbit in December 1997. In March 1999, we founded the Hanbit Shelter, Jeju's first temporary shelter for women and children victimized by domestic violence. I served as managing director of Hanbit Foundation from December 1997 to December 2006.

When did domestic violence surface as an issue in Jeju?

Domestic violence has been present for a long time within the patriarchal culture, but it was tacitly accepted by society and never revealed to the outside. Then, with the enactment of the Special Law on the Prevention of Domestic Violence in 1999, the problem started surfacing as an issue in Jeju as well.

What were the founding goals of Hanbit Shelter?

Our goal is to make a world for those who have lost hope, those dreaming of true freedom, where all lives can enjoy peace and freedom and share love. We provide services like counseling and psychological therapy for women and children victimized by domestic violence; legal and medical referrals; job training and placement; educational support for children; establishing a foundation for independence.

How many women have stayed at the shelter to date?

1,218 people between 1999 to 2009 (10 years); 679 mothers and 539 children, including pre-elementary children and elementary, middle, and high school students. We have a maximum capacity of 30 people; 20 people are currently staying here.

What is the age distribution of women coming to the shelter?

A wide range, from women in their 20s to those in their 70s; most are in their 30s and 40s.

What are the causes and influences behind domestic violence in Jeju?

As isolated phenomena, we see acts of violence resulting from alcoholism, pathological jealousy, extramarital affairs, gambling, and Internet addiction. But we cannot overlook the influences of a patriarchal culture.

Is it true that Jeju has the highest divorce rate in Korea? Is there any correlation between divorce and domestic violence?

Jeju's divorce rate is second after Incheon, and there is research showing that the divorce rate among Jeju women is high because they have strong earning capacities and independence, and thus do not depend on men. There is a strong correlation between domestic violence and divorce.

Even though they play a substantial role in the home economy, it seems like* haenyeo *may still accept the Confucian patriarchal system within the home. Is that true? Are there cases of older* haenyeo *who have experienced domestic violence?

It appears that the acceptance of the Confucian patriarchal system may stem from the advanced age of the *haenyeo* population (average age of 70 years). Even so, no *haenyeo* has ever come to the Hanbit Shelter for counseling or to stay.

Are there other similar shelters in Jeju besides the Hanbit Shelter?

In Jeju-si, there are two shelters for women victimized by domestic violence; four locations for women victimized by prostitution; one shelter for women who have suffered sexual assault; and 10 counseling sites for women victimized by domestic violence, sexual assault, and prostitution.

What has the reaction to the shelter's programs and services been like from the men and husbands in Jeju who are engaged in domestic violence, and from their family members? Has it been favorable or hostile? Have there been complaints?

We do counseling and education for husbands and couples counseling for the men who have assaulted the women in the shelter, as well as for ordinary husbands, when the wife and husband desire counseling. Everyone has expressed gratitude, and some of the husbands even become sponsors and volunteers. When we first started the shelter, the men were hostile about the fact that there was a shelter and the women were leaving home. Some called up the employees and made threats. It is now possible for wives and their children to leave for the mainland, where they will never be found, but I have heard that they are grateful to have a shelter instead, a trustworthy place that will look after the wives and children and keep them safe.

They are learning that they can be punished by the police for domestic violence, their wife can file a divorce suit, they can have their assets partitioned or lose custody of their children, and they may have to pay alimony and child support. So the views of the men in Jeju are gradually shifting.

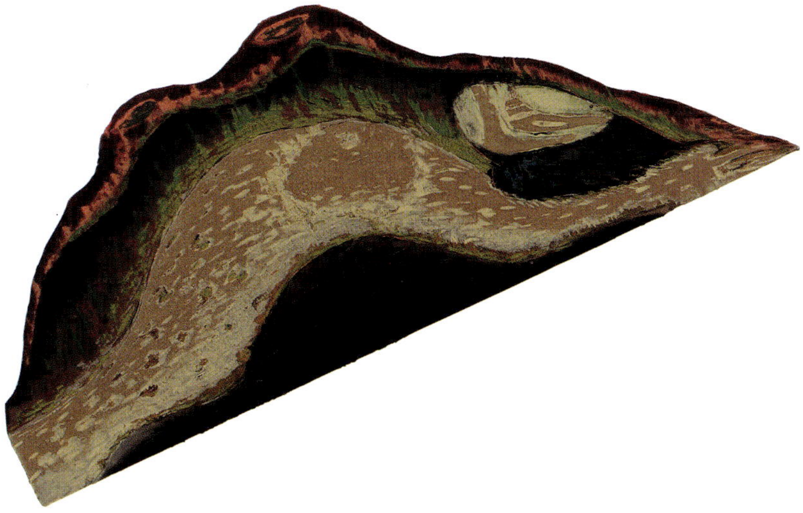

Mme. Okjo Transforms into Prison Guardian "Okhalmang"
53x46 cm | paint on backing paper | 2008

Sanbangdeok

(Goddess of Love)

Inside a cave at Mt. Sanbang, water drops from the ceiling, and the locals believe it represents the tears of the goddess Sanbangdeok, who guards the mountain. Legend has it that the beautiful daughter of Mt. Sanbang was in love with a young man named Goseong. However, a jealous official in town, who also had a crush on her, confiscated Goseong's property and cast him into exile. In despair, Sanbangdeok returned to the cave and turned herself into a rock. Visitors to the rock can hear her teardrops falling as they view the ocean from inside the grotto.

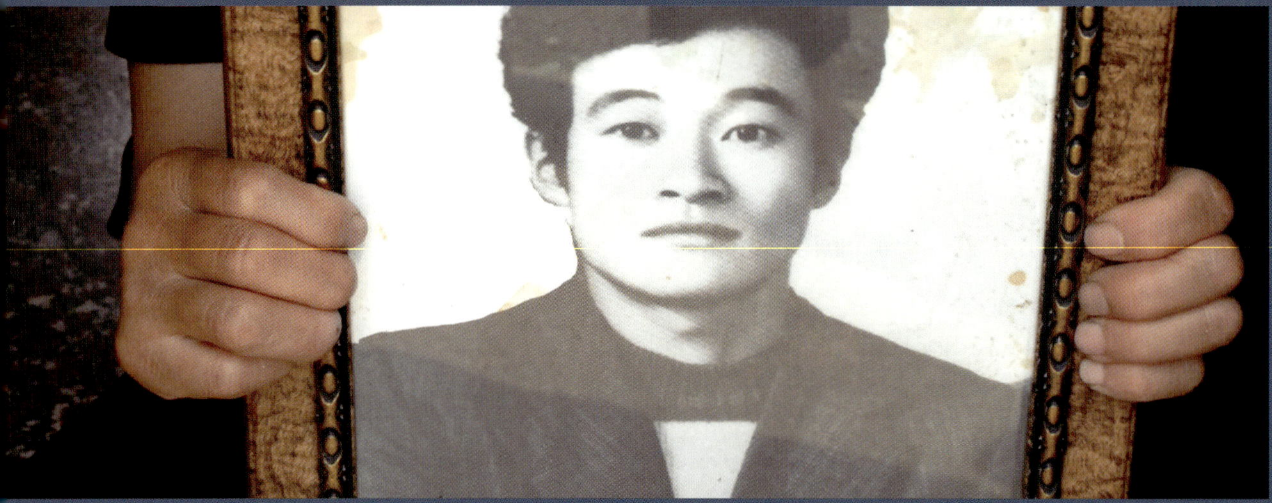

Chapter Six

LIFTED BY THE WIND & TIDES OF FAMILY

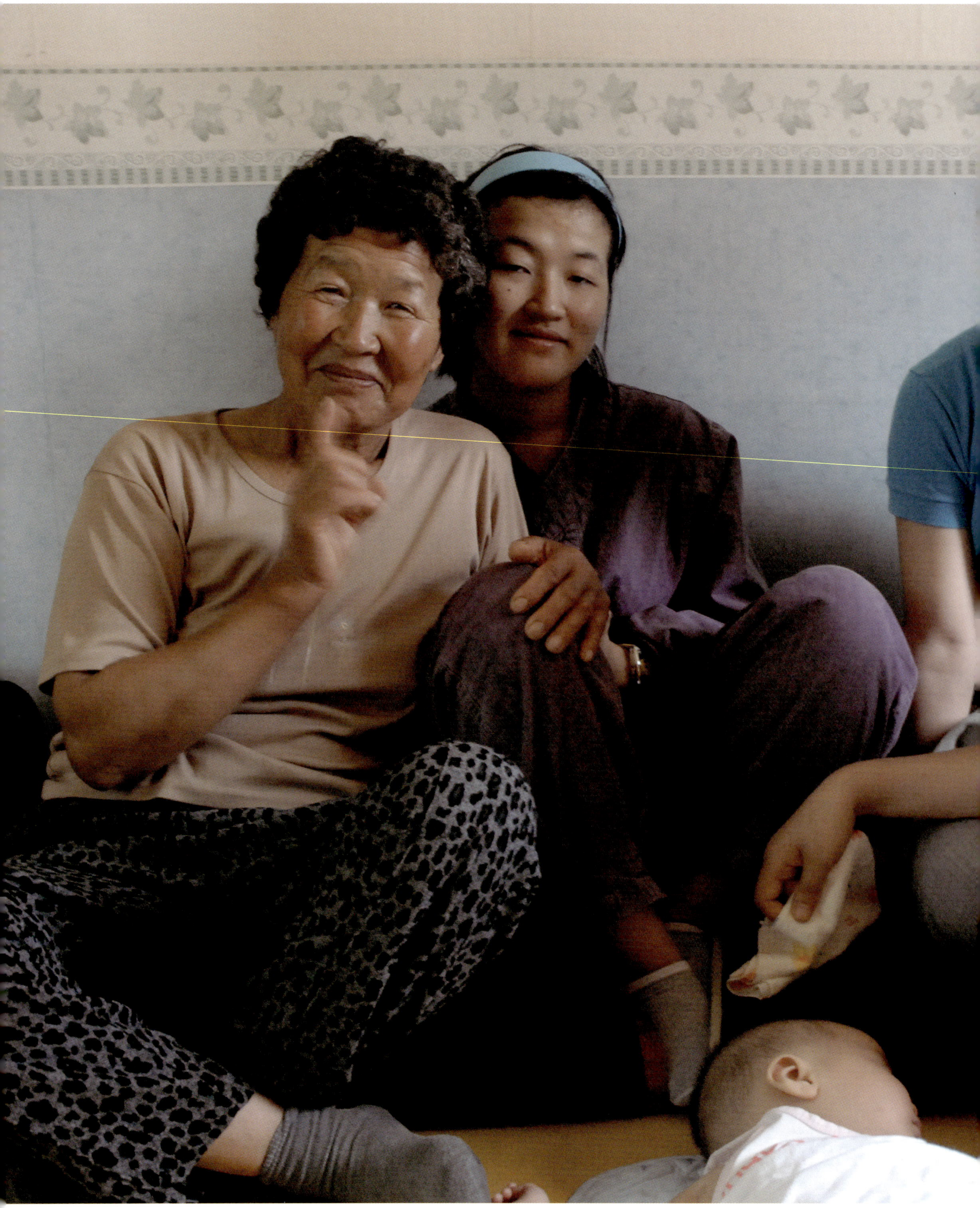

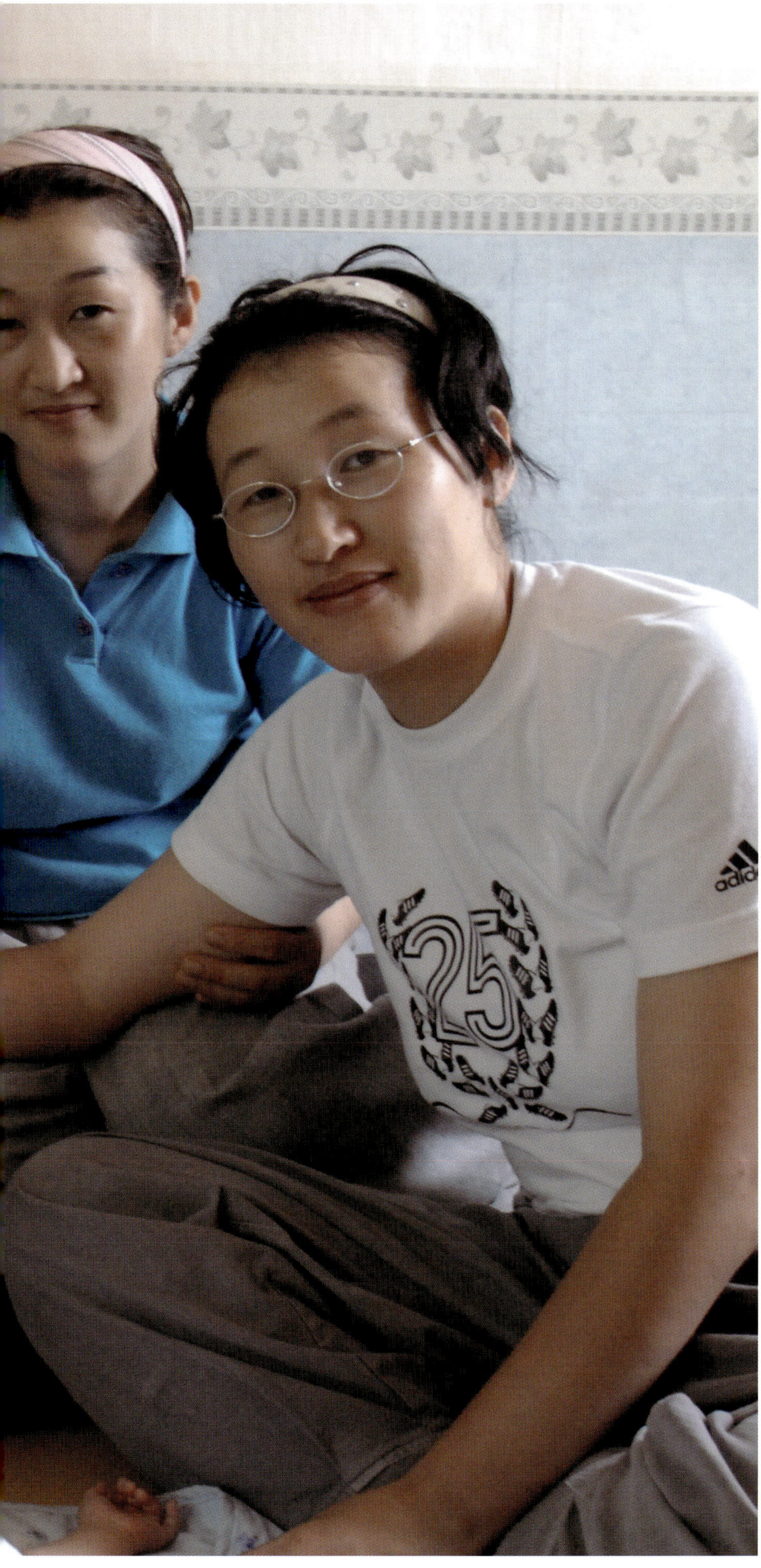

I once came across a quote that I unfortunately can only paraphrase: "When faced with suffering, respond with simplicity." It resonated deeply with me, especially after meeting Woo Sun-deok, the *haenyeo* interviewed in this section. A survivor of numerous sufferings, she conducts her life by the sea with the utmost simplicity and discipline. In between diving and farming, she mindfully walks more than five kilometers to Dongyeonsa, her nearby Buddhist temple. She encouraged me to bow to Buddha 108 times daily with a beaded mala. "This will cure every type of suffering," she said. Buddhists, she explained, believe that human beings experience 108 types of anguish in their lifetime. By prostrating oneself, an individual's mental anguish can be freed.

We met by chance on the road in Jongdal-ri. My friend Park Hyun-mi was actually driving me to interview another *haenyeo* whose only directions were "My house is near the tree." After circling around the village—lost—we encountered Woo Sun-deok walking along the road. "Excuse me," we asked, "can you help us find the home of Woo Bong-seon?" "Sure, I'll take you

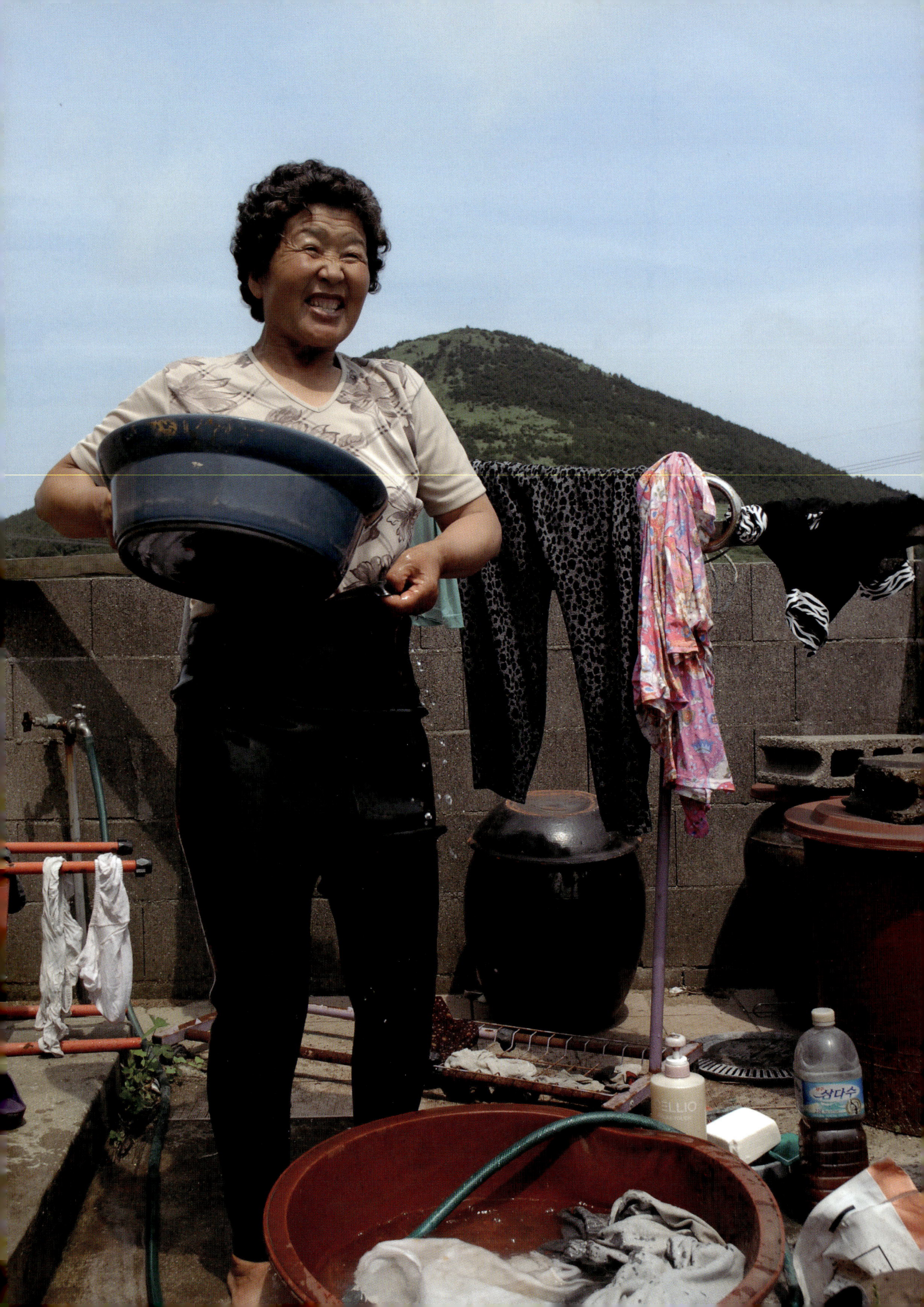
삼다수

there," she said, while getting in the car. We explained our mission, and why I had come to Jeju Island. Within a few minutes, she smiled and blurted out, "I think we were twins in a previous lifetime!" I was surprised by her unexpected intimacy, but flattered that we had made such a quick emotional connection as earthlings.

A week later, she invited me and Park Hyun-mi to attend her family's *jesa*. The ceremony was held at her local temple in memory of her husband, her eight-year-old son who drowned, two babies she had miscarried, and other ancestors. As three of her four daughters prepared food for the altar, she told me that she had cried in anticipation of my arrival. It baffled her because she didn't know whether they were tears of joy or sorrow. When I told her that I, too, had survived the death of a child, she seemed even more convinced that we were related in a previous lifetime. I was very touched to be included in her family's private commemoration, and to participate in the bowing ceremonies—albeit falling short of 108 prostrations.

My ongoing encounters with Woo Sun-deok were as serendipitous as the day we met. I could never reach her successfully by phone, or count on her keeping a pre-set appointment. After all, one lives according to the tides. But once I accepted this law of nature, I always found her by relying on my senses. One time, I listened for the *sumbisori*. With my Nikon telephoto lens, I once spotted Woo Sun-deok by her mismatched fins, one blue and one black. Another time, I found her gathering seaweed by the shore just as I happened to be driving by. Youngsook Han and I offered to come back the following day to help gather more seaweed. "Tomorrow? That's too late! Help me gather it now," she insisted. I couldn't refuse my twin.

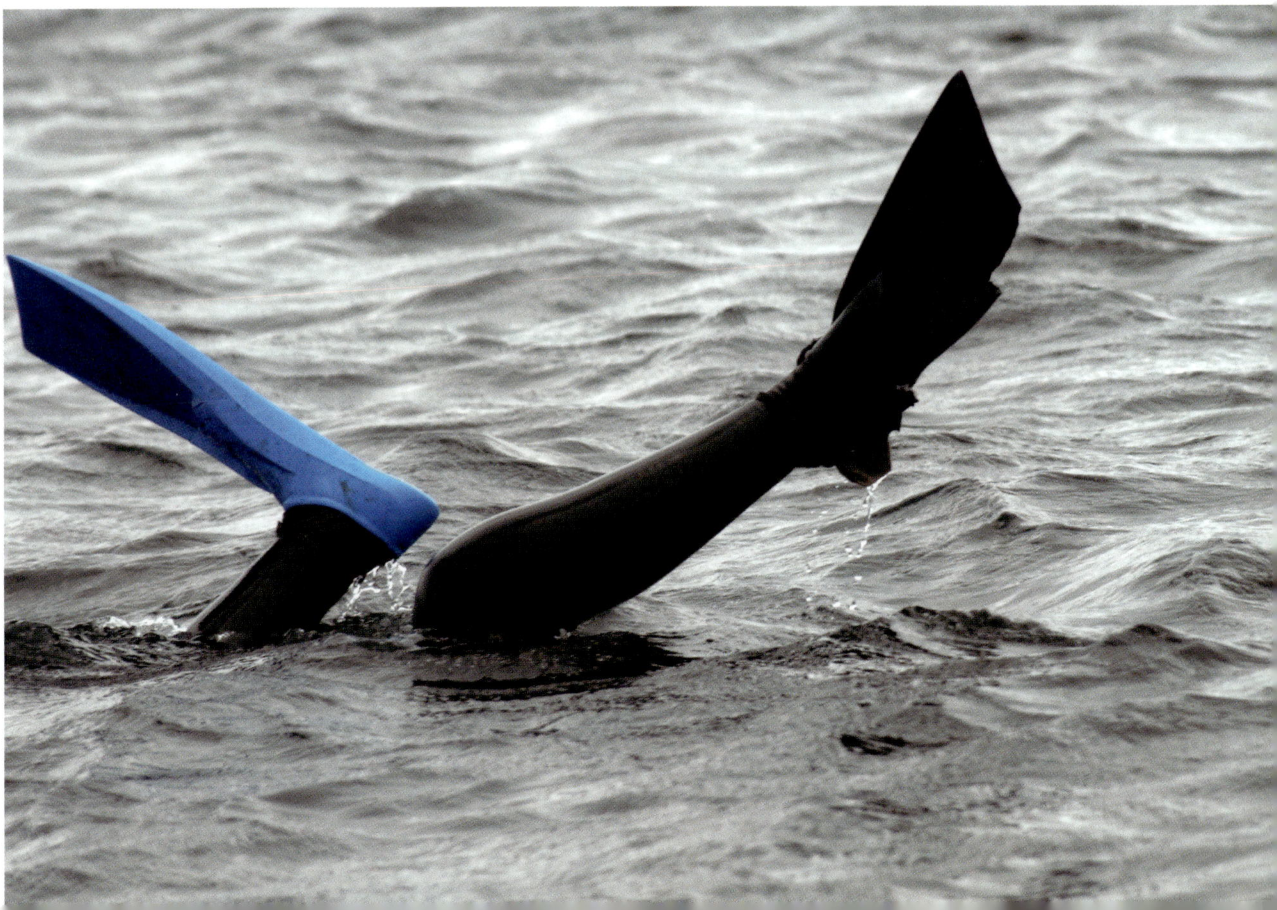

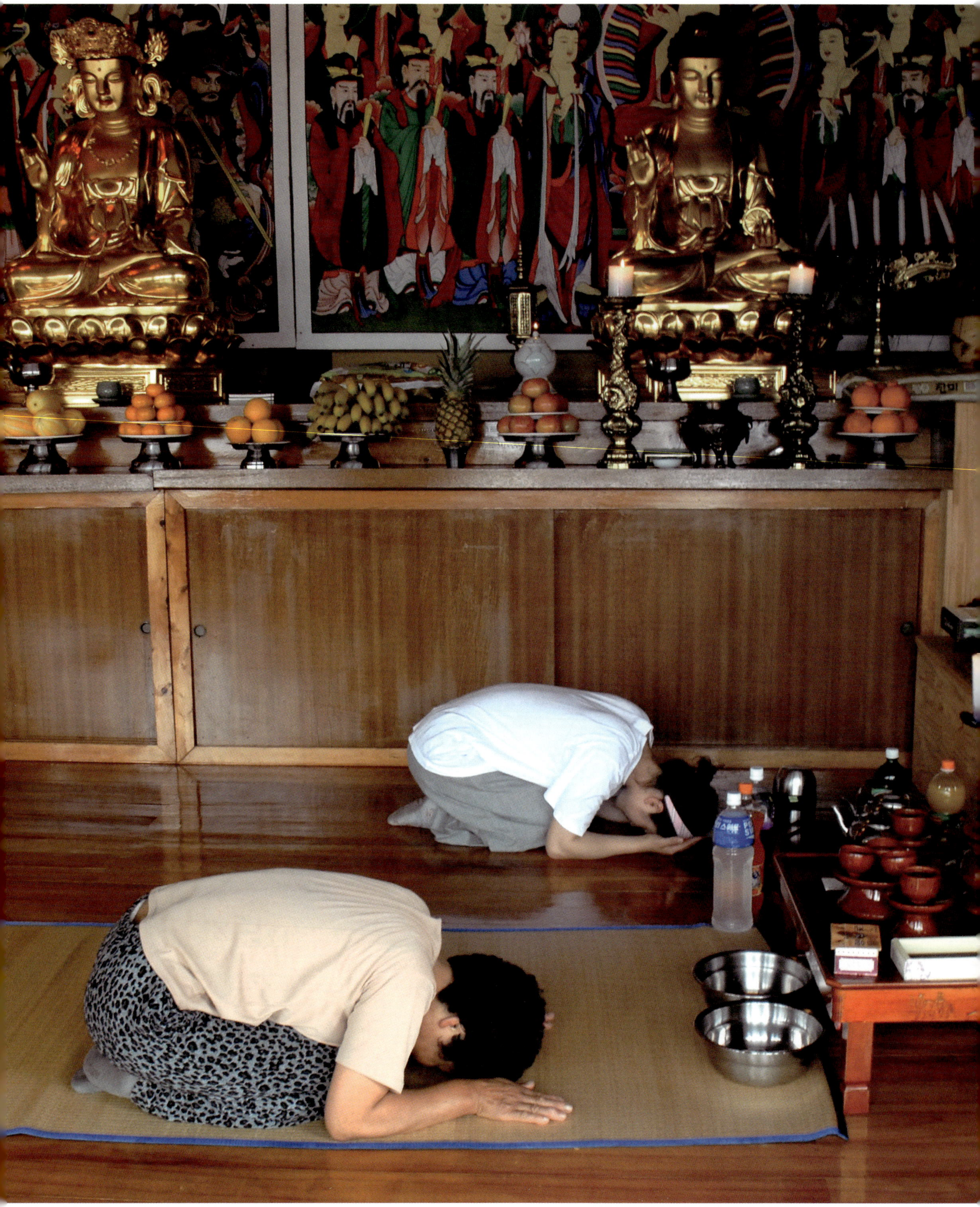

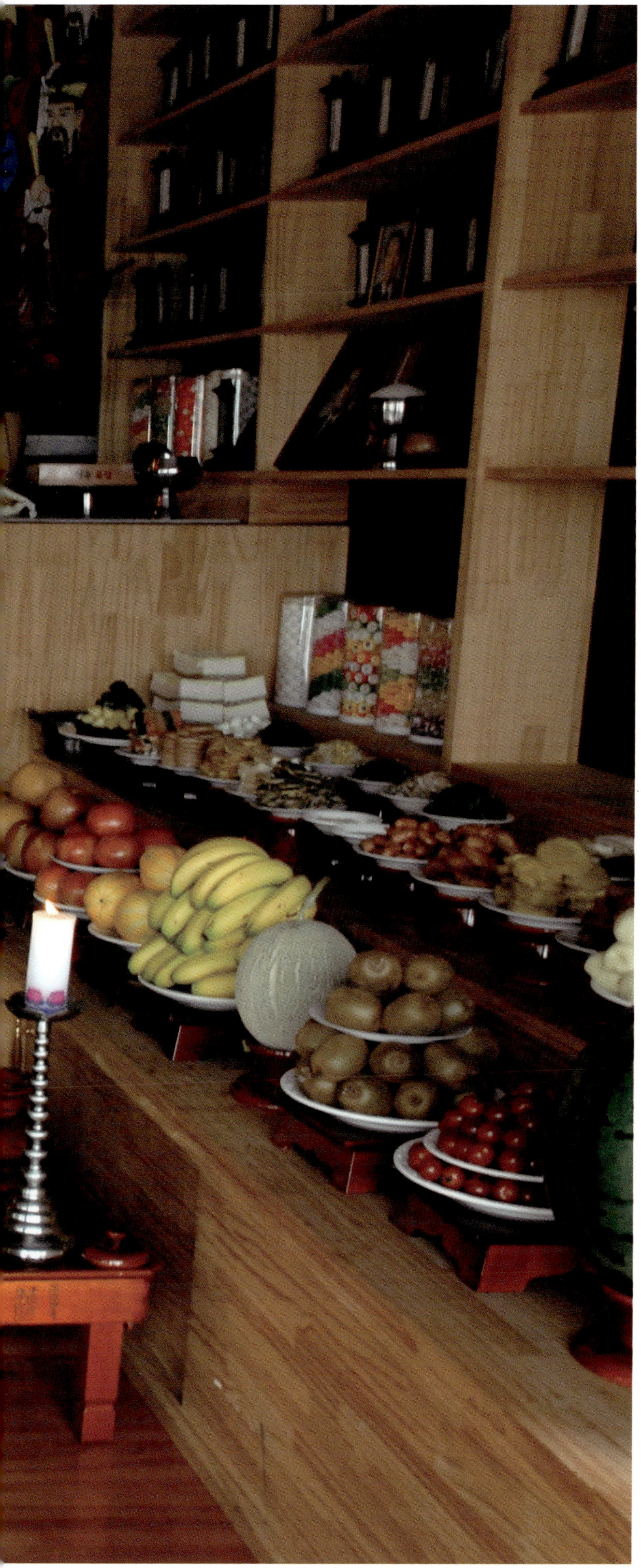

Jesa

The *jesa* is a memorial ceremony for a Korean family's ancestors. It is usually held on the anniversary of an ancestor's death. For the rituals, the wife of the eldest son usually prepares a variety of foods that are offered at the altar and then shared among the guests or neighbors. Bowing in respect is a common practice during *jesa*.

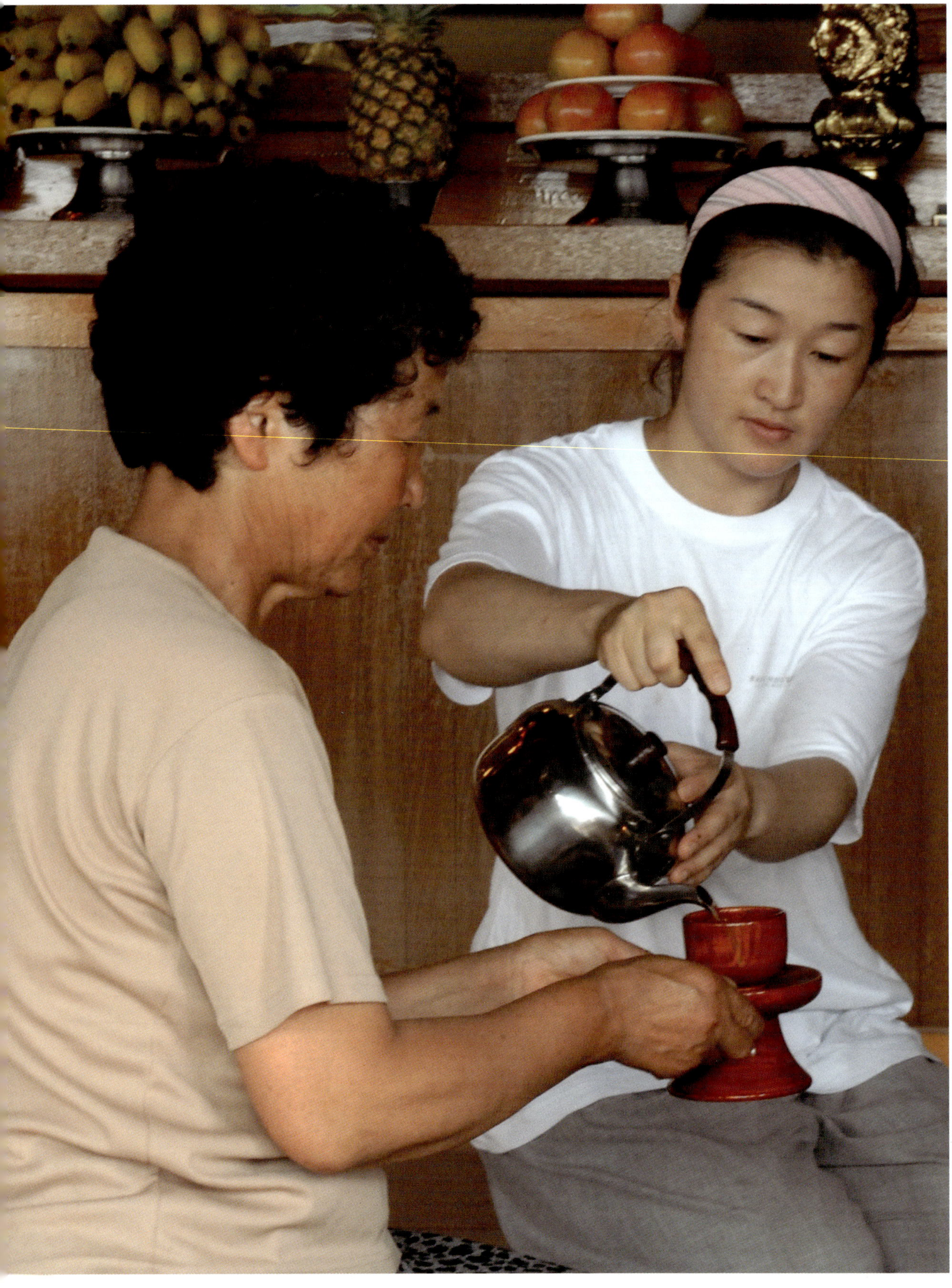

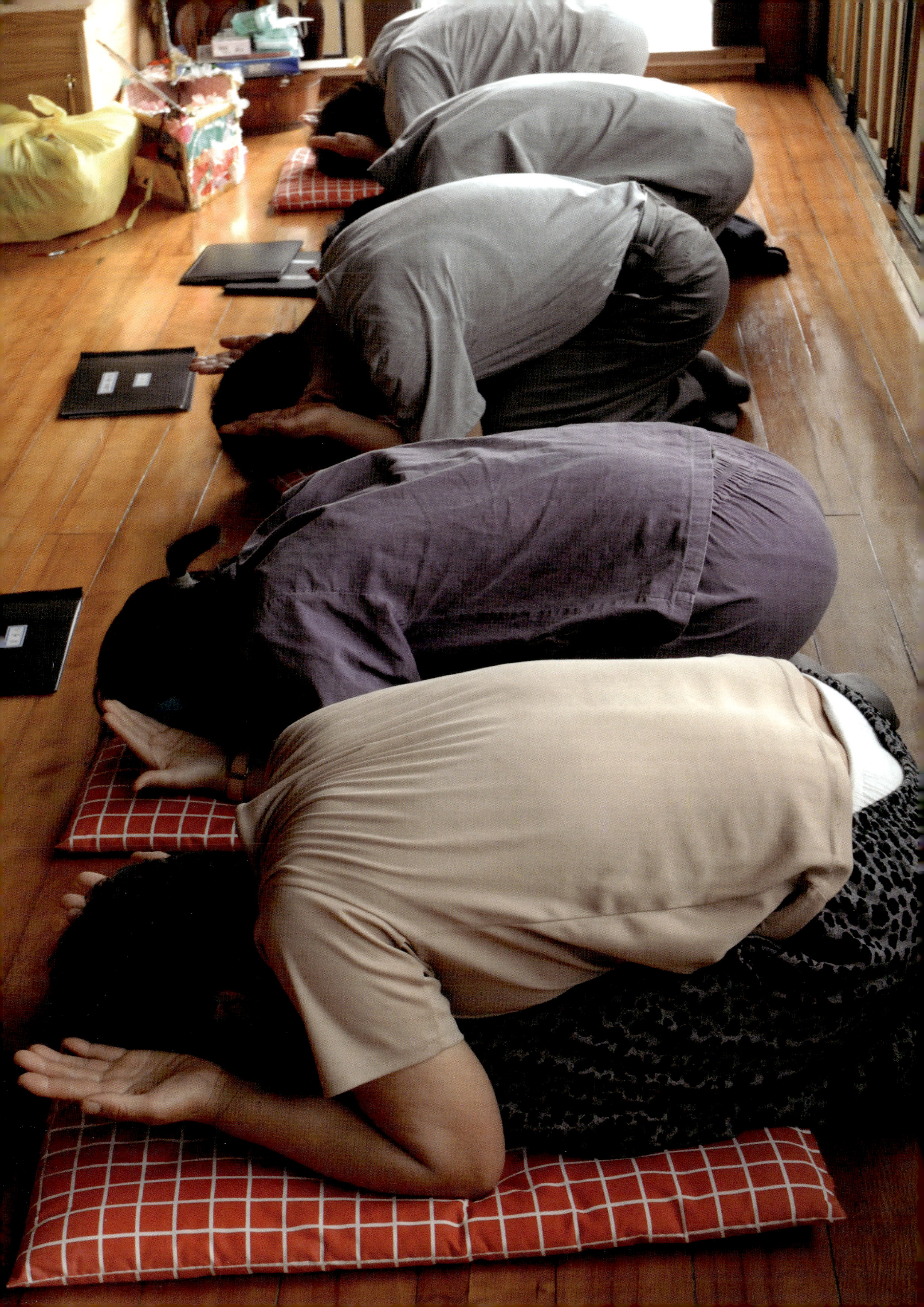

WOO SUN-DEOK

b.1953 | Jongdal-ri

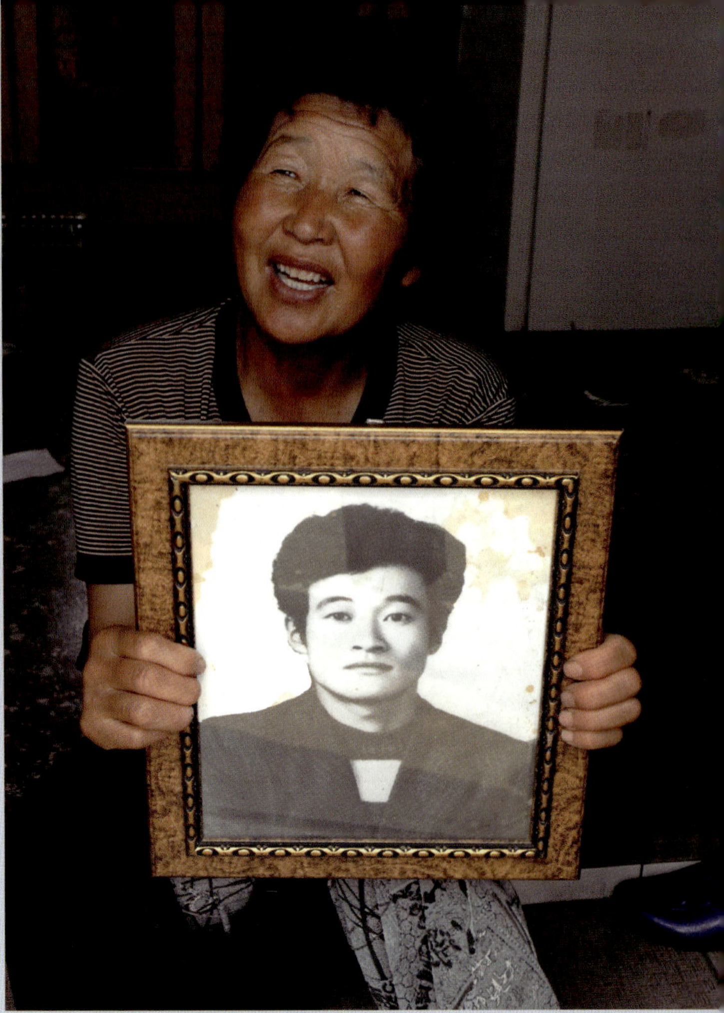

I never went to school, but I have gained wisdom from my life experiences. I learned how to read and write the Korean alphabet by myself. But after breathing carbon monoxide from the heating coals under the floor, I couldn't remember things well. Because I didn't attend school, I tried to let my children go and study, even though we had so many difficulties.

A fortuneteller once told me that I wouldn't live past the age of 31. But actually my husband died in his 30s. My youngest daughter had been crying, and he was angry and drunk. He mistakenly drank from a bottle of insecticide that he thought was *soju* (distilled alcohol from rice and other grains). I pray every day that he may be happy and peaceful in heaven.

We met when I was 26. My father forced me to marry him. I had to obey. I guess I liked him. He was very handsome and earned money by farming and fishing.

Every day I bow to Buddha 108 times in front of a small statue placed in my house. I began chanting prayers to Amitabha Buddha when I was six years old. I was born

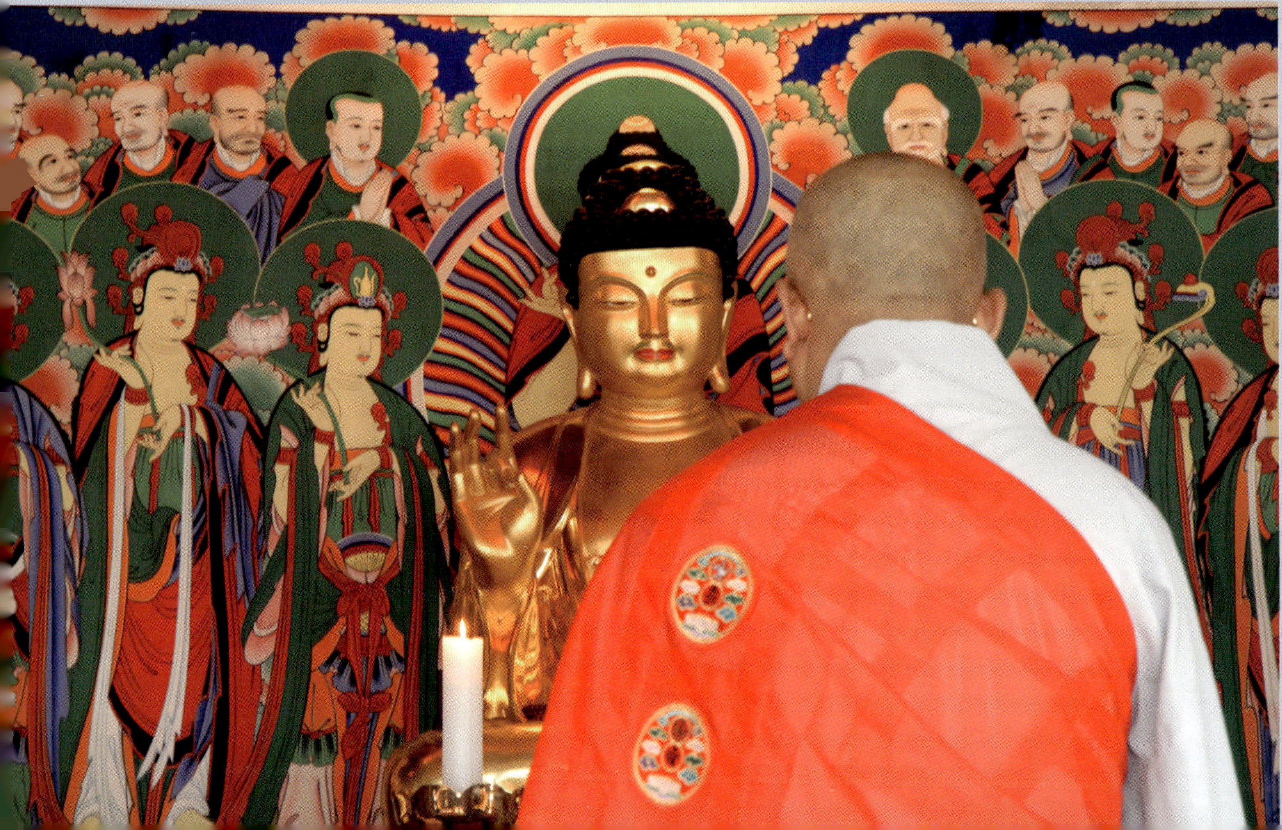

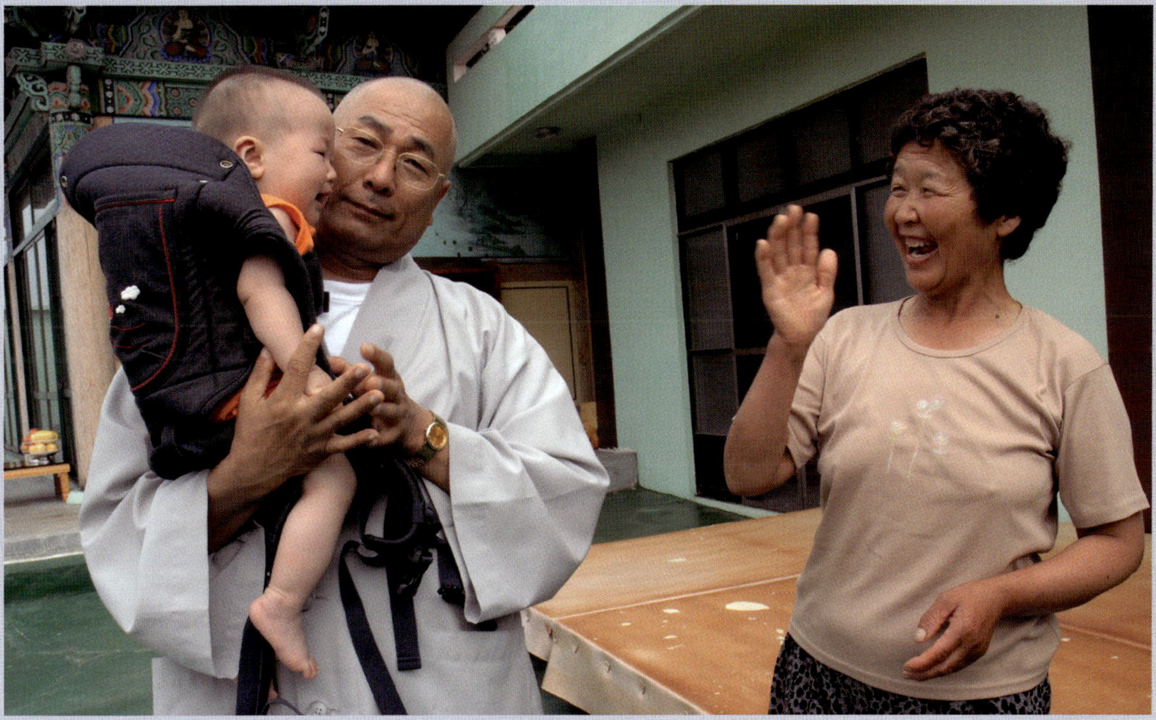

prematurely at seven months. Everybody thought I wouldn't survive, but I did. I also survived when I was once run over by a horse-driven carriage and hurt my head. I guess I fell asleep by the road. I have a big scar on my head. Several times, I nearly drowned when I was a teenager. I thank Buddha for surviving.

It wasn't until I became an adult that I really began to understand the true meaning of Buddhism. I used to believe in shamanism, but now I don't, although I still attend the shamanist *gut* in my village. I learned more about Buddhism by watching TV. I always try to be good-hearted. That way my children will be healthy and safe. I have four daughters and pray all the time for their safety and happiness. My son Yeong-ho died when he was eight years old. If he were alive now, he would be 32.

Whenever I observe *jesa*, I also include the food and objects that my son used to like. We have to comfort the spirits of the young ones, too. My husband and son are both buried at a graveyard nearby, but I don't go there. Why do I have to go there when I can meet them here at my nearby temple?

I don't preach Buddhism to others because some people think I'm already too much involved. For example, I always walk to the temple and help keep the grounds and altar clean. These days, I practice writing seven-letter phrases from the Buddhist scriptures. I do this in memory of my three other children, two of whom died before they were born. I also copy the scriptures to remove any bad karma. This practice is good for one's physical and mental health. I believe a bodhisattva will keep you safe and healthy. There isn't much time left for me. I am preparing for my next life.

Diving

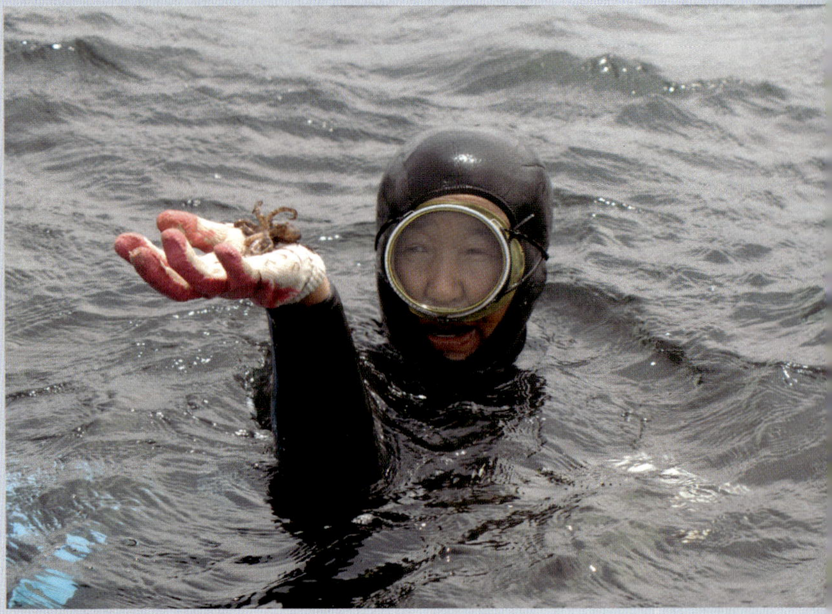

I went diving on the mainland when I was 16. Before that, I just went to the sea. Both my mother and grandmother were *haenyeo*, but I lived with my grandmother because there were seven children in our family. My younger brother died in the sea. We never recovered his body.

When I was young, my grandfather—who had two wives—wouldn't allow me to go to the sea. He wanted me to babysit my cousins. I remember rounding up cattle, carrying water from the well, weeding, and bringing lunch to him. Sometimes he beat me. When I was 10, I asked my grandmother why she used a cane when she walked. She said the cane is better than children and an uncle. Now I understand why she said that, because I can always rely on it.

In spite of my grandfather's restrictions, I went to the sea anyway and taught myself how to swim. The most dangerous experience I had was when a whale was swimming underneath me. Fortunately, I've never encountered a shark. I used to be able to remain underwater for three minutes. Now I can stay there for about one minute.

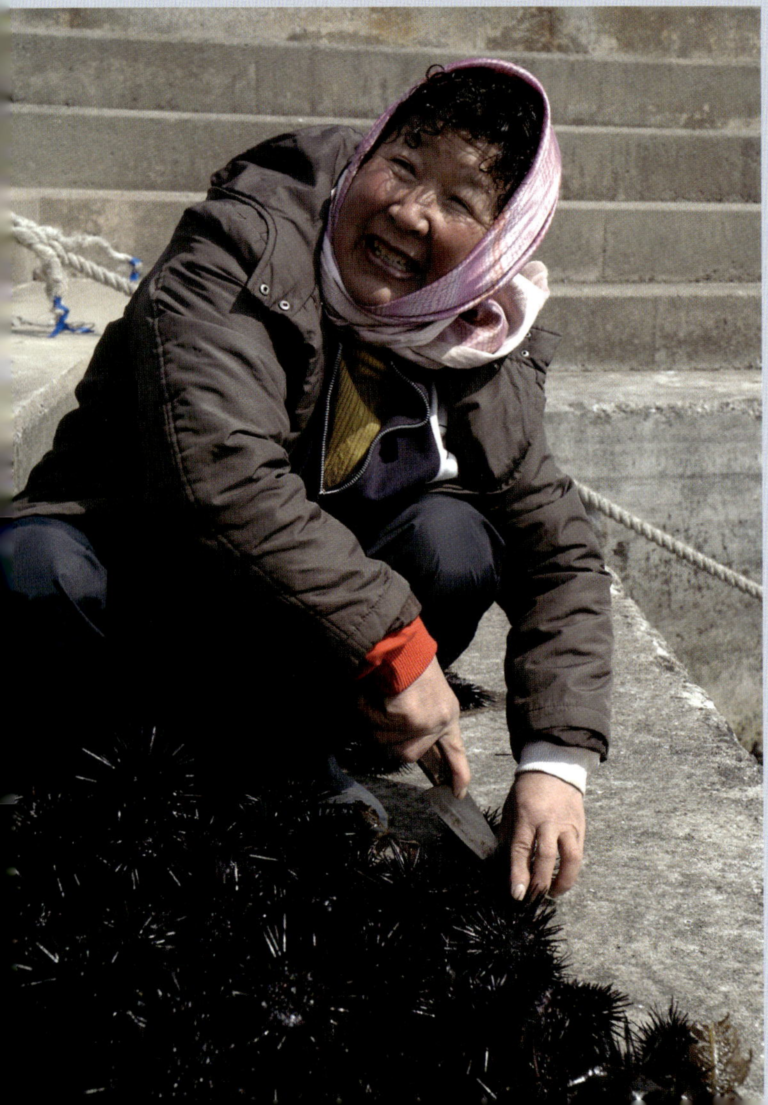

These days, I am careful not to be too greedy because there are fewer sea products than five years ago. Before, we could still catch a lot. I could even send my children to school based on that amount. But now we only collect the things that reach the shore by waves and wind.

(June, 2007)

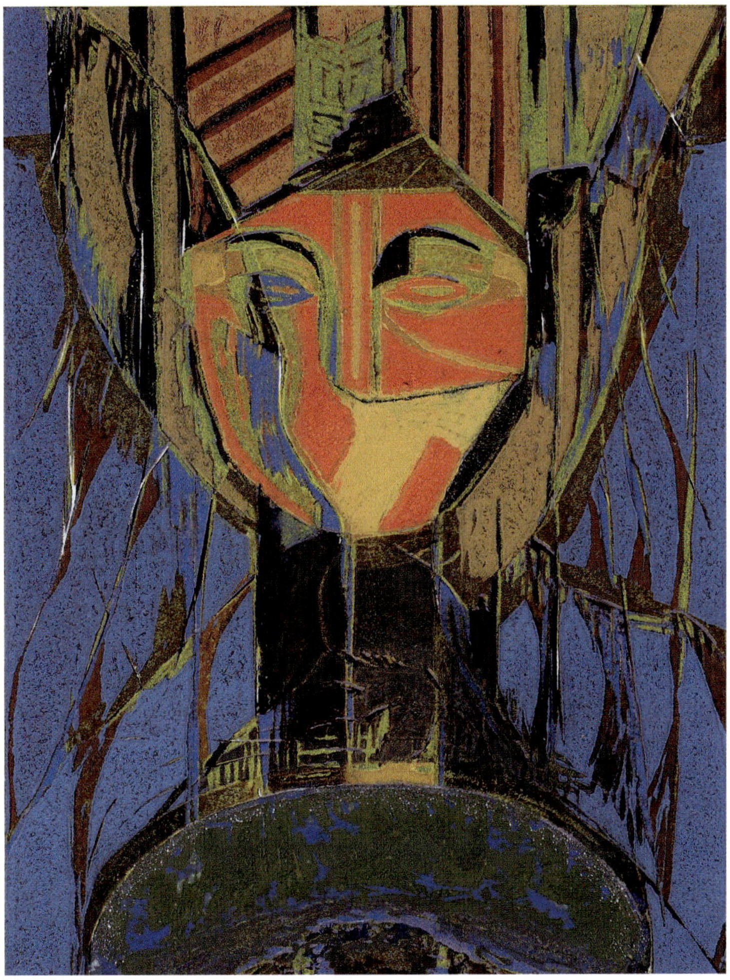

Halmanim 30x45 cm | woodcut | 2008

Samshin Halmang

(Goddess of Birth)

When a newborn is three days old, Korean believers set up a table on one side of the room for Samshin Halmang to pray for the longevity and health of a baby. When women become pregnant, they pray for themselves and their baby to also be protected from harm's way. Very often, images of Samshin Halmang show her holding a camellia branch at the prayer ceremony for a baby.

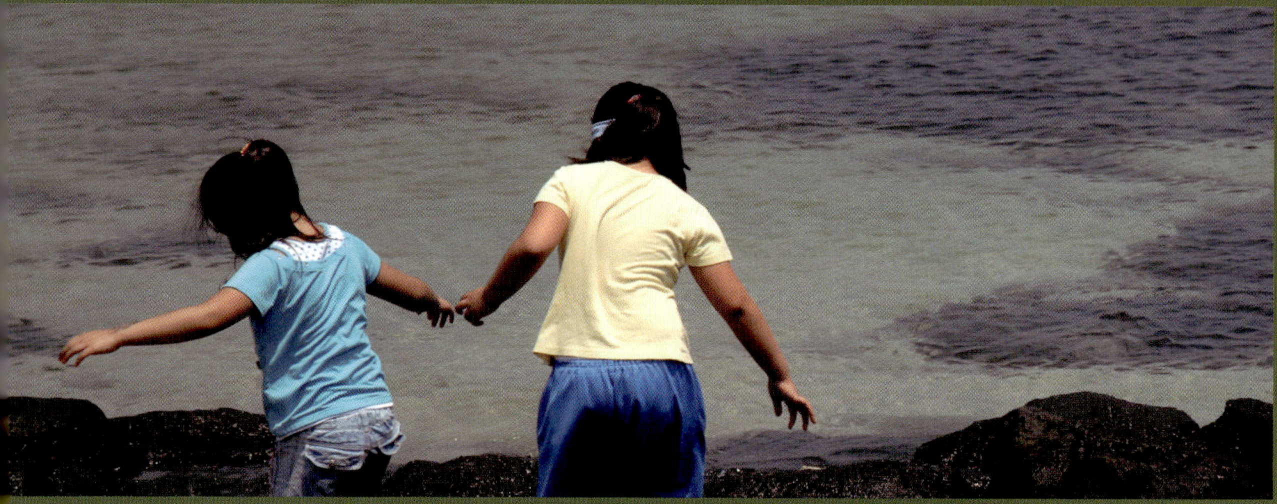

Chapter Seven

LIFTED BY THE WIND & TIDES OF the FUTURE

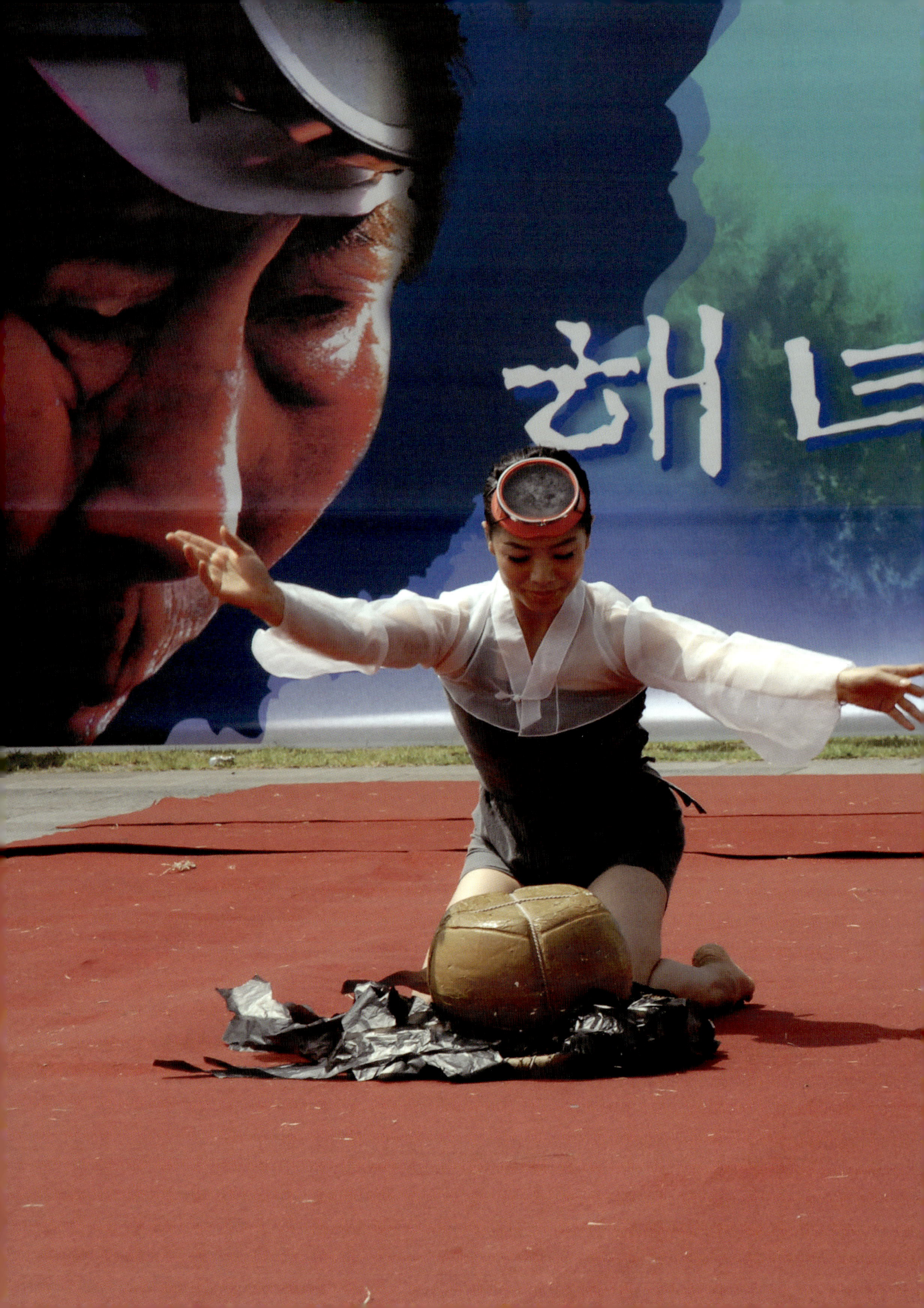
해

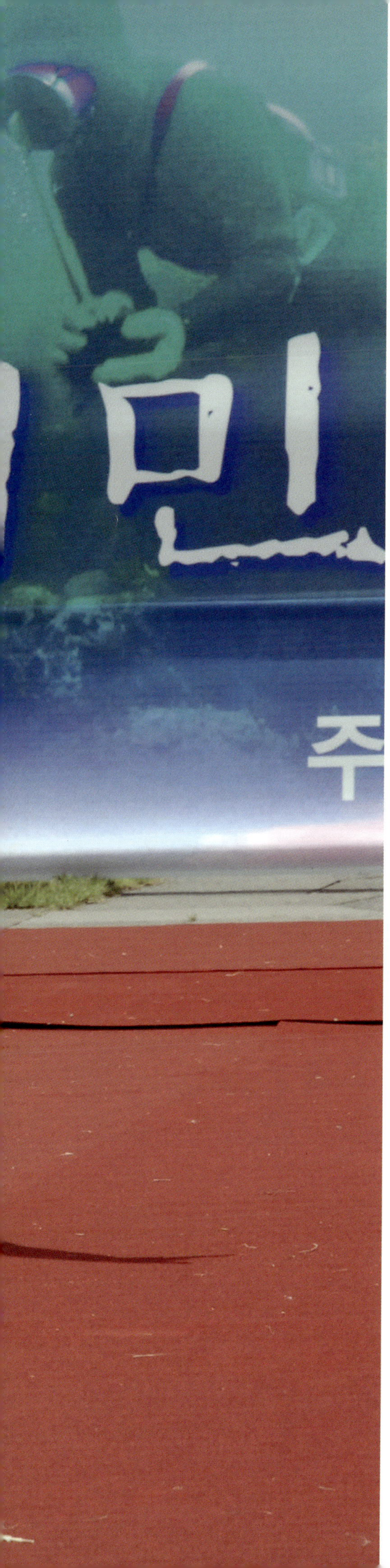

When visitors walk through the Haenyeo Museum in Hado-ri, they inevitably pass a wall displaying traditional and revealing Korean sayings:

"Would rather be born a cow than be born a woman";
"Never lay down babies and miyeok, however heavy they may be";
"Haenyeo work in the sea three days after childbirth."

These words remind visitors of the social and cultural context in which thousands of *haenyeo* have worked, suffered, and survived for centuries, says Choa Hye-gyung, museum researcher and *haenyeo* specialist. Today, the museum attracts many tourists, locals, and schoolchildren. The building—shaped like a volcanic cone and stone tower—was established in 2006 to honor the *haenyeo*, and to educate others about the island's fishermen and divers' history.

During my three visits to Jeju, Choa invited me to join her three-person research team that visited *haenyeo* village cooperatives in order to gather demographic, cultural, and social data. Their findings were later published in the Jemin Daily News, a Jeju-based newspaper.

Although the *haenyeo*'s legacy has been promoted by larger entities such as the museum, media, and the UNESCO, individuals have launched creative initiatives as well. In Gwideok-ri, for example, Im Myeong-ho established a Haenyeo School. "I wanted the young people to remember how the older generation of Jeju women lived. It's for their education, not job training," said Im, whose mother is a diver. The three-month weekly course runs from May to August. Once a year, locals attend a summer festival that includes diving demonstrations with and for the youth, *haenyeo* competitions, singing, dancing, interactive activities, and plenty of food.

Sherrin Hibbard, an Australian English language teacher, is one of the few foreigners to have completed the Haenyeo School's diving course. My friend Ahn Hye-kyoung introduced me to Sherrin, who lived in Jeju from 2004 to January 2011. A former commercial fisherwoman and boat builder, Sherrin shared my passion about *haenyeo* life. One day, while we were walking by the

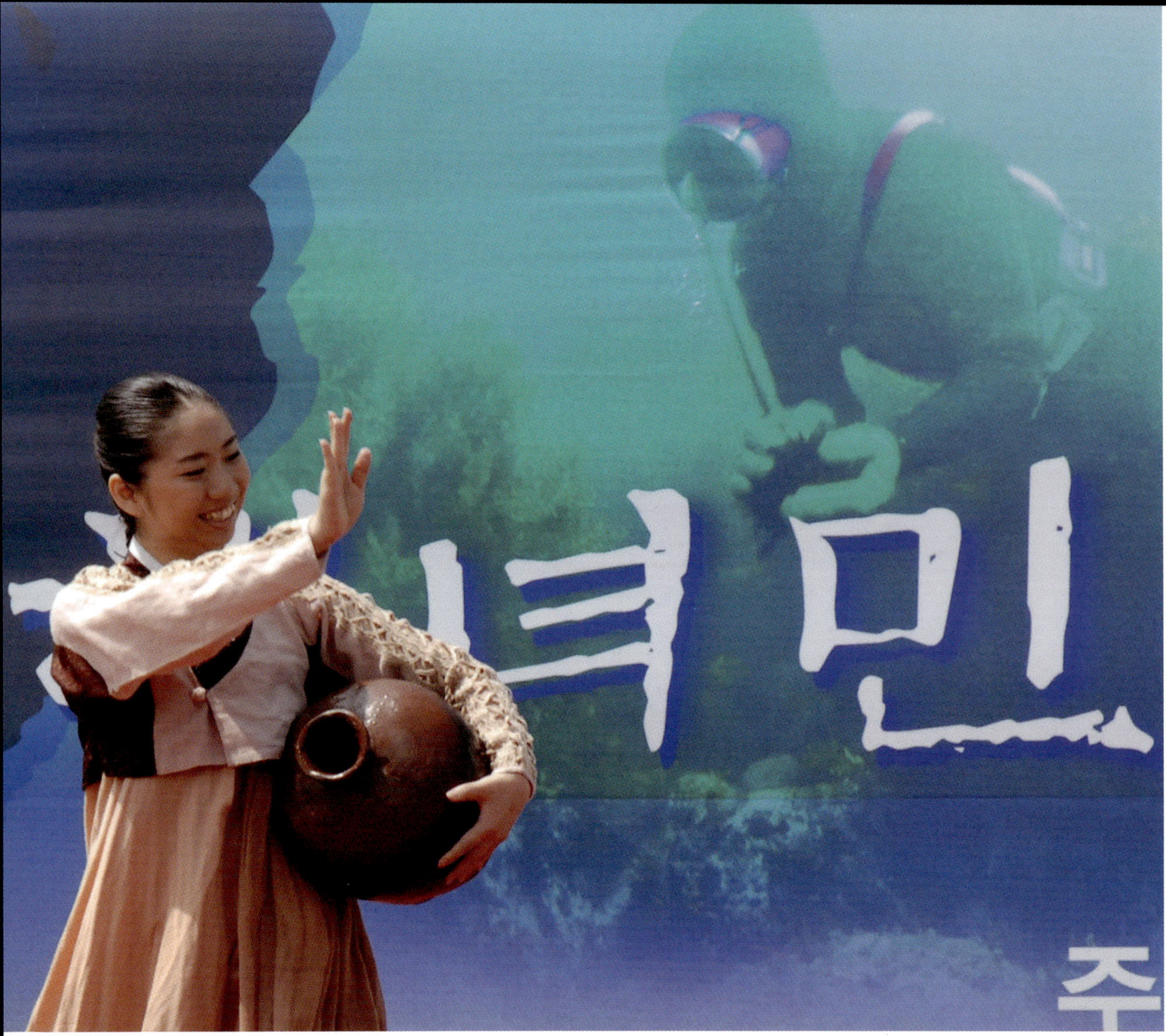

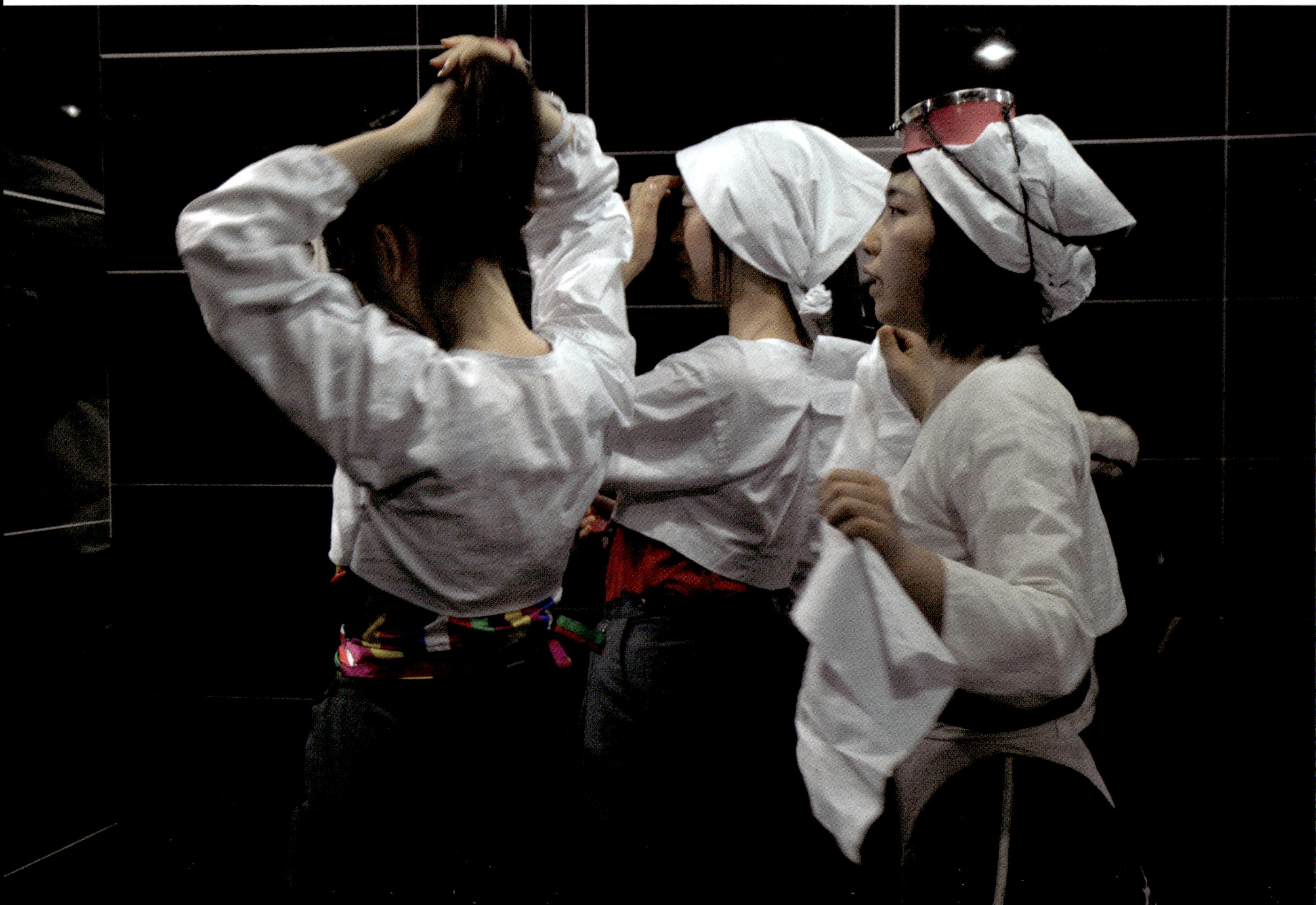

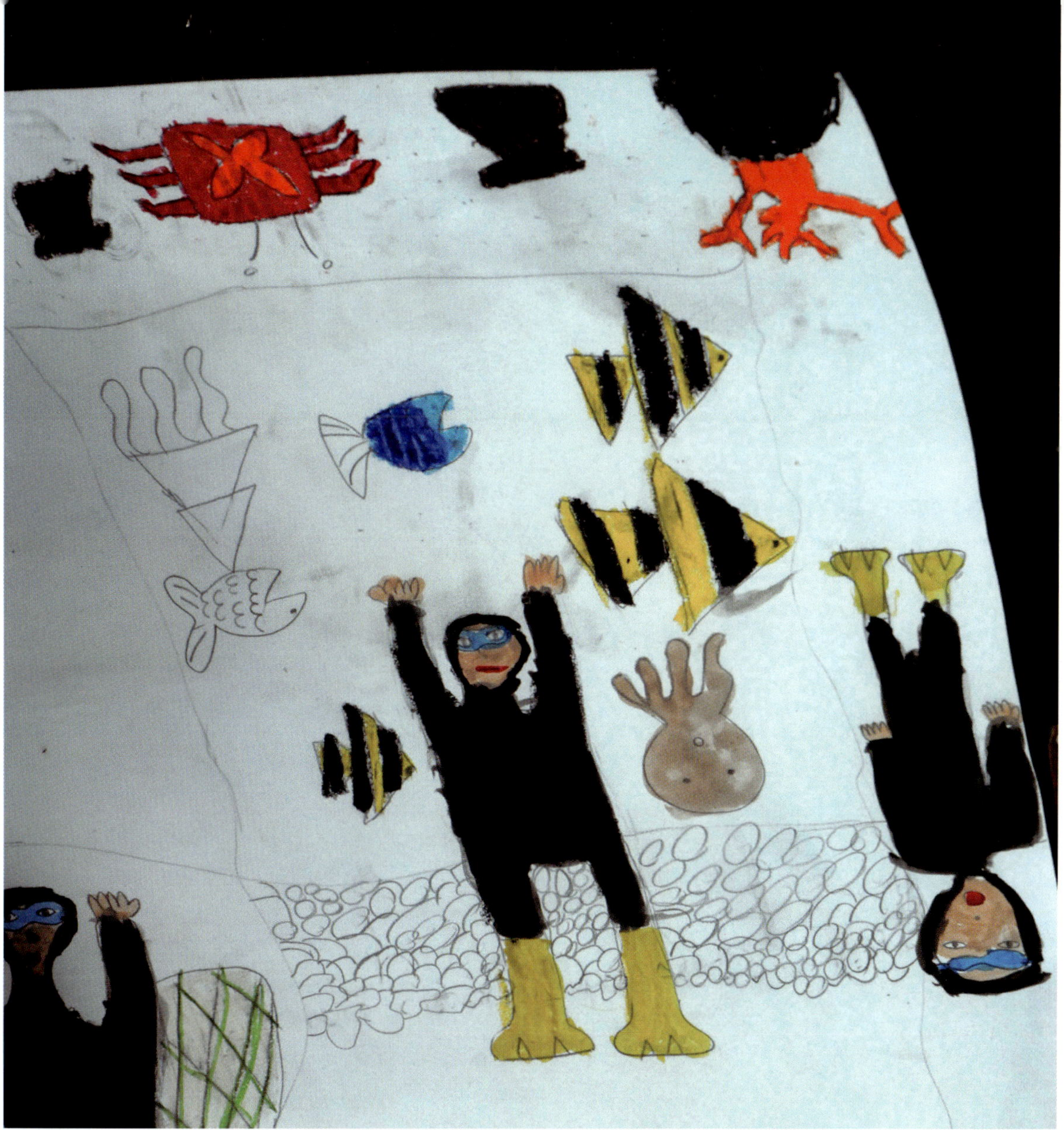

Picture painted by children at a local event honoring the *haenyeo*.

sea, she became upset as she pointed out plastic containers, beer cans, broken Styrofoam pieces, glass, and other litter carelessly left on the shore. She has since become a vocal advocate for the island's environmental protection.

From July 31 to August 26, 2010, as a part of the "World Environment and Peace Summer School," Sherrin swam nearly 200 kilometers around Jeju Island, accompanied by American Steve Oberhauser in a sea kayak, to inspire others that when it comes to the environment, every action counts.

When I began this project, I knew very little about the *haenyeo*'s culture and livelihood. I knew even less about Jeju's environmental issues. Now, whenever I think about this volcanic island off the southwestern coast of South Korea, I visualize the women diving in clean, unobstructed aquamarine waters—abundant with seaweed, turban shells, abalone, and healthy marine life. Jeju's 18,000 gods and goddesses may provide divine miracles. But without greater human effort, how will the *haenyeo*'s livelihood and legacy truly flourish as continuously and infinitely as the sea?

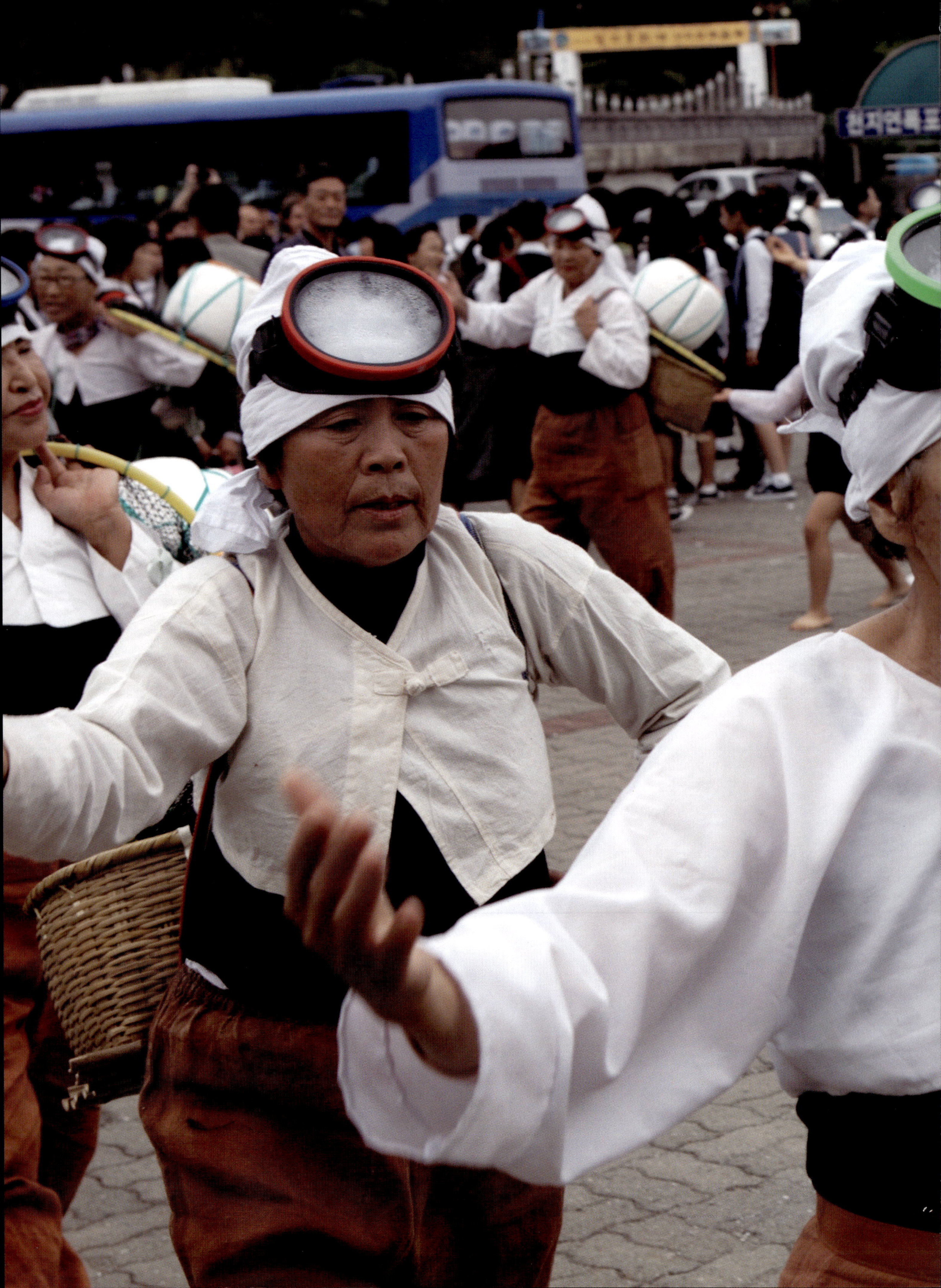

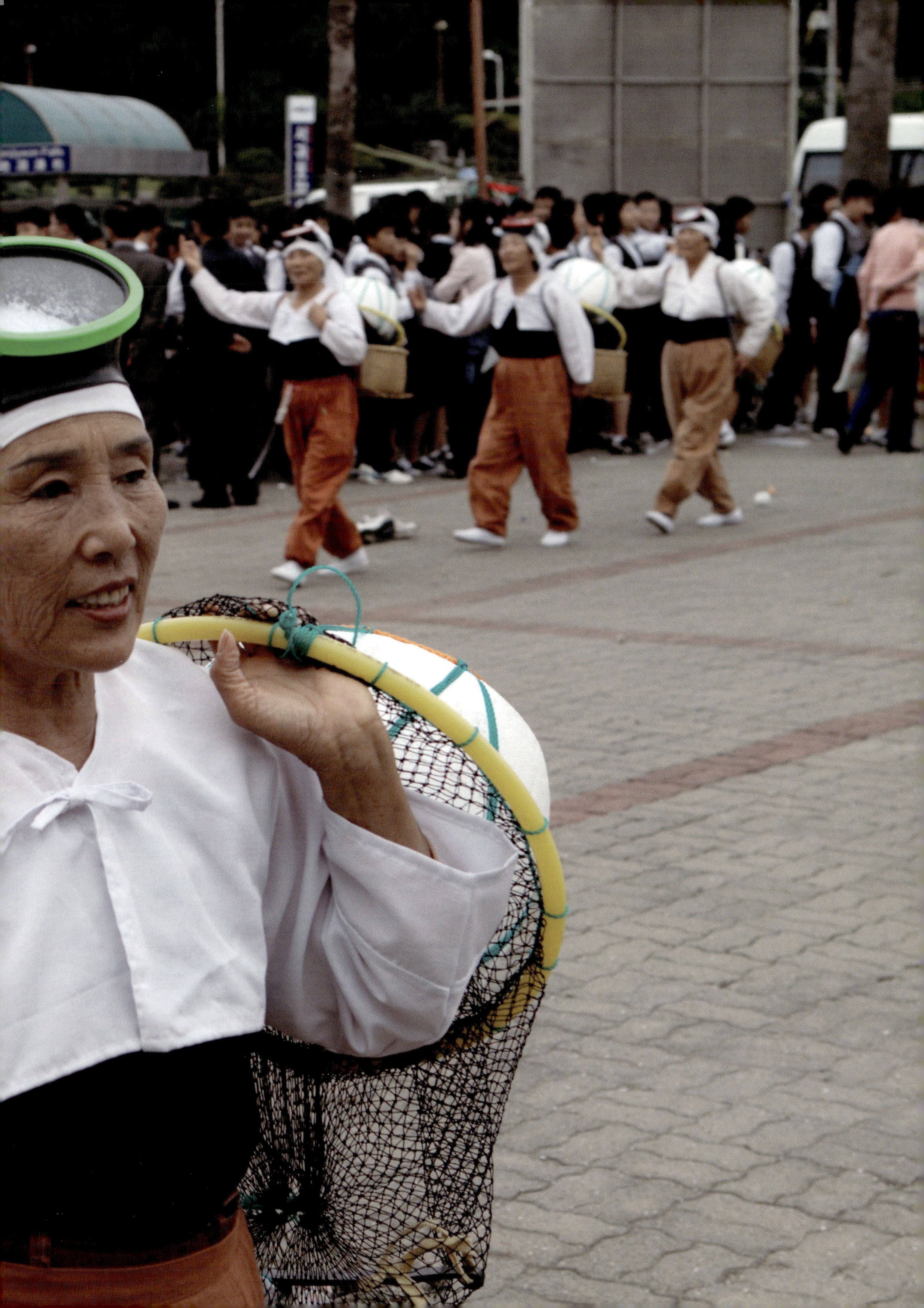

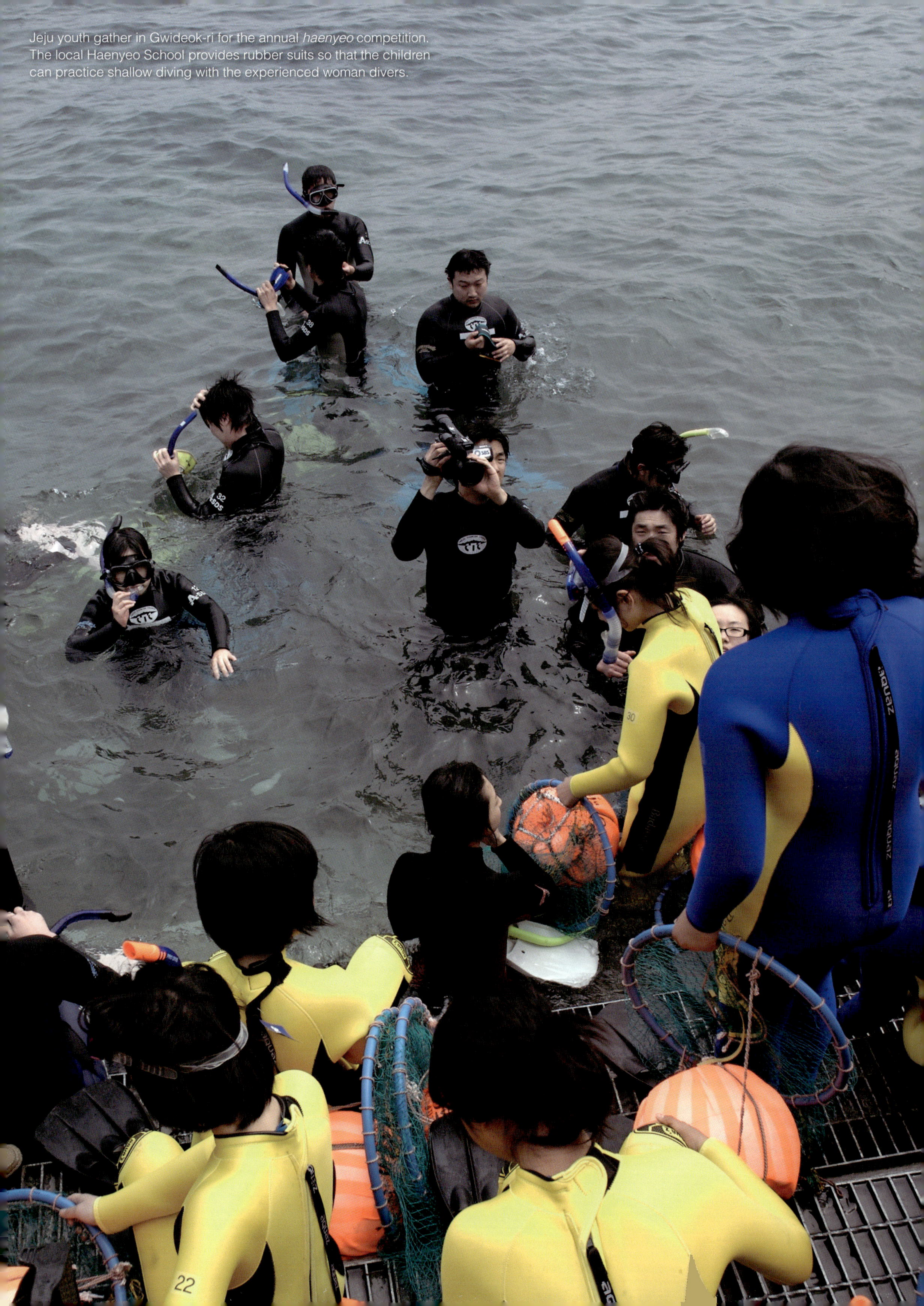

Jeju youth gather in Gwideok-ri for the annual *haenyeo* competition. The local Haenyeo School provides rubber suits so that the children can practice shallow diving with the experienced woman divers.

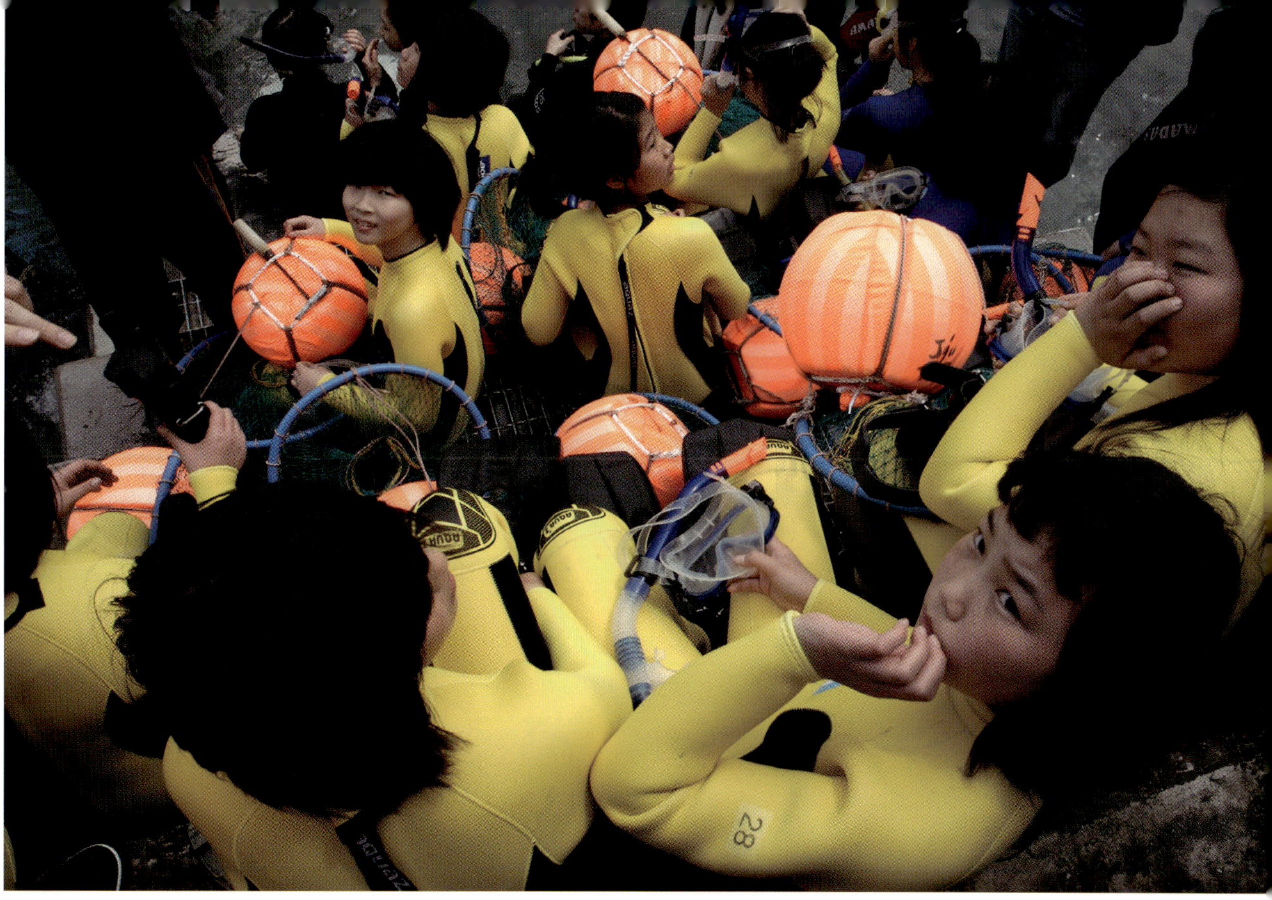

Haenyeo School

Although future generations of *haenyeo* are unlikely, one *haenyeo*'s son established a school in Gwideok-ri in order to honor and promote the history and culture of the *haenyeo*. Some locals and foreigners have enrolled in the three-month course, which offers a taste of the sea women's difficult and dangerous job.

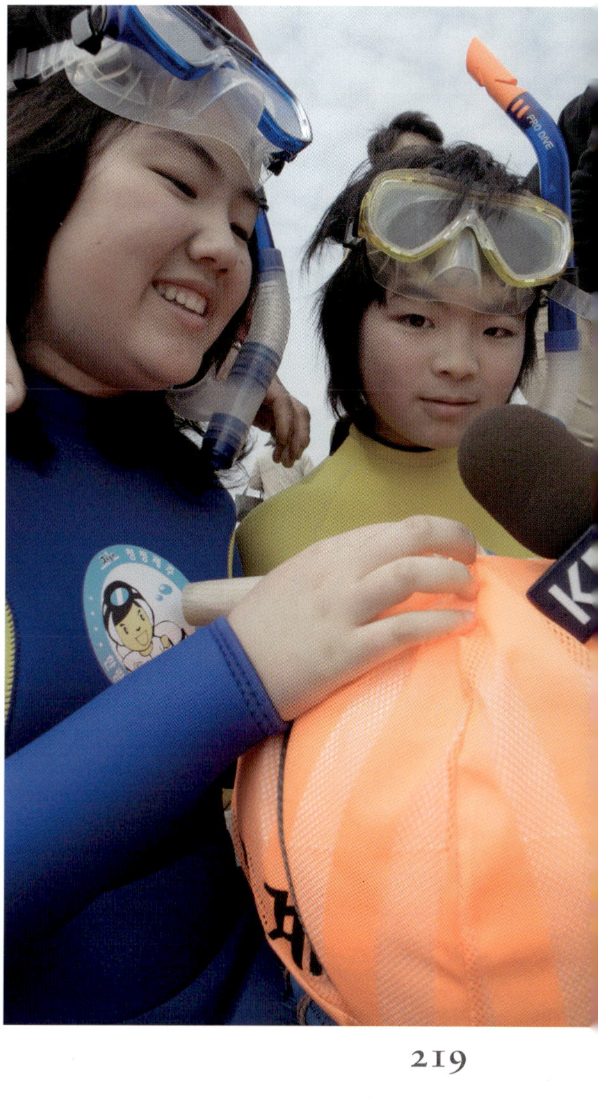

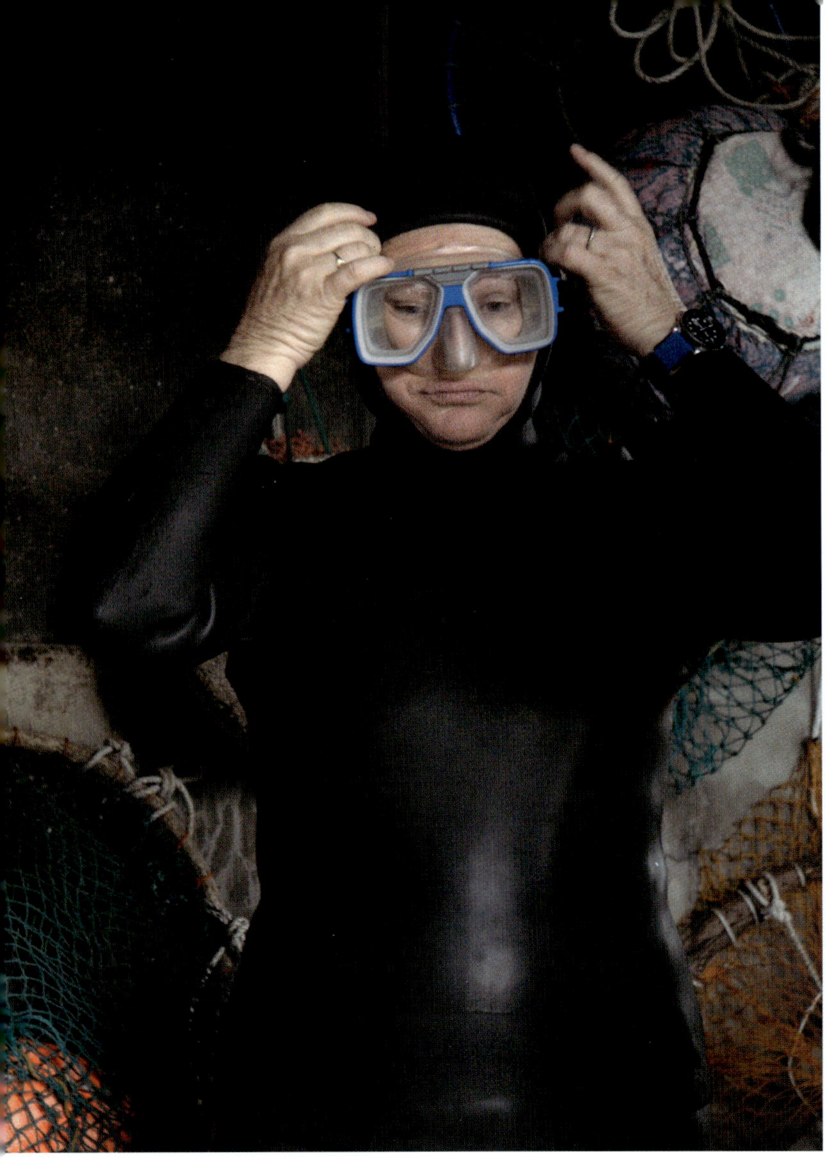

Australian Sherrin Hibbard—a former commercial fisherwoman and an ESL instructor in Jeju (2004–2011) —graduated from the three-month Haenyeo School diving program in 2009.

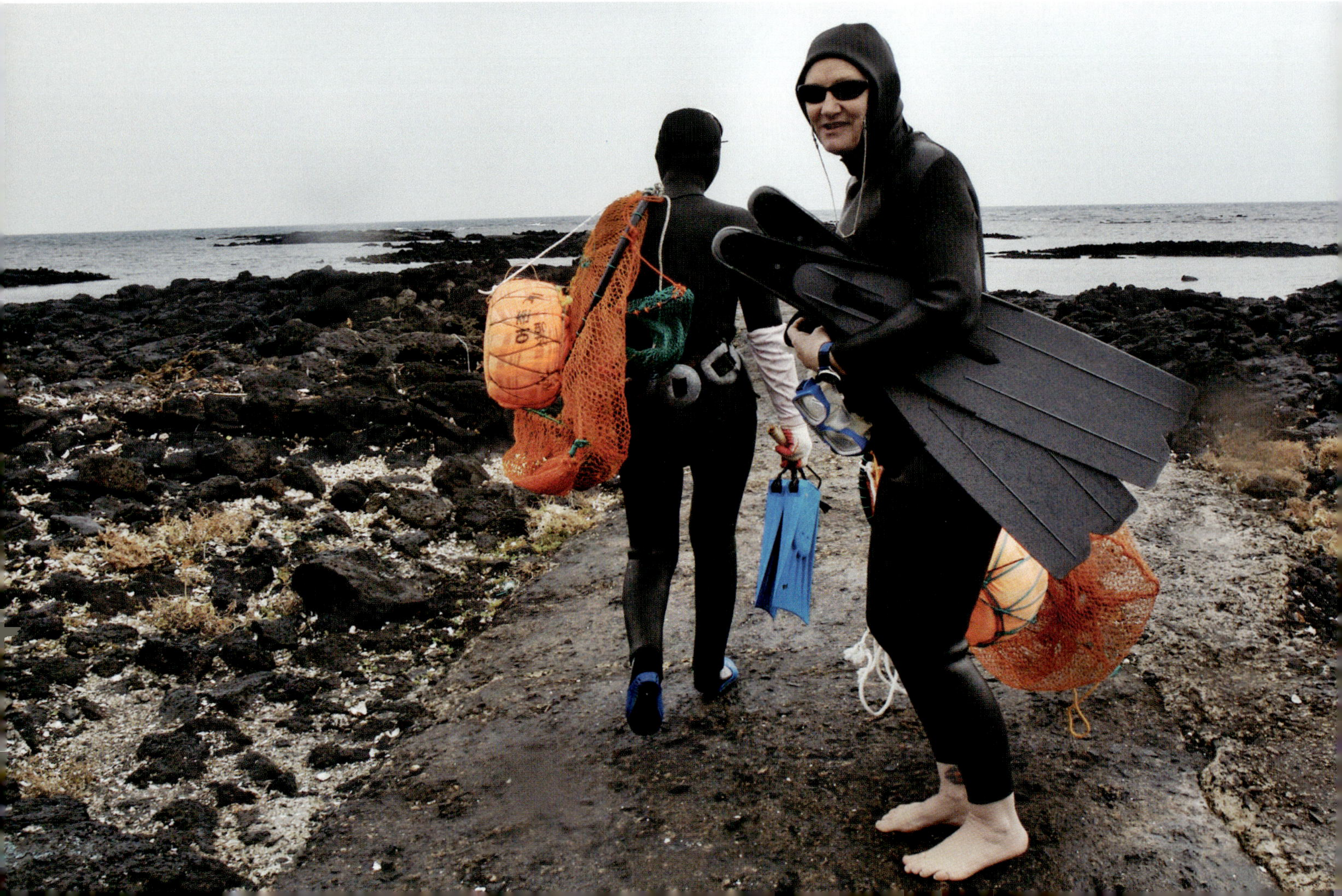

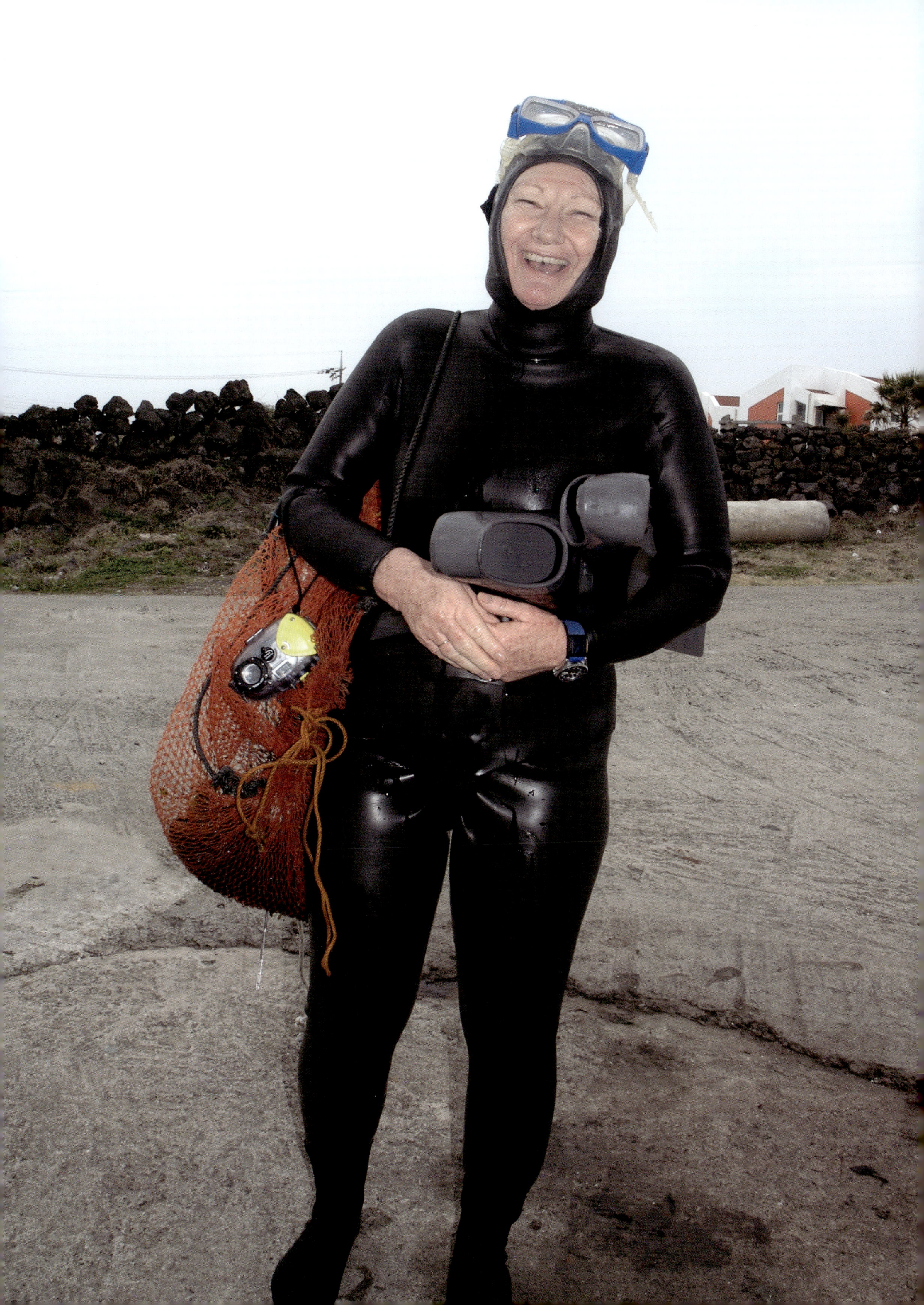

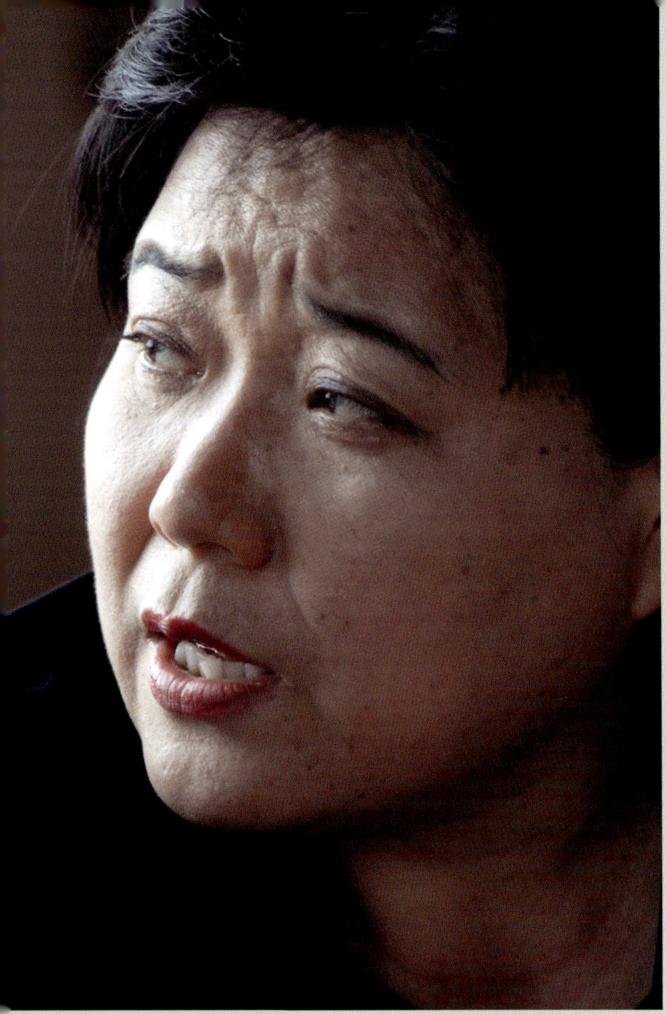

CHOA HYE-GYUNG

Researcher at Haenyeo Museum | Hado-ri

I've done research on Jeju folklore for a long time. Before arriving at the Haenyeo Museum, I taught at Jeju National University. I began studying the *haenyeo* in 2000 as I led a research team on "The Jeju Haenyeo's Value in Terms of the History of Marine Civilization." It took about five years, and I wrote a paper about it. I also translated a report written by a Japanese regarding Jeju *haenyeo*'s lives in Japan. I cried a lot while translating the report. Now, I do research at the museum to educate others about the *haenyeo*. Our museum opened in June 2006.

The museum began another research effort in 2006 in collaboration with the Jemin Daily News. There are 100 fishing cooperatives, and each one in each village has its own characteristics and culture. After the museum was built, we thought we needed to learn about the actual conditions of *haenyeo* society and document their remaining traditions.

Among the items we've collected are a lot of *haenyeo* songs. In fact, every Saturday afternoon, we offer a class to teach these traditional *haenyeo* songs. There are three *haenyeo* designated as Jeju's first Intangible Cultural Assets for preserving these traditional songs.

Types of Haenyeo

There are two categories of *haenyeo*: high-level *haenyeo* who go to the deep sea by boat, and low-level *haenyeo* who dive near the coast. If a *haenyeo* is high-level, she can make about 20 to 30 million won per year ($17,000–$26,000). A low-level *haenyeo* can make only 2 to 3 million won ($1,700–$2,600). Their wages fluctuate according to the current value of the Japanese yen, so they have to do both farming and diving in order to make a living. If a *haenyeo* is sick, she doesn't pay any penalty. If she doesn't work, she is simply not paid.

Procurement

After the *haenyeo* harvests her catch, she brings it back to the fishing cooperative, where the products are weighed. She receives a receipt for her catch and returns to the *eochongye* to get paid on a designated day soon after. She can sell any leftover products privately in the market. Nearly all of the agar-agar, *tot* (a kind of sea weed), and *gamtae* (laver) are exported to Japan.

Main Economic Provider

Before their families began cultivating cash crops such as tangerines and carrots, the *haenyeo*'s earnings were the main source of family income. These days, it's half and half—more coming also from the husbands. But in some places, like Udo Island, the *haenyeo* are still the main source of family income. Where the *haenyeo*'s earnings were, or still, are the main source of family income, the husbands have assisted by carrying the catch and taking care of the children. It depends on the family these days. Some of them farm. Others fish. Some are still sea assistants.

The *haenyeo*'s work is increasingly threatened by rising sea temperatures. Their catch has been reduced. Also, wastewater and chemicals used in fish farms have contaminated the sea.

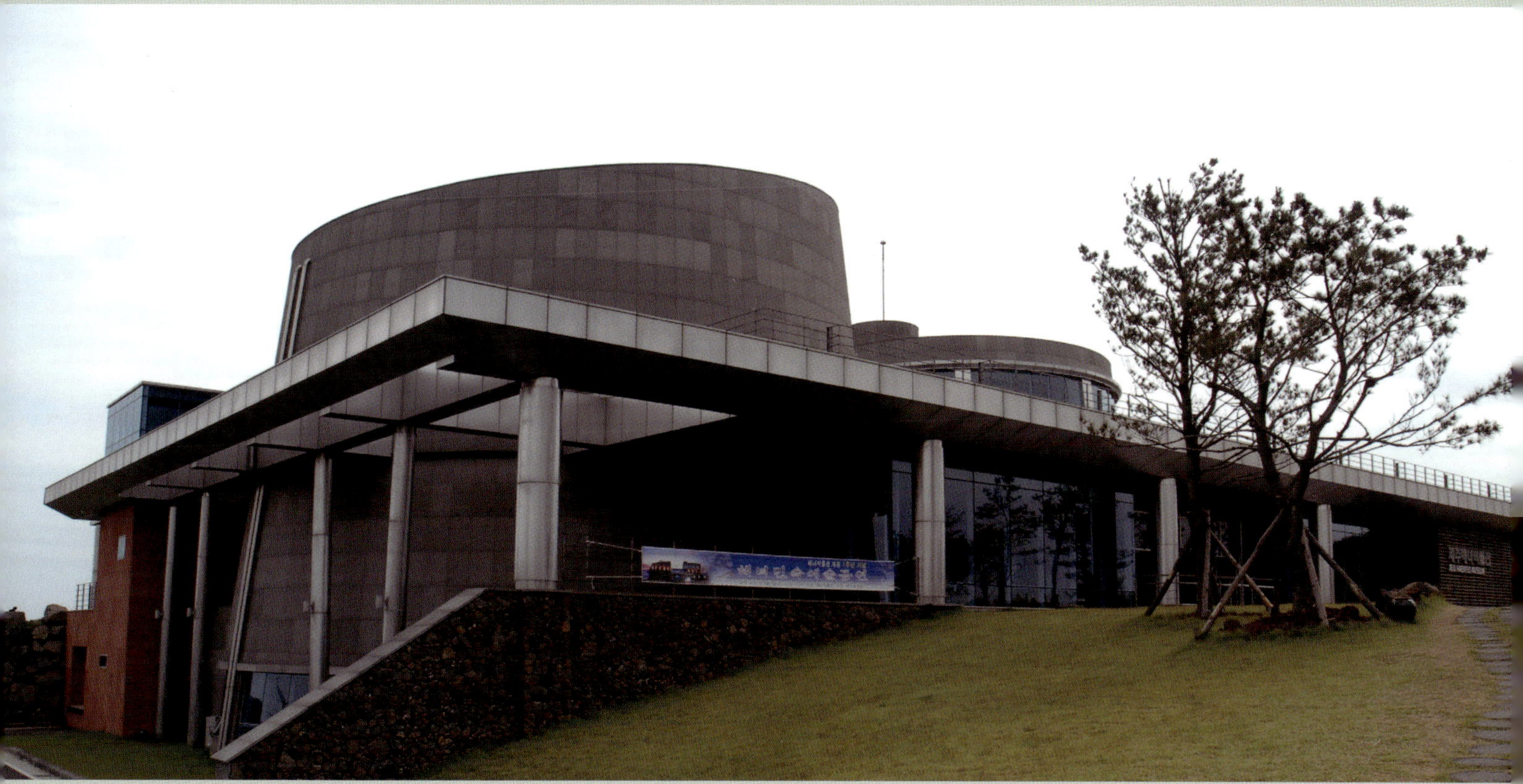

The Haenyeo Museum in Hado-ri pays tribute to the woman divers and fishermen of Jeju Island. One of the largest anti-Japanese protests during the colonial era (1910–1945) was led by women from the Hado-ri area.

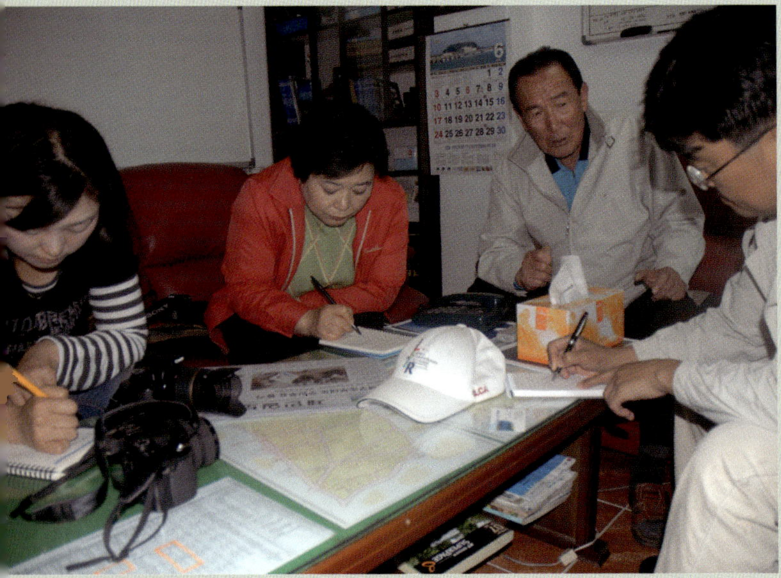

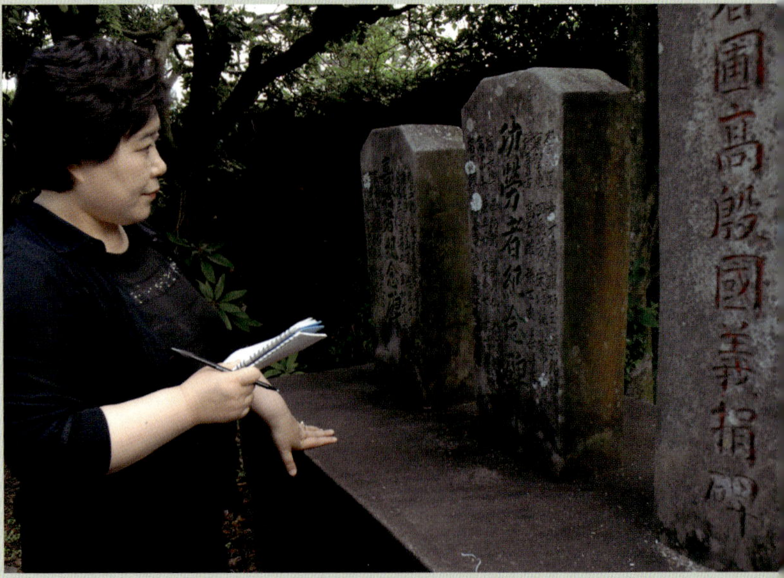

Ms. Choa documents the names of Onpyeong-ri heroines.

Anti-Japanese Protests

The *haenyeo*'s anti-Japanese activities began in Seongsan-ri in 1931 when they realized they were being cheated out of their deserved earnings. The protest then spread to the Gujwa area. In 1932, the biggest protest was held in Hado-ri. Many of the protests were stirred up by the local intellectuals, who were somewhat leftist in political orientation. They taught the *haenyeo* how to read and write as well as how to calculate the weight of their catch. Activities were also organized for them. About 17,000 people gathered together at that time. It was one of the biggest protests against Japan in the history of Korea's anti-Japanese movement. It was unique in that women spearheaded the protest.

Projection

I think *haenyeo* work may continue for another 20 to 30 years, as long as there are marine products to catch and one can earn enough. I guess even young women are likely to do the work. For example, I know of one young woman in her 30s from Jeju-si who had a hard time making a living. So she returned to her mother's village in Marado and began to do *haenyeo* work.

(August, 2007)

KO YEON-UK

b.1957 | Haengwon-ri

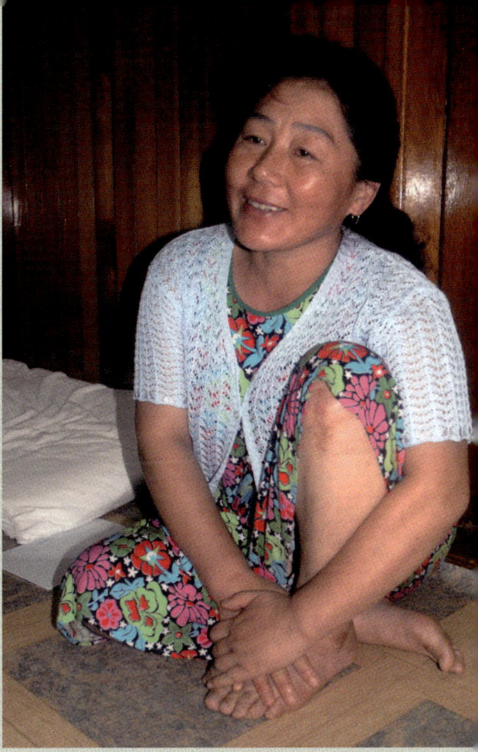

There are three women in Haengwon—including myself—who are designated as Jeju "Intangible Cultural Asset trainees" based on our knowledge of traditional *haenyeo* songs. I've been learning these songs for 10 years. We refer to them as "*haenyeo* rowing songs."

The local government authorities want us to teach these songs to the younger generation. In the old days, when the *haenyeo* rowed out to sea, they sang these songs, which expressed their hard lives and difficulties. Some of the words are more like utterances that are similar to the sound the *haenyeo* made while rowing across the rough ocean waters. The singing energized them and boosted their morale.

They sang about longing for their family and their hometown, and about their misery. These days, we don't sing because we're too preoccupied thinking about making money and don't face the same difficulties of the past. Years ago, the *haenyeo* sang on their way to sea and back because it was hard work, and they wanted to forget their woes. Nowadays, we have cars and motor boats that take us out to sea. We have much less hardship than before.

When I first began learning the songs, I didn't feel anything special. But after hearing the older *haenyeo* tell their stories about being separated from their families, I began to have the same feelings. Now I understand why they wrote and sang these songs.

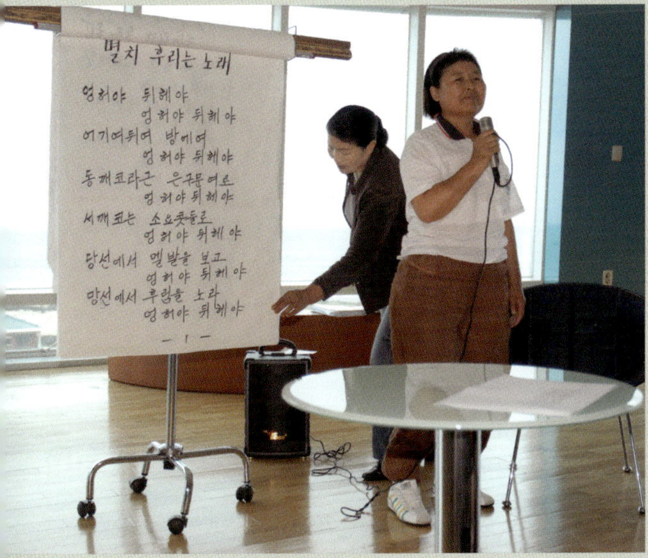

We teach these songs in the community hall of the Haenyeo Museum in Hado-ri, and we often perform several times a year. On stage, we wear cotton suits like the ones worn in the days before rubber suits.

Most young people don't want to learn the songs because they don't understand many of the words. They're written and sung in the Jeju dialect. Also, the songs don't make them feel happy. But I still think we should teach the children these songs so they appreciate the lives of their grandmothers and great-grandmothers. It's teaching history.

(August, 2007)

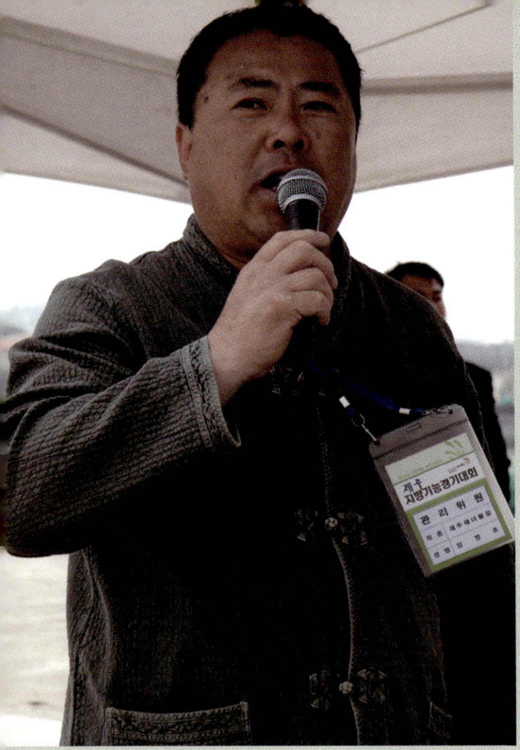

IM MYEONG-HO

b.1950 | Gwideok-ri

I conceived the idea of a Haenyeo School in 2007. The reason is that I wanted young people to remember how the older generation of Jeju women lived. It's for their education, not job training. In 20 years, 70 to 80 percent of the *haenyeo* will have disappeared from this island. I want to keep the tradition alive and give the public a chance to experience the *haenyeo*'s working life. *Haenyeo* are getting older and older, so they don't make as much money as before. By teaching people how to dive, some of them can earn extra money. I've persuaded the *haenyeo* in our village to reserve an area of the sea for outsiders. That way, others can experience diving and support the village by buying our sea products. That's my intention.

The local citizens' committee here in Hallim was seeking special projects for each village. My mother is a *haenyeo*, and I have memories of her peers earning money to send their children to school by selling their sea products. So that's how I came up with the idea of the school. My idea was accepted, and the project received 40 million won ($35,000) for its initial setup. We bought rubber suits and equipment for students who would be willing to pay for lessons. We still need a lot more money to have proper facilities. Several of us are working hard on this project without compensation.

The course runs every Friday from May to August. Hopefully, we can receive funding from the government so the students only have to pay a moderate fee. Those who aren't strong swimmers will practice first in shallow waters enclosed by a stone wall. Those who are more confident swimmers can dive outside the restricted area. We'll buy marine products such as turban shells and sea urchins and scatter them in the training area so that the students can actually collect something. Both women and men are welcome. During our first session, we taught 32 people, three men and 29 women. Eighty percent of them were in their 40s. The oldest was 53, and the youngest was in her early 30s.

(April, 2008)

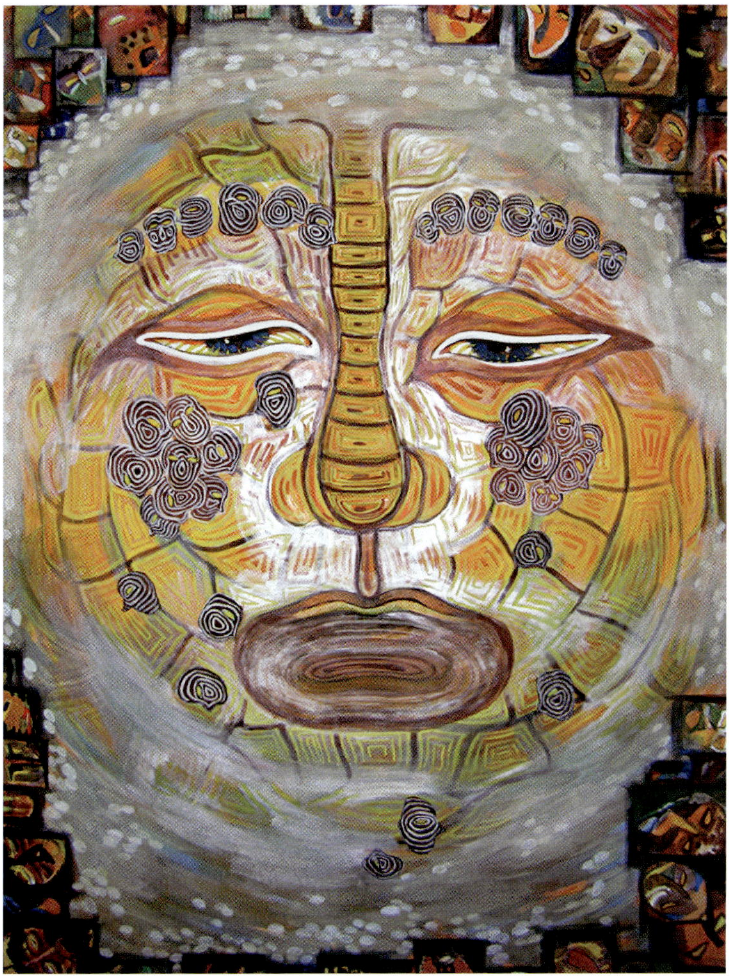

Seolmundae Halmang—Porridge Cauldron
112x163 cm | color on backing paper | 2009

Seolmundae Halmang

(Goddess of Creation)

The goddess Seolmundae Halmang is the creator of Jeju's landscape: the islands, mountains, valleys, hills, and rivers. There are several colorful tales of her—some of which are humorously graphic. Physically, she is always depicted as a giantess. One tale describes how she had difficulty finding clothes, and so she offered to build the locals a bridge to the mainland in exchange for making her underwear. Another tale describes how after eating millet porridge one day, she suffered from diarrhea, which subsequently created the 360 *oreum* (small mountains) of Jeju Island.

One of the most outrageous tales about her is about the day she made a cauldron of soup for her 500 sons. While cooking, she apparently fell asleep and toppled to her death in the broth. Returning home, her sons searched for their mother. The youngest son eventually discovered his mother's bones at the bottom of the cauldron. Distraught by their grief, the sons were petrified into stones. Today, these stones, known as the 500 Generals, are the outcrops rocks at Yeongsil, a spot on Mt. Halla.

Haenyo Song: Harvest

Sonnet L'Abbe *

We cull the island's most spectacular fields.

Cheju's long grasses have always belonged to the women.
Like our inland sisters who crouch in flocks
along the roadsides, cutting and tying
tall stalks into bundles,
we too wrap our heads in white towels, and bend to trim
the jagged underwater lawns.

We envy the sun's long arms, its deep reach.
Crystals of light collect on the sea floor,
settle like sugar in a glass of water, dusting
the greedy seaweed fronds. When stirred by our fins,
light disperses, dissolves. It clings to our bodies.
We swim, pollinating the watery garden.

Other crops move in the wet meadow: we hunt
mobile vegetables - cucumbers with fingers,
flowers with feet!
The urchin flees, millimeters per minute,
on its single, toothed paw. The sola retreats
into its white turban, tries to pass for one of us.
The conch shies from the hand, curls into itself
As a bud cringes before it is picked.

Eighty-nine fires lit on Halla.

Nagasaki, Hiroshima: dropped
casually as pebbles into a pond,
but the ripples lashed our shores for years.

Spread on their dissection table,
Korea was a little rabbit
On a stranger's map,
dangling in China's paw or snagged
in the hind claw of Russia

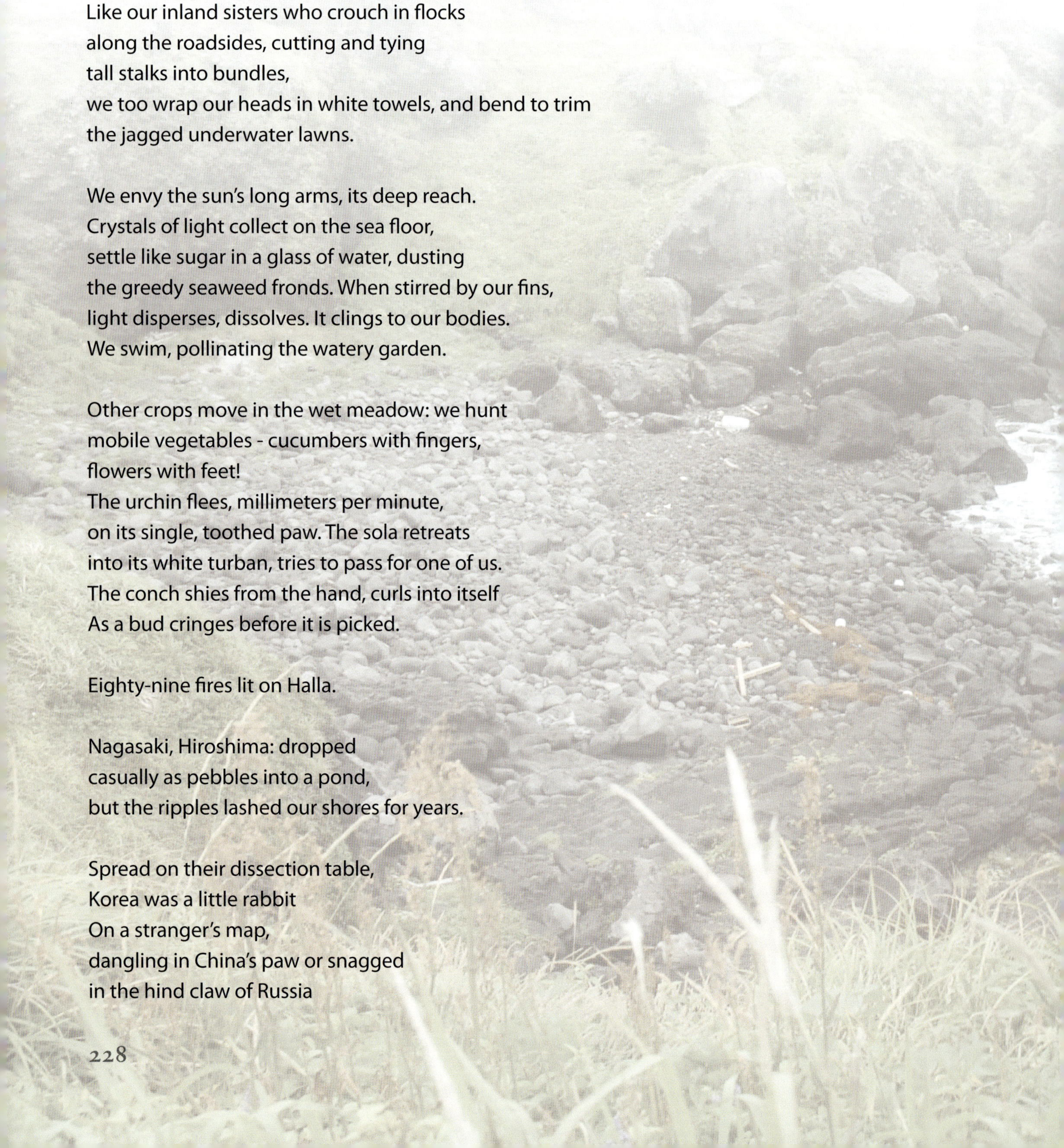

if they had cared to look:
they performed their secret operation
blindfolded, in a far-away room,
the paper decision to sever its head
as easy as unpinning
a drawn donkey's tail.

Everyone forgets islands
but the armies.
Cheju, both their rabbit's foot
and a dropping at its heel.
Our own country
gnawing us off at the ankle
to escape.

What the sea gave back freely moved us first:

He bobbed up, his pale back
a bullet-pitted coral,
shreds of skin around the wounds
like the red blooms of anemone's flower.

His mute body told the whole story,
the exact cost of silence -
two blunt stumps announced his lost thumbs;
his tongue, waterlogged and swollen with secrets,
tumbled from the cave of his mouth,
dumbstruck as the long-hidden survivor
who emerges from the dark shelter, and stumbles
into the sober, devastated day.

My own child in the basket
beside the water, among the fishbaskets
and waterjugs. A boy disguised
as part of our harvest.

The stories from Orari, from Bucholi,
of black stones soaked in red.
Suddenly, blood oranges.

Hatred is a crafty child,
who finds even in a farmer's field
torture's toybox:
who needs weapons when at hand
is onion's green-tailed whip,
the rape-efficient orange root,
and the killstone of white radish?

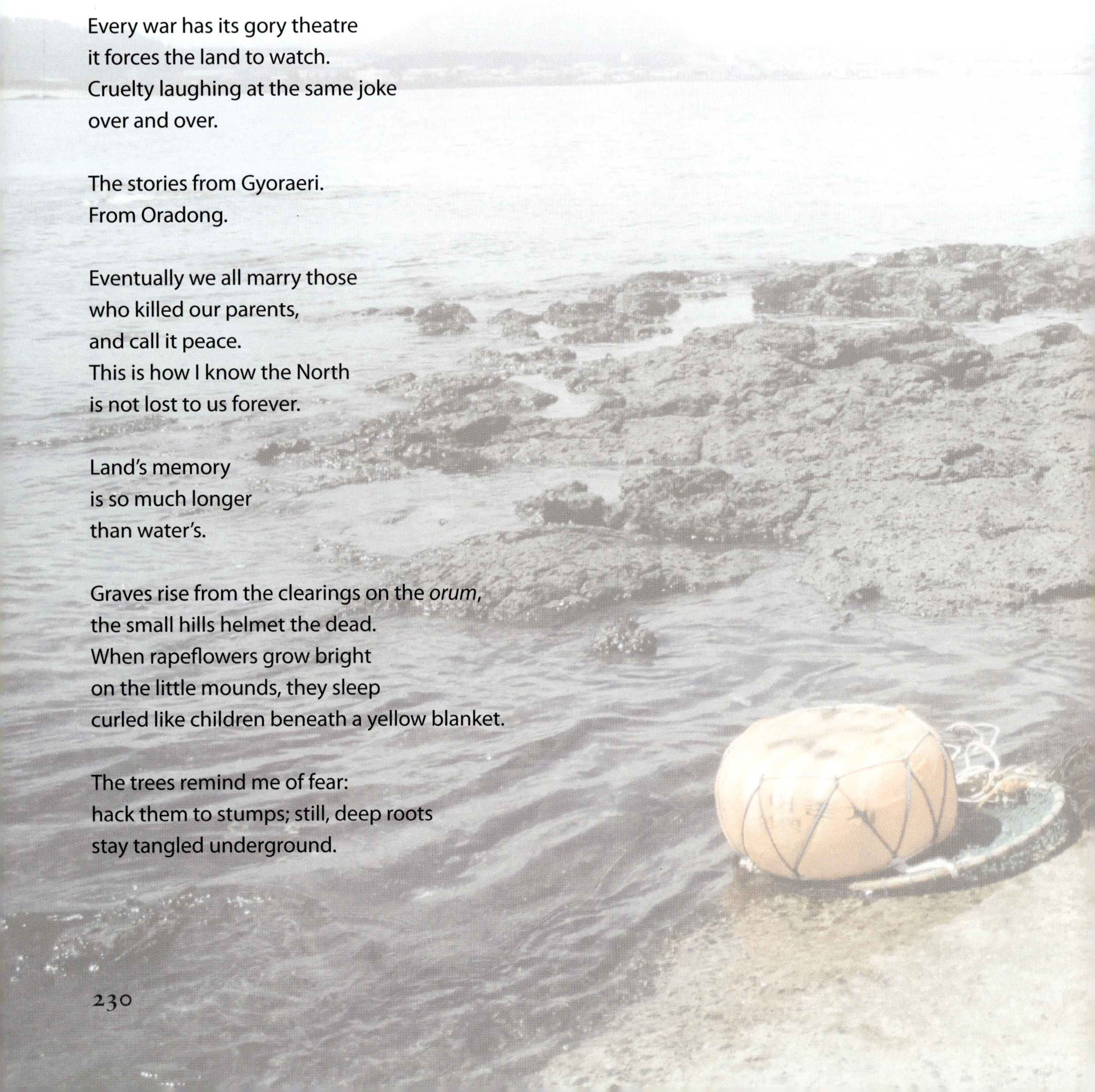

Every war has its gory theatre
it forces the land to watch.
Cruelty laughing at the same joke
over and over.

The stories from Gyoraeri.
From Oradong.

Eventually we all marry those
who killed our parents,
and call it peace.
This is how I know the North
is not lost to us forever.

Land's memory
is so much longer
than water's.

Graves rise from the clearings on the *orum*,
the small hills helmet the dead.
When rapeflowers grow bright
on the little mounds, they sleep
curled like children beneath a yellow blanket.

The trees remind me of fear:
hack them to stumps; still, deep roots
stay tangled underground.

The sun says nothing, recedes
as if its radiant face would offend.
We walk in the timid light
that filters through the gauze of cloud
bandaging the island, afraid
to scar the soil with our shadows.

I am full of lessons I cannot share:

How wild grass cleans the mist
from goggles, how to hook the sicklecord
at the elbow, not the neck.

How to find feather stars and sea lilies.
The flounce of scallops, sea slugs' ruffled skirts,
the split gourd shape of cuttlefish.

How the concave of the abalone,
its hard slick of colour, is like the skin
of gasoline on water, a liquid prism.

The depth of crab and shrimp. The myth
of sand dollars, the bottle mouths of sea squirts.
Polyps. Molluscs.

How we learned to respect
their reluctance to leave, the bravado of shells.
How we learned to love
the sea's slow resistance.

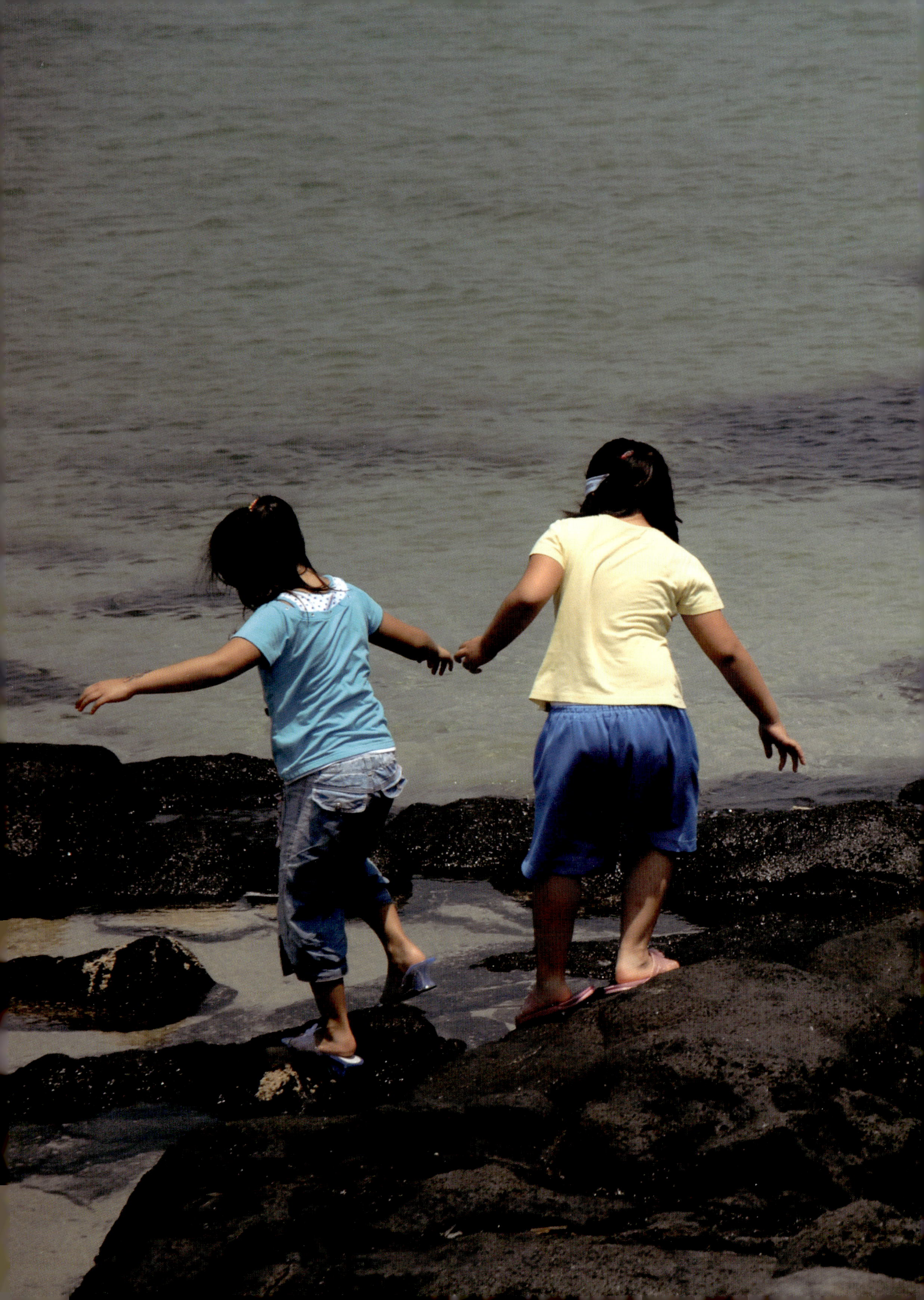

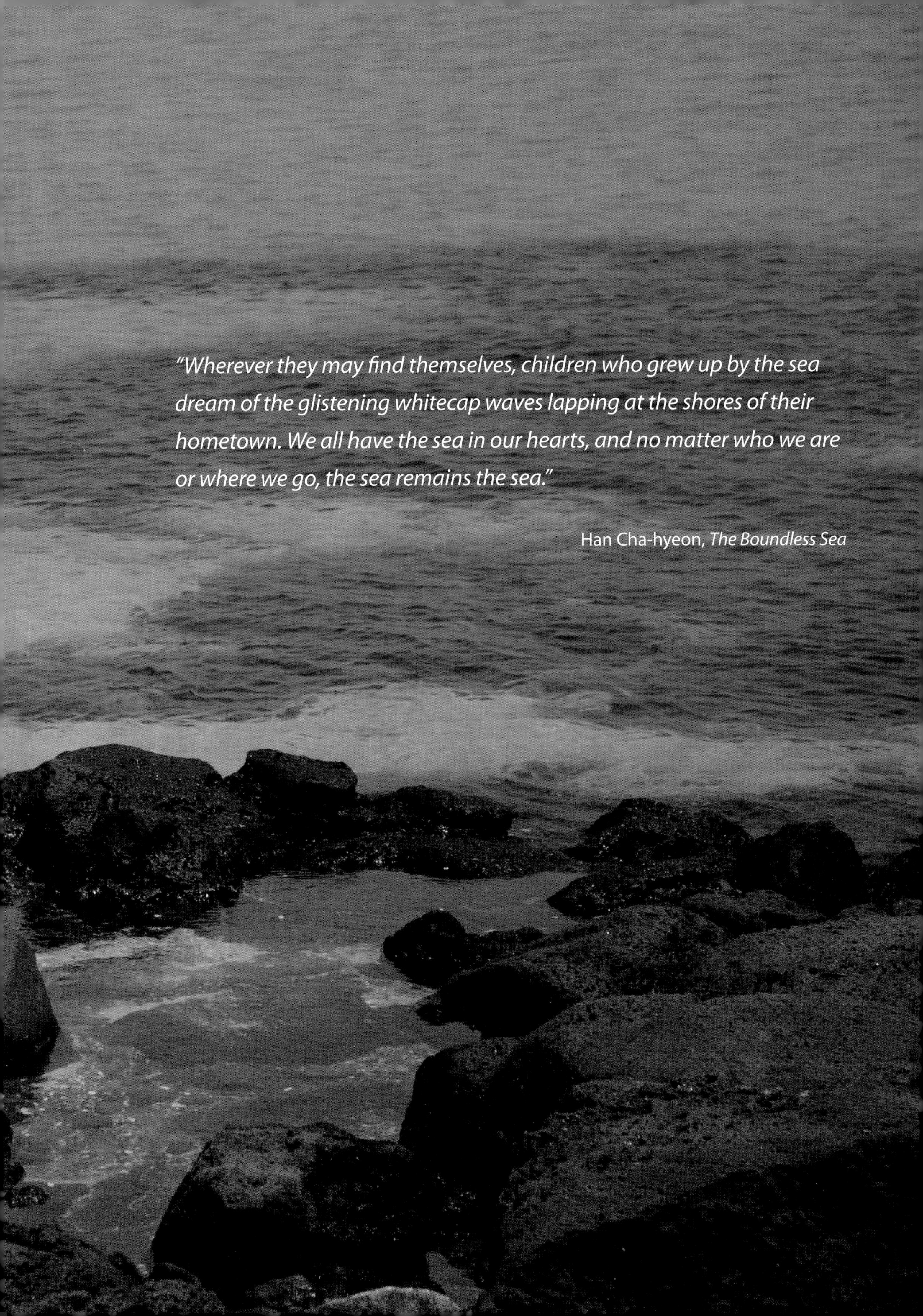

"Wherever they may find themselves, children who grew up by the sea dream of the glistening whitecap waves lapping at the shores of their hometown. We all have the sea in our hearts, and no matter who we are or where we go, the sea remains the sea."

Han Cha-hyeon, *The Boundless Sea*

ACKNOWLEDGEMENTS

When I first arrived in Jeju to embark on my writing project, I didn't know anyone. I was given a couple of referrals, but for the most part I didn't know a soul. Somehow, through the grace of Jeju's 18,000 gods and goddesses, I was blessed to meet extraordinary people along the way. Each one played a significant role in bringing this book to fruition. My gratitude to Lee Samuel (former Secretary-General of Korean National Commission for UNESCO) and his wife Son Duk-soo for their support and for introducing me to Yang Soon-ja, who not only fed me well but also introduced me to many facets of Jeju culture, including the healing and fashionable powers of young persimmons.

My gratitude to Moon Sook-hee and Kim Gi-yong for my first homestay in 2007, and to Martha and James Chun for their frequent hospitality and translations when I needed them upon my first arrival to Jeju.

By the time I left, I had made many new friends. My gratitude to: Ahn Hye-kyoung of Art Space C for her enthusiasm, generosity of spirit, resourcefulness, and key introductions; Kim Myeong-sin for sharing the experiences of her *haenyeo* family; Sherrin Hibbard for an illuminating awareness about the island's environment; Chris Evans for writer chats, Aussie vocabulary, and ice cream cone bets; Sung Chang-mo for his Café 1263 hospitality; Lee Jae-hyun for being more than a taxi driver (photographer, humorist, scavenger, and all-around guardian); Fred Dustin for urging me to rent the house in Gimnyeong-ri and for including me in several staff lunches and dinners; Park Hyun-mi for translations, and for bailing me out with immigration authorities when my visa expired; Stella Chun for her kindness and hospitality whenever I headed south to Seogwipo; Laurie Kim and her husband for meals and fun times with the children of Gimnyeong-ri; Im Yeong-ae for introducing me to the Gimnyeong-ri women's volunteer group; my cousin Peggy Choy and Jeju artist Koh Gil-chun for bringing Sa Sam (the Massacre of April 3, 1948) to my attention; Ko Young-lim for stimulating discussions about feminism and politics; and Ko Kyung-hee for the book that inspired me to imagine faces carved in Jeju volcanic rocks.

I could not have understood much of the historical and political context of *haenyeo* life without the research and published works of many Korean academicians and professionals—several of whom are listed in the back of the book. Special gratitude, however, is extended to: Choa Hye-gyung, researcher at the Haenyeo Museum in Hado-ri; Kim Hyong-hoon and Kwon Mi-soon (then Haenyeo Museum research team members); painter Kang Yo-bae; and Lee Eun-Joo of the Sa Sam Massacre research committee.

I would like to express special gratitude to the Jeju *simbang* whom I interviewed and photographed during their special gut ceremonies for the *haenyeo* and fishermen: Suh Sun-sil (Gimnyeong-ri), Lee Yong-ok (Jocheon-ri), and Ko Sun-an (Hado-ri)

There would be no book without the *haenyeo* and several husbands whom I met and interviewed with my translator/interpreter, Youngsook Han. We are both indebted to the *haenyeo* for their generosity of time, recollections, and patience in sharing their stories. We also thank them for all the fresh shellfish they gave us straight from the sea. From Gimnyeong-ri: Kim Ui-ja and Shin Seong-sun, Oh Yeong-geum, Kang Do-saeng, Jung Dong-seon, Yoon Sun-nyeo, Hwang Chun-seok, Ko Chun-hwa, and the women from the Gimnyeong-ri Women's Volunteer Group. From Jongdal-ri: Woo Bong-seon, Woo Sun-deok, and Kim Do-won. From Woljeong-ri: Oh Jeong-ja and Kim Du-seok. From Haengwon-ri: An Mi-seon and Ko Yeon-uk. From Onpyeong-ri: Cho Sun-wol. From Gwideok-ri: Ko Myeong-ho. From Hallim-eup: Kang Deok-kyo. From Sagye-ri: Cho Jeong-sun and Im Seong-ok. From Bukchon-ri: Hong Suk-ja. Special thanks also to former teacher Ko Chang-ho of Onpyeong-ri and boat driver Kim Gye-ho of Ojo-ri. My gratitude also extends to Lim Baek-yeon, divers' association leader, who arranged for me to go underwater with a local haenyeo in Hado-ri; Im Myeong-ho of Hallim-eup, who established the Haenyeo School; poet Yang Min-sook; artist Hong Jin-suk for her work on Jeju goddesses; and Han Ki-ok and Kang Mi-kyoung of the Hanbit Shelter for Women.

No writer can produce a book without the support of family, longtime friends, and fellow writers. My gratitude to Lady Borton in Hanoi, Vietnam, for writing advice in the early stages of my book; Dr. Harold Sunoo, my 94-year-old father-in-law who translated many news articles about the *haenyeo* from Korean to English and who helped subsidize my stay in Jeju; my cousin Gail Whang for constant feedback at all stages of my project; and my girlfriends who came to Jeju during my stays: Jai Lee Wong, Ton Thi Thu Nguyet, and Thu Kim Vu ("Yaeji").

At home in California, my gratitude to my neighbor and fellow writer Kathy Lazarus for reminding me to speak in my own voice.

I am most indebted to my publisher Kim Hyung-geun (Hank) and his wife Yang Mi-soon for their faith in me to write another book with Seoul Selection. They are my family whenever I am in Seoul. I consider them one of Korea's best unofficial ambassadors through their bookstore across the street from Gyeongbokgung Palace. My gratitude to Seoul Selection staff for their meticulous, thoughtful, and beautiful contributions to the book: Chung Kyung-a, Jung Hyun-young, Lee Jin-hyuk, and Kim Eugene.

To my son, David, and grandchildren—Jolena and Devon—my deepest appreciation for kick-starting my '60s. Most of all, I thank my husband, Jan, for his patience, generosity, and tolerance whenever I dream up a project that requires an occasional escape. You are a writer and photojournalist's dream spouse.

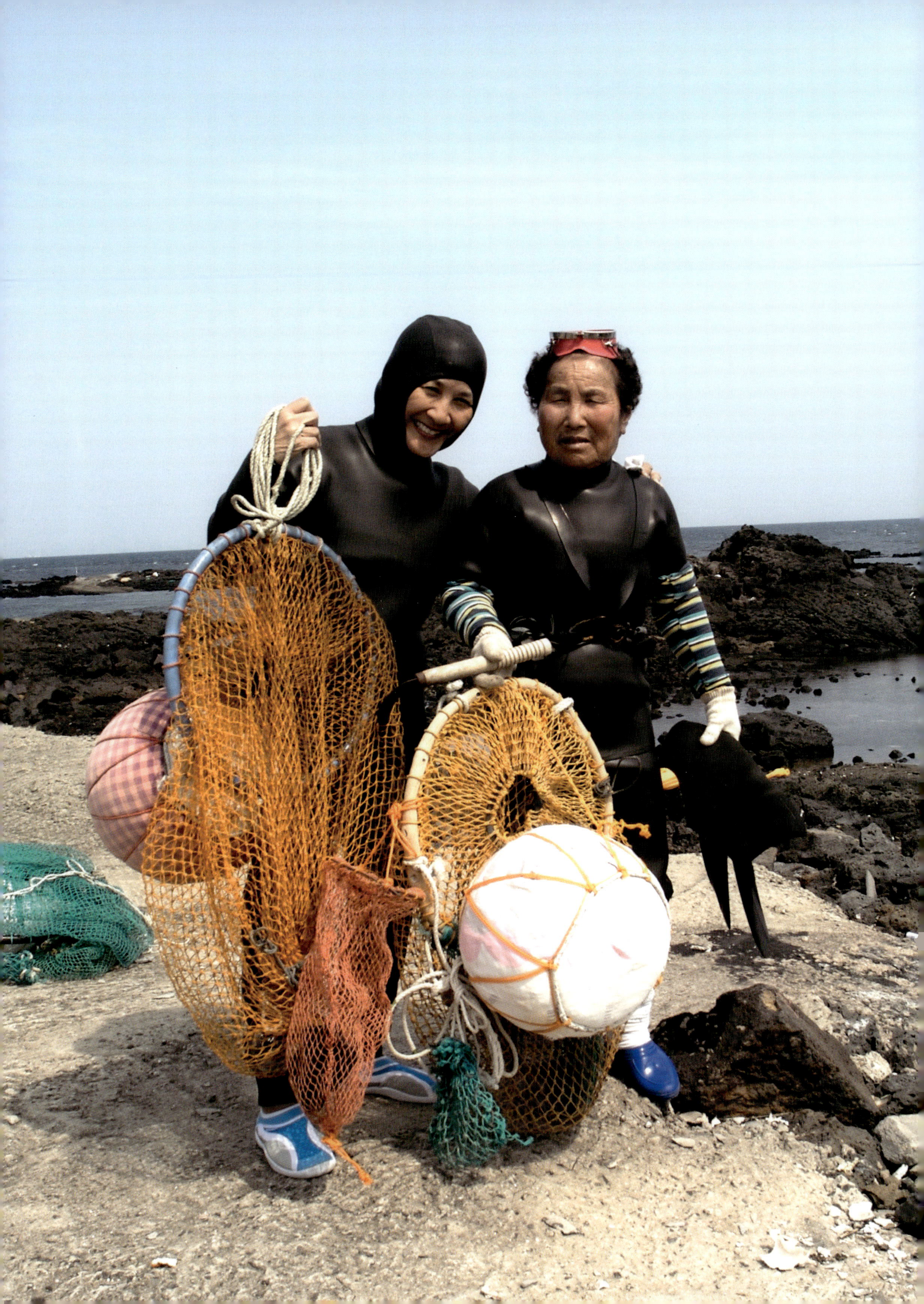

REFERENCE

Books

Hyun, Ki-young. *Aunt Suni*. Gak. 2008.

Kim, Soonie & Hahn, Jenie. *The Goddesses of Jeju Island*. Jeju Munhwa. 2002.

Lee, Sung-eun. *Sumbisory: Haenyeo in U-do*. NOONBIT. 2007.

Kendall, Laurel. *Life and Hard Times of a Korean Shaman: Of Tales and Telling of Tales*. University of Hawaii Press. 1988.

Jeju Haenyeo Museum. *Mother of the Sea, the Jeju Haenyeo*. Jejucom. 2007.

Jeju Haenyeo Museum. *The Work and Cultural Asset of Jeju Women Sea Divers*. hana CNC. 2009.

Paek, Ja-won. *Young-shil, the Spirit Grove*. Mok-Suk-Won Publishing. 1991.

Online Academic Resources

Cumings, Bruce. "American Involvement in April 3 Cheju Massacre." Excerpts from his speech in April 4, 1998.

Gwon, Gwi-sook. "Changing Labor Processes of Women's Work: The Haenyeo of Jeju Island." *Korean Studies*, Vol. 29, 2005: 114-136.

Han, Rim-wha & Kim, Soon Hee. "Jeju Women's Lives in the Context of April 3rd Uprising." *Asian Women*, Vol. 17, 2003: 21-37.

Hong, Gui-young. "Becoming a "Legitimate" Ancestor: A Sociocultural Understanding of a Sonless Jamnyeo's Life Story." *Forum: Qualitative Social Research*, Vol.5, No.3, 2004.

Kim, Jennifer Jung, "Kisaeng and Haenyeo: Modern Working Women in Colonial Korea." Association for Asian Studies Annual Meeting, 2004.

Kim, Seong-nae. "Sexual Politics of State Violence: On the Cheju April Third Massacre of 1948." in *Race, Panic and the Memory of Migration*. Meaghan Morris & Brett De Bary. (eds.). Hong Kong: Hong Kong University Press. 2002.

Kim, Seong-nae. "Shamanic Epics and Narrative Construction of Identity on Cheju Island." *Asian Folklore Studies*, Vol. 63, 2004: 57-78.

Kim, Soon-hee. "Jeju Island Womens Divers' Association in South Korea: A Source of Social Capital." Asian Journal of Women's Studies, Vol. 9, No. 1, 2003:37-59.

Korean Films and DVDs

My Mother, the Mermaid [*Ineogongju*] (Park Heung-sik, 2004): a feature film about a *haenyeo* and her daughter

Between [*Sai-e-seo*] (Lee Chang-jae, 2006): a documentary film on Korean shaman Kim Hye-kyoung

Weblinks

- http://english.visitkorea.or.kr/enu/index.kto (Official Korea tourism guide)
- http://english.jejusamda.com/?category=F (Overview of Jeju)
- http://www.jejuweekly.com (Local Jeju news)
- http://www.jejulife.net (Archive of Jeju features and news)
- http://www.artjs.com (Artwork of Jeju goddesses by Hong Jin-suk)
- http://www.jejuolle.org:8080/eng/ (Jeju walking tours)
- http://www.jejumaze.com (Maze Park in legendary Gimnyeong-ri)
- http://library.thinkquest.org/28170/45.html (Diving)
- http://home.hiwaay.net/~krcool/Astro/moon/moontides/ (Tides)
- http://www.youtube.com/watch?v=sBb-oofr-7A (Short documentary cut)
- http://www.youtube.com/watch?v=BZ1uIR0e0bo (Jeju Chilmeoridang Yeongdeunggut)
- http://www.gloriachang.com/radio/seawomen.htm (Radio show by Gloria Chang on "Sea Women")
- http://www.oscars.org/video/watch/36saa_lastmermaids.html (Short documentary "The Last Mermaids" by Liz Chae)
- http://ijtj.oxfordjournals.org/content/3/3/406.full ("Seeking Truth after 50 Years: The National Committee for Investigation of the Truth about the Jeju 4.3 Events" by Hunjoon Kim)
- http://wolcottwheeler2.blogspot.com/2007/01/1948-cheju-do-civil-war-how-to-kill.html ("The 1948 Cheju-do Civil War: How to Kill over 35,000 Innocent People in One Year and Get Away with It." by Wolcott Wheeler)
- http://www.japanfocus.org/-Khiem___Kim_Sung_soo/2848 ("Crimes, Concealment and South Korea's Truth and Reconciliation Commission" by Do Khiem and Kim Sung-soo)
- http://www.japanfocus.org/-Heonik-Kwon/3172 ("Healing the Wounds of War: New Ancestral Shrines in Korea" by Heonik Kwon)
- http://www.thejejubigswim.blogspot.com/ (The Jeju Big Swim)
- http://www.jejuweekly.com/news/articleView.html?idxno=719 (Article by J.M.G. Le Clezio, which appeared in France's GEO magazine)
- http://hosted.ap.org/specials/interactives/_international/korean_divers/index.html?SITE=NCAGW (Associated Press article on Jeju's Haenyeo School)
- http://www.imow.org (Website of the International Museum of Women that features photos from Moon Tides that won Community Choice Awards, 2010)

CREDITS

Photographer & Interviewer	Brenda Paik Sunoo
Translator & Interpreter	Youngsook Han
Publisher	Kim Hyung-geun
Editor	Lee Jin-hyuk
Assisting Editor	Kim Eugene
Copy Editor	Colin A. Mouat
Proofreader	Ben Jackson
Designer	Jung Hyun-young

* Additional photographs on pages 20, 44 by Sherrin Hibbard, 237 by Lee Jae-hyun, and 192, 216 by Yun Dae-gyun
* Additional translation for interview with Han Ki-ok by Colin A. Mouat, Ko Chun-hwa by Ko Young-lim, Woo Bong-seon and Woo Sun-deok by Park Hyun-mi, Yoon Man-sun by Yang Soon-ja.
* Woodcuts on page 81, 123, 149, 175, 195, 209, 227, courtesy of the artist Hong Jin-suk
* Paintings on page 138, 140, courtesy of the artist Kang Yo-bae